Progress in Inflammation Research

Series Editor

Prof. Michael J. Parnham PhD
Senior Scientific Advisor
GSK Research Centre Zagreb Ltd.
Prilaz baruna Filipovića 29
HR-10000 Zagreb
Croatia

Advisory Board

G. Z. Feuerstein (Wyeth Research, Collegeville, PA, USA)
M. Pairet (Boehringer Ingelheim Pharma KG, Biberach a. d. Riss, Germany)
W. van Eden (Universiteit Utrecht, Utrecht, The Netherlands)

Adhesion Molecules: Function and Inhibition

Klaus Ley

Editor

Birkhäuser
Basel · Boston · Berlin

Editor

Klaus Ley
La Jolla Institute for Allergy & Immunology
9420 Athena Circle
La Jolla, CA 92037
USA

(valid from November 2007)

Library of Congress Control Number: 2007929896

Bibliographic information published by Die Deutsche Bibliothek
Die Deutsche Bibliothek lists this publication in the Deutsche Nationalbibliografie;
detailed bibliographic data is available in the internet at http://dnb.ddb.de

ISBN 978-3-7643-7974-2 Birkhäuser Verlag AG, Basel – Boston – Berlin

© 2007 Birkhäuser Verlag AG, P.O. Box 133, CH-4010 Basel, Switzerland
Part of Springer Science+Business Media
Printed on acid-free paper produced from chlorine-free pulp. TCF ∞
Cover design: Markus Etterich, Basel
Cover illustration: see page 275. With friendly permission of Eric A. Severson
Printed in Germany
ISBN 978-3-7643-7974-2   e-ISBN 978-3-7643-7975-9

9 8 7 6 5 4 3 2 1   www.birkhauser.ch

Contents

Transendothelial migration

List of contributors

Scott D. Auerbach, Center for Excellence in Vascular Biology, Departments of Pathology, Brigham and Women's Hospital and Harvard Medical School, 77 Avenue Louis Pasteur, Boston, MA 02115, USA; e-mail: sauerbach@rics.bwh.harvard.edu

Daniel C. Bullard, Department of Genetics, Kaul Building 640A, 720 South 20th Street, University of Alabama at Birmingham, Birmingham, AL 35294, USA; e-mail: dcbullard@genetics.uab.edu

Stefan Butz, Max-Planck-Institute for Molecular Biomedicine, Röntgenstraße 20, 48149 Münster, Germany; e-mail: butzs@uni-muenster.de

Rochelle M. Conway, Department of Biological Sciences, Box 413, University of Wisconsin-Milwaukee, Milwaukee, WI 53201, USA; e-mail: rmconway@uwm.edu

Myron I. Cybulsky, Toronto General Research Institute, 200 Elizabeth Street, Eaton-11, Toronto, Ontario, M5G 2C4, Canada; e-mail: myron.cybulsky@utoronto.ca

Britta Engelhardt, Theodor Kocher Institute, University of Bern, Freiestrasse 1, 3012 Bern, Switzerland; e-mail: bengel@tki.unibe.ch

Jamison J. Grailer, Department of Biological Sciences, Box 413, University of Wisconsin-Milwaukee, Milwaukee, WI 53201, USA; e-mail: jgrailer@uwm.edu

John M. Harlan, Division of Hematology, Department of Medicine, University of Washington, Seattle, WA 98195, USA; e-mail: jharlan@u.washington.edu

Sharon J. Hyduk, Toronto General Research Institute, 200 Elizabeth Street, Toronto, Ontario, M5G 2C4, Canada; e-mail: sharon.hyduk@utoronto.ca

Sirpa Jalkanen, Department of Bacterial and Inflammatory Diseases, National Public Health Institute, 20520 Turku, Finland; e-mail:

Minsoo Kim, Department of Surgery, Division of Surgical Research, Rhode Island Hospital and Brown University School of Medicine, Providence, RI 02903, USA; e-mail: minsoo_kim@brown.edu

Francis W. Luscinskas, Brigham and Women's Hospital, 77 Avenue Louis Pasteur, NRB 7 Rm. 752, Boston, MA 02115, USA; e-mail: fluscinskas@rics.bwh.harvard.edu

Rodger P. McEver, Cardiovascular Biology Research Program, Oklahoma Medical Research Foundation,825 N.E. 13th Street, Oklahoma City, OK 73104, USA; e-mail: rodger-mcever@omrf.org

William A. Muller, Department of Pathology, Northwestern University Feinberg School of Medicine, 303 East Chicago Avenue, Chicago, IL 60611, USA; e-mail: wamuller@northwestern.edu

Sussan Nourshargh, Centre for Microvascular Research, Williams Harvey Research Institute, Queen Mary and Westfield College, Charterhouse Square, EC1M 6BQ, London, UK; e-mail: s.nourshargh@qmul.ac.uk

Charles A. Parkos, Epithelial Pathobiology Research Unit, Department of Pathology and Laboratory Medicine, Emory University School of Medicine, Whitehead Biomedical Building, Room 105B, Atlanta, GA 30322, USA; e-mail: cparkos@emory.edu

Marko Salmi, MediCity Research Laboratory, Tykistökatu 6A, 20520 Turku, Finland; e-mail: marko.salmi@utu.fi

Alan R. Schenkel, Department of Microbiology, Immunology and Pathology, College of Veterinary Medicine and Biomedical Sciences, Colorado State University, Fort Collins, CO 80523, USA; e-mail: alan.schenkel@colostate.edu

Eric A. Severson, Epithelial Pathobiology Research Unit, Department of Pathology and Laboratory Medicine, Emory University School of Medicine, Whitehead Biomedical Building, Room 115, Atlanta, GA 30322, USA; e-mail: esevers@Learnlink.emory.edu

Douglas A. Steeber, Department of Biological Sciences, Box 413, University of Wisconsin-Milwaukee, Milwaukee, WI 53201, USA; e-mail: steeber@uwm.edu

Traci J. Storey, Department of Biological Sciences, Box 413, University of Wisconsin-Milwaukee, Milwaukee, WI 53201, USA; e-mail: tjstorey@uwm.edu

Hariharan Subramanian, Department of Biological Sciences, Box 413, University of Wisconsin-Milwaukee, Milwaukee, WI 53201, USA; e-mail: subrama3@uwm.edu

Dietmar Vestweber, Max-Planck-Institute for Molecular Biomedicine, Röntgenstraße 20, 48149 Münster, Germany; e-mail: vestweb@mpi-muenster.mpg.de (vestweb@uni-muenster.de)

Mathieu-Benoit Voisin, Centre for Microvascular Research, Williams Harvey Research Institute, Queen Mary and Westfield College, Charterhouse Square, EC1M 6BQ, London, UK; e-mail: m.voisin@qmul.ac.uk

Lin Yang, Center for Excellence in Vascular Biology, Departments of Pathology, Brigham and Women's Hospital and Harvard Medical School, 77 Avenue Louis Pasteur, Boston, MA 02115, USA

Karyn Yonekawa, Division of Nephrology, Department of Pediatrics, University of Washington, Seattle, WA 98195, USA; e-mail: karyny@u.washington.edu

Preface

Endothelial and leukocyte adhesion molecules guide the trafficking of all leukocytes to sites of inflammation, of lymphocytes to primary, secondary and tertiary lymphatic organs, and of monocytes to almost all tissues of the body where they become macrophages and dendritic cells. Some of these same adhesion molecules are also involved in leukocyte and endothelial cell development and differentiation, as well as leukocyte egress from the bone marrow, aspects that are not covered in this book. Instead, we focus on the role of adhesion molecules in inflammation.

The first leukocyte-endothelial adhesion molecule, L-selectin, was discovered in 1983 [1]. In the quarter century following this and many other important discoveries, we have learned a lot about the physiology and pathophysiology of inflammatory cell recruitment. This book covers the 16 most important leukocyte adhesion molecules in detail, organized into 12 chapters written by leading specialists who, in some cases, are also the discoverers of the respective molecule. The book is organized according to the leukocyte adhesion cascade, which starts with capture/tethering, followed by rolling, firm adhesion and transendothelial migration [2].

The first chapter is by Rodger McEver, the discoverer of P-selectin Glycoprotein Ligand-1 (PSGL-1) [3], the most studied selectin ligand. Doug Steeber covers L-selectin, the most important molecule in lymphocyte homing to lymph nodes and Peyer's patches, but also critically involved in inflammation [4]. The endothelial E- and P-selectins are described and discussed by Dan Bullard, who constructed some of the earliest selectin knockout mice [5].

In the section on firm adhesion, Francis "Bill" Luscinskas writes about Intercellular Adhesion Molecule-1 (ICAM-1), a molecule whose function he helped define [6]. Britta Engelhardt discusses the structure and function of $\alpha_4\beta_1$ and $\alpha_4\beta_7$ integrins. She contributed important studies on the role of α_4 integrins in inflammatory disorders of the central nervous system [7]. The main ligand of $\alpha_4\beta_1$ integrin, vascular cell adhesion molecule-1 (VCAM-1), is covered in a chapter by Drs. Hyduk and Cybulsky. Dr. Cybulsky is a pioneer in the adhesion molecule field and defined much of the biological functions of VCAM-1 [8]. The next chapter focuses on the two most important of the four known β_2 integrins, LFA-1 (CD11a/CD18) and Mac-1 (CD11b/CD18). The authors are Alan Schenkel, who discovered Mac-1-dependent migration of myeloid cells inside blood vessels [9], and Min Soo Kim, who contributed groundbreaking studies to our understanding of conformational LFA-1 activation [10].

In recent years, transendothelial migration has received tremendous attention. William "Bill" Muller is a pioneer in this area and has studied Platelet Endothelial Cell Adhesion Molecule-1 for two decades [11]. Sussan Nourshargh, another leading investigator in transmigration, discusses the role of $\alpha_6\beta_1$ integrin in leukocyte adhesion and transmigration [12]. Marko Salmi and Sirpa Jalkanen worked together to explain the functions of Vascular adhesion protein-1 (VAP-1), a molecule they discovered [13]. The role of ESAM in neutrophil emigration into inflamed tissues is presented by Dietmar Vestweber and Stefan Butz, who discovered key functions of this molecule involved in transmigration [14]. Eric Severson and Charles "Chuck" Parkos review the structure and function of JAM proteins, a family of endothelial and epithelial molecules whose function they helped elucidate [15].

The book concludes with a chapter by Karyn Yonekawa and John Harlan, who evaluate the promises and limitations of targeting adhesion molecules for therapy. Harlan made many outstanding contributions to the field of leukocyte adhesion molecules, including the production and generous distribution of one of the first useful antibodies to the β_2 integrins, mAb 60.3 [16]. Their chapter is particularly interesting, because it shows how far leukocyte adhesion molecules have come in the last 25 years and how far they have yet to go. In a very critical analysis, Yonekawa and Harlan are certainly not overly exuberant, but it is quite clear that interfering with adhesion molecule function by antibodies, with their expression by small molecules, or, in the case of selectin ligands, with their glycosylation holds enormous potential for a variety of therapeutic interventions. The current stars of adhesion molecule-directed therapy are antibodies to α_4 integrins and to LFA-1, which are approved for the treatment of relapsing-remitting multiple sclerosis and chronic moderate to severe psoriasis, respectively. It is safe to predict that more adhesion molecule-based therapies will make it to market and help countless patients in the years to come.

This book would not have been possible without the dedication and hard work of each of the authors, who provided chapters of the highest quality and at the cutting edge of inflammatory adhesion molecule research. I thank each of them. I would also like to thank Detlef Klueber, the publisher's representative, and Becky Ellwood, who helped me keep track of the authors and the various versions of their chapters. Finally I hope that you, the reader, will find this little book both enjoyable and useful.

May 2007   Klaus Ley

References

1   Gallatin WM, Weissman IL, Butcher EC (1983) A cell-surface molecule involved in organ-specific homing of lymphocytes. *Nature* 304: 30–34

2   Butcher EC (1991) Leukocyte-endothelial cell recognition - Three (or more) steps to specificity and diversity. *Cell* 67: 1033–1036

3   Norgard KE, Moore KL, Diaz S, Stults NL, Ushiyama S, McEver RP, Cummings RD, Varki A (1993) Characterization of a specific ligand for P-selectin on myeloid cells. A minor glycoprotein with sialylated O- linked oligosaccharides. *J Biol Chem* 268: 12764–12774

4   Steeber DA, Campbell MA, Basit A, Ley K, Tedder TF (1998) Optimal selectin-mediated rolling of leukocytes during inflammation *in vivo* requires intercellular adhesion molecule-1 expression. *Proc Natl Acad Sci USA* 95: 7562–7567

5. Bullard DC, Kunkel EJ, Kubo H, Hicks MJ, Lorenzo I, Doyle NA, Doerschuk CM, Ley K, Beaudet AL (1996) Infectious susceptibility and severe deficiency of leukocyte rolling and recruitment in E-selectin and P-selectin double mutant mice. *J Exp Med* 183: 2329–2336

6   Luscinskas FW, Cybulsky MI, Kiely JM, Peckins CS, Davis VM, Gimbrone MA Jr (1991) Cytokine-activated human endothelial monolayers support enhanced neutrophil transmigration *via* a mechanism involving both Endothelial-Leukocyte Adhesion Molecule-1 (ELAM-1) and Intercellular Adhesion-Molecule-1 (ICAM-1). *J Immunol* 146: 1617–1625

7   Vajkoczy P, Laschinger M, Engelhardt B (2001) Alpha4-integrin-VCAM-1 binding mediates G protein-independent capture of encephalitogenic T cell blasts to CNS white matter microvessels. *J Clin Invest* 108: 557–565

8   Cybulsky MI, Gimbrone MA Jr (1991) Endothelial expression of a mononuclear leukocyte adhesion molecule during atherogenesis. *Science* 251: 788–791

9   Schenkel AR, Mamdouh Z, Muller WA (2004) Locomotion of monocytes on endothelium is a critical step during extravasation. *Nat Immunol* 5: 393–400

10 Kim M, Carman CV, Springer TA (2003) Bidirectional transmembrane signaling by cytoplasmic domain separation in integrins. *Science* 301: 1720–1725

11 Muller WA, Ratti CM, McDonnell SL, Cohn ZA (1989) A human endothelial cell-restricted, externally disposed plasmalemmal protein enriched in intercellular junctions. *J Exp Med* 170: 399–414

12 Dangerfield J, Larbi KY, Huang MT, Dewar A, Nourshargh S (2002) PECAM-1 (CD31) homophilic interaction up-regulates alpha6beta1 on transmigrated neutrophils *in vivo* and plays a functional role in the ability of alpha6 integrins to mediate leukocyte migration through the perivascular basement membrane. *J Exp Med* 196: 1201–1211

13 Salmi M, Jalkanen S (1992) A 90-kilodalton endothelial cell molecule mediating lymphocyte binding in humans. *Science* 257: 1407–1409

14 Wegmann F, Petri B, Khandoga AG, Moser C, Khandoga A, Volkery S, Li H, Nasdala I,

Brandau O, Fassler R et al (2006) ESAM supports neutrophil extravasation, activation of Rho, and VEGF-induced vascular permeability. *J Exp Med* 203: 1671–1677

15   Parkos CA, Colgan SP, Diamond MS, Nusrat A, Liang TW, Springer TA, Madara JL (1996) Expression and polarization of intercellular adhesion molecule-1 on human intestinal epithelia – consequences for cd11b/cd18-mediated interactions with neutrophils. *Molecular Medicine* 2: 489–505

16   Harlan JM, Killen PD, Senecal FM, Schwartz BR, Yee EK, Taylor RF, Beatty PG, Price TH, Ochs HD (1985) The role of neutrophil membrane glycoprotein GP 150 in neutrophil adherence to endothelium *in vitro*. *Blood* 66: 167–178

Capture and rolling

P-selectin glycoprotein ligand-1 (PSGL-1)

Rodger P. McEver

Cardiovascular Biology Research Program, Oklahoma Medical Research Foundation, 825 N.E. 13th Street, Oklahoma City, OK 73104, USA

Introduction

The selectins are type 1 membrane glycoproteins that have an N-terminal C-type lectin domain, followed by an epidermal growth factor (EGF)-like motif, a series of short consensus repeats, a transmembrane domain, and a cytoplasmic tail (see Chapters 2 and 3). Leukocytes express L-selectin, whereas activated platelets express P-selectin and activated endothelial cells express P-selectin and E-selectin. Each selectin mediates leukocyte rolling to vascular surfaces through Ca^{2+}-dependent interactions of the lectin domain with cell-surface glycoconjugates. All selectins bind with low affinity to glycans with terminal components that include α2-3-linked sialic acid and α1-3-linked fucose, typified by the sialyl Lewis x (sLe^x) determinant (NeuAcα2-3Galβ1-4[Fucα1-3]GlcNAcβ1-R) [1, 2]. Crystal structures of sLe^x bound to the lectin domains of P- and E-selectin reveal a network of interactions between the fucose, a single Ca^{2+} ion, and several amino acids, including those that coordinate the Ca^{2+} [3]. This explains the Ca^{2+} requirement for binding. Targeted disruption of the α1-3-fucosyltransferases Fuc-TIV and Fuc-TVII in mice eliminates selectin-mediated leukocyte trafficking [4, 5]. This demonstrates that physiologically relevant selectin ligands require α1-3-linked fucose. However, there is abundant evidence that selectins bind better to some glycoproteins modified with sLe^x-capped glycans than to others [1, 2]. This chapter focuses on the most thoroughly characterized of these glycoproteins: P-selectin glycoprotein ligand-1 (PSGL-1), which mediates important biological functions through interactions with each of the selectins.

Structure and tissue distribution of PSGL-1

PSGL-1 was initially described as the dominant glycoprotein ligand for P-selectin on human myeloid cells [6]. Biochemical characterization indicated that PSGL-1 is a disulfide-linked homodimer with two 120-kDa subunits as determined by SDS-PAGE. Each subunit has two to three N-glycans that are dispensable for binding to P-selectin. In contrast, it has many clustered, sialylated O-glycans that render it susceptible to cleavage with O-sialoglycoprotein endopeptidase [7] (Fig. 1A). The amino acid sequence of human PSGL-1 derived from cDNA and genomic cloning reveals that it is a type 1 membrane protein of 402 or 412 amino acids [8, 9]. A signal peptide is followed by a propeptide that is cleaved by paired basic-amino acid-converting enzymes. The extracellular domain of the mature protein is rich in serines, threonines, and prolines, and includes 15 or 16 decameric repeats, all hallmarks of mucins. Electron micrographs of rotary-shadowed PSGL-1 demonstrated that it is a highly extended protein, consistent with the presence of multiple serine/threonine-linked O-glycans that straighten the polypeptide backbone of a mucin [10]. Three N-terminal tyrosines at residues 5, 7, and 10 of the mature protein are located in an anionic consensus sequence that favors tyrosine sulfation [8]. A single extracellular cysteine at the junction with the transmembrane domain mediates disulfide-linked dimerization [11]. Cross-linking studies demonstrated that PSGL-1 forms noncovalent homodimers through interactions of the transmembrane domains even when this cysteine is mutated [12]. This suggests that newly synthesized PSGL-1 initially forms a noncovalent homodimer in the endoplasmic reticulum that is then stabilized by a disulfide bond. The cytoplasmic domain of human PSGL-1 has 69 amino acids with no consensus sequences for phosphorylation [8].

The primary sequence of murine PSGL-1 suggests mucin-like features like those of the human protein [13]. The extracellular sequence has an anionic N-terminal region with two rather than three tyrosines; it also has many serines, threonines, and prolines, and a single cysteine near the transmembrane domain. Overall, there is very little similarity between the human and murine extracellular sequences. In contrast, the sequences of the transmembrane and cytoplasmic domains are highly conserved, and this conservation is evident in the sequences of several other vertebrate species now available in databases. This suggests that these domains have important conserved functions. A single exon encodes the entire open reading frame in both the human and murine genes for PSGL-1 [9, 13].

PSGL-1 is primarily expressed in hematopoietic cells. In bone marrow it is found on some CD34+ stem/progenitor cells and on myeloid cells at many stages of maturation, but not on erythroid cells, megakaryocytes, or platelets [14, 15]. It has been suggested that platelets express PSGL-1 [16], but this finding has not been confirmed; if present on these cells, the levels must be very low. In peripheral blood, PSGL-1 is present on virtually all leukocytes, although at lower levels on B cells [15]. PSGL-1 is expressed on circulating dendritic cells, on tissue monocyte-derived dendritic cells, and on some dendritic cells in lymphoid organs. As determined by immunohistochemistry of human tissues, most vascular endothelial cells do not express PSGL-1 except at a few sites of chronic inflammation [15]. Mice express PSGL-1 in venular endothelial cells of the small intestine, which may have significance for human disease [17] (see below). Human umbilical vein endothelial cells have been reported to express PSGL-1 at low levels [18]. These data suggest that

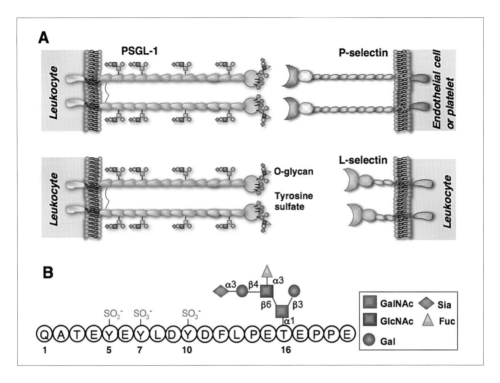

Figure 1

Interactions of PSGL-1 with P-selectin and L-selectin.

(A) PSGL-1 has an extended extracellular domain containing many O-glycans with hetero-geneous structures that are attached to serine or threonine residues. Only a small number of representative O-glycans are depicted. The extracellular domain is followed by a transmem-brane domain and cytosolic tail. PSGL-1 forms homodimers through transmembrane-domain interactions that are stabilized by a single disulfide bond just outside the membrane. Both P- and L-selectin have an N-terminal C-type lectin domain, followed by an EGF-like domain, a series of consensus repeats, a transmembrane domain, and a cytosolic tail. P-selectin forms noncovalent dimers, probably through transmembrane-domain interactions. There is no evi-dence that L-selectin dimerizes. The lectin domains of P- and L-selectin interact with a mem-brane-distal region of PSGL-1 that is modified with sulfated tyrosines and a specific O-glycan structure. (B) The N-terminal region of PSGL-1 that binds to P- and L-selectin. A synthetic glycosulfopeptide with this structure binds to P- and L-selectin with the same affinity as native PSGL-1. The amino acid sequence, represented by single-letter code, represents that of the mature protein after proteolytic cleavage of the signal peptide and propeptide during biosynthesis. The O-glycan has a short core 2 branch with N-acetylglucosamine in β1-6 link-age to N-acetylgalactosamine attached to the threonine. The core 2 branch comprises the tetrasaccharide sLe^x. Sia, sialic acid; Gal, galactose; GalNAc, N-acetylgalactosamine; GlcNAc, N-acetylglucosamine; Fuc, fucose.

endothelial cells in some environments can express PSGL-1. As discussed below, PSGL-1 must have appropriate post-translational modifications to bind to selectins. Thus, mere expression of the polypeptide backbone may not confer biological function.

Post-translational modifications of PSGL-1 required for binding to selectins

Like all mucins, PSGL-1 has O-glycans attached to many serine and threonine residues in the extracellular domain. The structures of the O-glycans have been determined for PSGL-1 derived from the human myeloid HL-60 cell line [19]. Most have relatively short, sialylated core 1 or branched core 2 structures. None are detectably sulfated. Remarkably, even though all PSGL-1 molecules from HL-60 cells bind to P-selectin, only a minority of the O-glycans are fucosylated [19]. Fucosylation of O-glycans on murine leukocytes appears to be even more rare [20]. This suggests that α1-3-fucosyltransferases have site-specific preferences for modifying O-glycans on PSGL-1. It further implies that most of the O-glycans function only to extend the polypeptide backbone.

The first suggestion of a site-specific location of a critical fucosylated O-glycan arose from the discovery that mAbs with epitopes at the extreme N-terminal region of human and murine PSGL-1 block binding to both P- and L-selectin [13, 14, 21–25]. The mAbs to the N-terminal region of human and murine PSGL-1 that block binding to P- and L-selectin in biochemical assays also block rolling of leukocytes on P- or L-selectin under flow conditions. Such studies first demonstrated that PSGL-1 is the dominant ligand for leukocyte rolling on these two selectins. *In vitro*, leukocytes use interactions of PSGL-1 with L-selectin to roll on adherent leukocytes or to initiate leukocyte aggregation [21, 24]. These leukocyte-leukocyte interactions lead to secondary tethering of leukocytes to a P- or E-selectin surface [26], a potential mechanism to amplify leukocyte recruitment to the vessel wall.

In human PSGL-1, the binding site for P- and L-selectin comprises a peptide sequence containing three tyrosine sulfate residues near a threonine to which a specific O-glycan is attached (Fig. 1B). Biochemical assays and expression of recombinant forms of PSGL-1 have demonstrated that optimal binding to P-selectin and L-selectin requires sulfation of the tyrosines and addition of a branched, core 2 O-glycan capped with sLe^x to the threonine [27–32]. Targeted deletion of the murine gene encoding core2GlcNAcT-I, the major core 2 β1-6-N-acetylglucosaminyltransferase in leukocytes, eliminates binding of leukocytes to P-selectin [33]. This suggests that core2GlcNAcT-I has the essential role in constructing core 2 O-glycans on PSGL-1. There are two tyrosyl protein sulfotransferases that catalyze addition of sulfate esters to tyrosine residues. Targeted deletion of the gene for either enzyme does not impair leukocyte binding to selectins [34, 35], suggesting that both enzymes contribute to tyrosine sulfation of PSGL-1.

The structural requirements for binding P- and L-selectin have been more definitively mapped by synthesis of glycosulfopeptides modeled after the N-terminal region of PSGL-1 [36–38] (Fig. 1B). Each sulfate contributes to binding affinity, and the peptide itself confers weak binding. The fucose moiety is essential for binding, whereas the sialic acid plays a lesser role. The position of the O-glycan in relation to the tyrosine sulfates is critical. Even the core backbone of the O-glycan is important: a short core 2 O-glycan capped with sLe^x supports binding, whereas an isomeric core 1 O-glycan does not [37]. Furthermore, a glycosulfopeptide with an extended core 2 branch containing fucosylated polylactosamine capped with sLe^x binds poorly [39]. These studies reveal specific stereochemical requirements for optimal binding of PSGL-1 to P- and L-selectin. Relative to P-selectin, L-selectin binds to PSGL-1 with lower affinity and more rapid dissociation kinetics [36–39]. Therefore, there are subtle but important differences in specificity that may be dictated by specific residues in the lectin domain of each selectin.

X-ray crystallography was used to solve the structure of the lectin and EGF domains of P-selectin bound to a recombinant N-terminal fragment of PSGL-1 [3]. The PSGL-1 fragment is sulfated on all three tyrosines, and the sequence of the core 2 O-glycan is nearly identical to the one in the synthetic glycosulfopeptide that confers optimal binding to P-selectin (Fig. 1B). The PSGL-1 fragment binds to a large but shallow surface on the lectin domain, opposite to where the EGF domain is attached. The fucose has multiple binding interactions, some of which participate in coordinating the single Ca^{2+} ion in the lectin domain. The galactose and sialic acid residues make fewer contacts. Other regions of the lectin domain contact certain amino acids plus the sulfates of the middle and C-terminal tyrosines of PSGL-1. Monomeric P-selectin binds with equivalent affinity to native or recombinant PSGL-1 and to the synthetic glycosulfopeptides, with dissociation constants estimated at ~80–800 nM [3, 36, 37, 40, 41]. In contrast, P-selectin binds to carbohydrates containing only sLe^x with dissociation constants of 1–10 mM [42–44]. The multiple interactions of carbohydrate, amino acids, and sulfate residues with the lectin domain explain why PSGL-1 binds to P-selectin with much higher affinity than do peptide-free oligosaccharides that contain sLe^x. The shallow binding site on P-selectin may contribute to the rapid binding kinetics. The N-terminal tyrosine sulfate is not visualized in the co-crystal structure, which is surprising since a recombinant form of PSGL-1 or a glycosulfopeptide with sulfation restricted to this tyrosine still binds to P-selectin with appreciable affinity [31, 32, 37]. Interestingly, the N-terminal region of murine PSGL-1, which can bind to both human and murine P-selectin, has only two tyrosines that might be sulfated [13]. Compared to the human sequence, these tyrosines are positioned much closer to the two threonines that are the best candidates for attachment of a fucosylated O-glycan. Mutagenesis studies suggest that only one tyrosine and one threonine contribute to binding [45]. Thus, there may be additional ways in which P-selectin can bind to both sLe^x and a sulfated peptide segment.

PSGL-1 binds differently to E-selectin than to P- and L-selectin. Sialylation and α1-3-fucosylation of glycans are required for binding, and at least some of these modifications appear to be on core 2 O-glycans [30]. However, tyrosine sulfation of PSGL-1 is not required for binding to E-selectin [28, 29]. Because mAbs to the N terminus of PSGL-1 do not block interactions with E-selectin [46], there may be other binding sites on PSGL-1 [47, 48], although the limited number of fucosylated O-glycans should restrict the number of these sites [19].

All myeloid cells bind to all three selectins, suggesting that they express the full complement of glycosyltransferases required to construct selectin ligands on PSGL-1 and other glycoproteins [14]. In contrast, most circulating T and B cells do not bind to selectins [14, 49]. During the transition to the effector cell phenotype, T cells acquire selectin-ligand binding function, which appears to be primarily due to up-regulated expression of core2GlcNAcT-I and Fuc-TVII [50]. *In vitro*, cytokines such as TNF-α induce endothelial cells to express glycosyltransferases that modify PSGL-1 so that it can bind selectins [18]. *In vivo*, however, available evidence does not support a general increase in expression of functional PSGL-1 on endothelial cells at all sites of inflammation.

Unlike most leukocytes, some subsets of natural killer cells and dendritic cells express glycoforms of PSGL-1 that have glycans with sulfate esters attached to the C-6 position of *N*-acetylglucosamine residues [51, 52]. The functional significance of these modifications is not clear, although PSGL-1 on natural killer cells has been reported to bind L-selectin [51]. This might be due to attachment of a sulfate ester to the C-6 position of the *N*-acetylglucosamine residue of sLe^x to produce 6-sulfo-sLe^x, a well-characterized binding determinant for L-selectin on the glycans of mucins from lymph node endothelial cells [53].

Mechanisms for leukocyte rolling through PSGL-1

Leukocytes must adhere to vascular surfaces under the hydrodynamic environment of the circulation. The relative motion between the flowing cell and the vascular surface limits the time for a cell adhesion molecule to bind its ligand. This imposes kinetic constraints on selectin-ligand interactions. Fast association and dissociation rates for bonds between PSGL-1 and P-selectin have been suggested as requirements for leukocytes to roll on P-selectin in flow [54]. The kinetics may be even faster for interactions of L-selectin with its ligands, including PSGL-1. Surface plasmon resonance studies have confirmed rapid binding kinetics for unstressed bonds between PSGL-1 and P-selectin. The measured dissociation rate is ~1 s^{-1}, and the estimated association rate is a remarkably high 4.4×10^6 M^{-1} s^{-1} [40]. There are ~25 000 PSGL-1 molecules on each leukocyte [55]. Leukocytes roll on purified immobilized P-selectin at densities as low as 10–25 sites/μm^2 [14, 48]. Cultured human endothelial cells stimulated with thrombin or histamine express ~25–50 P-selectin molecules/μm^2 [56]. The density of P-selectin on activated endothelial cells *in vivo* is unknown, but immunohistochemical studies suggest that it is much lower than that of E-selectin. By contrast, the density of P-selectin on activated platelets is much higher: ~350 sites/μm^2 [57–59].

The hydrodynamic environment also imposes mechanical constraints on selectin-ligand interactions. Adhesive bonds are subjected to force, which affects their dissociation rates. The most intuitive response to force is an increase in dissociation rate that shortens bond lifetime. This phenomenon has been termed a "slip bond." The first experimental demonstration of slip bonds was for interactions between P-selectin and PSGL-1. The lifetimes of transient tethers of neutrophils or PSGL-1-transfected cells on limiting densities of P-selectin were measured as a function of wall shear stress, or force on the tether [32, 60]. The lifetimes appear to obey first-order kinetics, consistent with the tethers representing "quantal units" that may represent single bonds. The lifetimes shorten only gradually as force on the tether is increased. If the tether represents only one bond, the bond must have high mechanical strength. However, some transient tethers have more than one bond, even though they have lifetimes that appear to obey first-order dissociation kinetics [61]. It is possible that tethers at higher wall shear stresses require more bonds, which would distribute the force and thus lower the average force per bond. In this scenario, the strength of individual bonds need not be high.

A less intuitive response to force is a decrease in dissociation rate that lengthens bond lifetime. This phenomenon has been termed a "catch bond." Although catch bonds were hypothesized many years ago [62], they were only recently demonstrated experimentally, again for interactions between P-selectin and PSGL-1 [63]. The catch bonds were observed at lower levels of force than those previously employed, explaining why they were not initially detected. As force initially increases, bond lifetimes lengthen, a characteristic of catch bonds, until the lifetimes reach an optimal level. As force increases further, bond lifetimes shorten, a characteristic of slip bonds. Thus, depending on the force level, interactions between P-selectin and PSGL-1 behave as either catch bonds or slip bonds. This biphasic response to force introduces a new dimension into how cell adhesion molecules might act in mechanically stressful environments such as the circulation.

L-selectin-dependent rolling develops only above a certain shear threshold; below this shear threshold, leukocytes roll unstably and detach [64, 65]. Such flow-enhanced rolling may prevent spontaneous aggregation of free-flowing leukocytes that express both L-selectin and PSGL-1. Transitions from catch bonds to slip bonds have also been observed between L-selectin and PSGL-1 [66]. The force range that elicits catch bonds between L-selectin and PSGL-1 is higher than for catch bonds between P-selectin and PSGL-1, and catch bonds have been shown to govern L-selectin-dependent cell rolling at threshold shear [67]. Catch bonds enable increasing force to convert short-lived tethers into longer-lived tethers, which decrease rolling velocities and increase the regularity of rolling steps as shear rises from the threshold to an optimal value. As shear increases above the optimum, transitions to slip bonds shorten tether lifetimes, which increase rolling velocities and decrease rolling regularity. Thus, force-dependent alterations of bond lifetimes govern L-selectin-dependent cell adhesion below and above the shear optimum.

The structural mechanisms that enable catch bonds constitute an intriguing puzzle. Substituting a residue in the EGF domain of L-selectin with the corresponding residue from P-selectin, which was predicted to increase the flexibility of the interdomain hinge of L-selectin, reduces the shear threshold for adhesion *via* two mechanisms [68]. One affects the on-rate by increasing tethering through greater rotational diffusion. The other affects the off-rate by strengthening rolling through augmented catch bonds with longer lifetimes at small forces. Thus, allosteric changes remote from the ligand-binding interface regulate both bond formation and dissociation. As the interdomain hinge straightens, force might cause the interacting molecules to slide across the binding interface, causing new interactions to form that compensate for the breakage of other interactions, a mechanism for catch bonds [68]. Force might also propagate conformational changes from the lectin-EGF domain interface to the ligand-binding interface to enhance binding affinity [69].

Microspheres decorated with PSGL-1 or with glycosulfopeptides modeled after the P-binding domain of PSGL-1 roll on immobilized P-selectin [70]. Similarly, microspheres bearing L-selectin roll on immobilized PSGL-1 [67]. This demonstrates that intrinsic molecular interactions are sufficient to confer rolling adhesion under flow. The cell surface organizations of P-selectin and PSGL-1 further contribute to the efficiency of leukocyte tethering and rolling in flow. PSGL-1 is concentrated on the tips of microvilli, which enhances tethering rates [14, 71, 72]. Both P-selectin and PSGL-1 are long molecules that extend their binding sites well above the membrane [10, 55]. P-selectin must have a minimal length to support leukocyte rolling *in vitro* [73], and this may also be true for PSGL-1 [70]. Molecular extension enhances bond formation and may reduce repulsion between the glycocalyces of apposing cells [74]. Adhesive forces applied through P-selectin–PSGL-1 bonds cause rolling leukocytes to extrude thin membrane tethers at the trailing edge of the cell; these tethers reduce the force required to balance the torque applied to the cell by fluid shear and prolong the time required for bond dissociation [75, 76]. Tether extension and retraction is dynamic, rapidly increasing as wall shear stress increases and then rapidly decreasing as wall shear stress decreases [77]. Clustering of PSGL-1 in microvilli and of P-selectin in clathrin-coated pits may also delay bond dissociation [78]. Indeed, differential signaling of endothelial cells regulates the degree of clustering of P-selectin in clathrin-coated pits [79]. The cytoplasmic domain of PSGL-1 has a positively charged juxtamembrane segment that binds to members of the ezrin/radixin/moesin (ERM) family *in vitro* [80, 81]. Because ERM proteins connect the cytosolic tails of some membrane proteins to actin filaments, they might contribute to the location of PSGL-1 in microvilli of resting cells and to the redistribution of PSGL-1 to the uropods of polarized, activated cells [80]. Based on experiments with transfected cells, the cytoplasmic tail of PSGL-1 was reported to be essential for rolling on P-selectin [81]. However, our unpublished data indicate that deletion of the cytoplasmic domain of PSGL-1 causes at most modest defects in cell rolling. Interactions of other membrane proteins with the cortical cytoskeleton may regulate tether extensions even if the cytoplasmic domain of PSGL-1 is deleted [82].

Dimerization through molecular self-association contributes to rolling efficiency. Mutation of the cysteine that covalently links two subunits of PSGL-1 into homodimers was reported to eliminate PSGL-1-dependent cell rolling on P-selectin [11], but this has not been confirmed [12]. However, cells expressing a chimeric, monomeric form of PSGL-1 that contains the transmembrane domain of CD43 roll less stably on P-selectin than cells expressing wild-type dimeric PSGL-1 [61]. P-selectin forms dimers and oligomers, probably through interactions with the transmembrane domains [55, 83]. All these changes delay dissociation of the last bonds at the trailing edge of the cell, where force is applied. This allows more time for new bonds to form at the leading edge of the cell. The result is slower and more regular rolling motions. As wall shear stress is increased, the velocities of leukocytes rolling on selectins plateau rather than continue to increase. It has been suggested that intrinsic features of the selectin-ligand bond contribute to this "automatic braking system" [84]. However, fixed cells or rigid microspheres displaying PSGL-1 do not stabilize their rolling velocities as wall shear stress increases [70]. This indicates that fixation-sensitive cellular features are critical for this phenomenon. Extension of thin membrane tethers will reduce the force on bonds [76], and enlargement of the contact area through shear-induced cellular deformation may increase the probability of forming bonds [85]. A portion of PSGL-1 is concentrated in lipid rafts through an unknown mechanism [86, 87]. Cholesterol chelators that disrupt rafts impair leukocyte rolling on P-selectin [88]. Although this finding was interpreted as evidence that raft localization of PSGL-1 enhances rolling, the chelators stiffen the cell membrane, which as noted above can also impair rolling [70]. Whether rafts affect the rolling function of PSGL-1 requires further study.

Signaling through PSGL-1

There are a large number of reports that engagement of PSGL-1 transduces signals into leukocytes [89–104]. The bulk of these studies suggest that signaling through PSGL-1 requires integration with signaling through receptors for chemokines or other mediators to elicit effector responses such as integrin activation, chemokine or cytokine synthesis, production of oxygen radicals, or generation of a procoagulant cell surface. In at least some instances, partial activation of leukocytes during the isolation procedure may have accounted for the apparent ability of PSGL-1 to propagate effector responses in the absence of a second agonist [105]. A limitation of many papers is that signaling was induced through cross-linking of PSGL-1 with antibodies or selectins over a time period of minutes to hours. Such studies might be relevant for adhesion of activated platelets to leukocytes through P-selectin–PSGL-1 interactions, where such adhesion may be sustained because of the higher P-selectin density on platelets. The contribution of signaling during the rapidly reversible interactions of PSGL-1 with selectins during rolling adhesion has been less well documented. Despite these limitations, the available data suggest that signaling through PSGL-1 may, depending on the context, be biologically important.

The cytoplasmic domain of PSGL-1 does not have canonical sequences for tyrosine phosphorylation or for binding to SH2 or SH3 domains of cytosolic adaptors, kinases, or phosphatases. Engagement of PSGL-1 does not detectably cause phosphorylation of PSGL-1 on either tyrosine or serine/threonine residues [93]. Nevertheless, cross-linking of PSGL-1 causes rapid tyrosine phosphorylation of several cytoplasmic proteins, including the ERK family of mitogen-activated protein kinases [93]. Another of these tyrosine-phosphorylated proteins may be the kinase Syk. Cross-linking of PSGL-1 on myeloid and lymphoid cells activates Syk [98]. It was suggested that the cytoplasmic tail of PSGL-1 binds to and activates ERM proteins, enabling tyrosine phosphorylation of an ITAM motif on the ERM proteins that then binds to and activates Syk. Co-precipitation of PSGL-1 and ERM proteins and of ERM proteins and Syk were demonstrated. The authors did not demonstrate a ternary complex of the three proteins or confirm tyrosine phosphorylation of the ERM proteins to prove that they were activated. Nevertheless, Syk activation is a plausible mechanism for downstream signaling events such as gene transcription and integrin activation, which are known to occur after ligand binding to receptors with ITAM motifs in their cytoplasmic domains [106]. Inhibition of Syk reportedly inhibits rolling of neutrophils and transfected cells on P-selectin, although no mechanism for how Syk might modulate rolling was proposed [88]. Src kinases might be the proximal activators of ERM proteins, since at least a portion of PSGL-1 is in rafts where Src kinases are concentrated [86, 87]. Stirred neutrophils and activated platelets form aggregates through binding of P-selectin to PSGL-1; binding of β_2 integrins to platelet ligands then stabilizes the aggregates [107]. Stable adhesion requires the action of Src kinases that promote outside-in signaling through integrins, but a contribution to inside-out activation of integrins following PSGL-1 engagement cannot be excluded. PSGL-1-mediated rolling on P-selectin enhances β_2 integrin-dependent adhesion of Th1 cells on ICAM-1. Inhibitors of protein kinase C block this effect, and cross-linking of PSGL-1 activates several protein kinase C isoforms [102]. The possible intermediate steps between selectin binding to PSGL-1, protein kinase C activation, Syk activation, and inside-out signaling of integrins have not been resolved.

Physiological and pathological functions of PSGL-1–selectin interactions

The contributions of PSGL-1 to leukocyte adhesion *in vivo* have been documented in numerous studies, mostly in murine models. Studies with blocking mAbs to the N-terminal region of PSGL-1 initially demonstrated that PSGL-1 is the dominant ligand for mediating leukocyte rolling on P-selectin on inflamed endothelial cells *in vivo* [108, 109] (Fig. 2). Targeted disruption of the gene encoding murine PSGL-1 confirmed these observations; very few PSGL-1-deficient leukocytes roll on P-selectin on activated venules, and those that do, roll very rapidly and irregularly [110, 111]. PSGL-1-deficient leukocytes exhibit reduced tethering to E-selectin in cytokine-activated venules; however, those leukocytes that do tether to E-selectin roll with velocities equivalent to those of wild-type leukocytes [111]. Thus, PSGL-1 contributes to tethering to but not rolling on E-selectin, demonstrating that other E-selectin ligands are required for rolling (Fig. 2). One of the latter may be CD44 [112]. Studies with blocking mAbs to PSGL-1 and with PSGL-1-deficient mice revealed that PSGL-1 is the dominant ligand that mediates L-selectin-dependent rolling on inflamed venules [113]. Bone marrow transplantation experiments demonstrated that leukocyte-expressed PSGL-1 serves as the main L-selectin ligand. The majority of observed L-selectin-dependent leukocyte rolling is between free-flowing leukocytes and already adherent leukocytes or possibly leukocyte fragments, followed by E-selectin-dependent leukocyte rolling along the endothelium [113] (Fig. 2). L-selectin binding to PSGL-1 initiates tethering events that enable L-selectin-independent leukocyte-endothelial interactions. These studies suggest that leukocyte-derived PSGL-1 is primarily responsible for L-selectin-dependent adhesion during inflammation. However, mAbs to PSGL-1 reduce intestinal inflammation in a murine model of Crohn's disease, where endothelial cells were shown to express PSGL-1 [17]. Thus PSGL-1 expressed on endothelial cells may support L-selectin-dependent leukocyte rolling in some types of inflammation (Fig. 2). Platelets have been reported to express PSGL-1 that mediates direct rolling of platelets on P-selectin expressed on endothelial cells *in vivo* [16]. Nonetheless, whether platelets express PSGL-1 is controversial. Because activated platelets also adhere to leukocytes through P-selectin–PSGL-1 interactions, platelets might use this mechanism to roll on leukocyte fragments that have previously adhered to the endothelial surface, as noted above for some L-selectin-dependent interactions [113].

Consistent with the observed contributions of PSGL-1 to leukocyte adhesion to each of the selectins, blocking PSGL-1 function or targeted deletion of PSGL-1 reduces inflammation or thrombosis in models that were previously shown to depend on one or more of the selectins. In particular, the effects of PSGL-1 deficiency largely resemble those of P-selectin deficiency in P-selectin-dependent models of inflammation. PSGL-1 also contributes to E-selectin-dependent inflammation [114]. The reduced inflammation in PSGL-1-deficient mice is not as pronounced as in E-selectin-deficient mice, consistent with PSGL-1 not being the only E-selectin

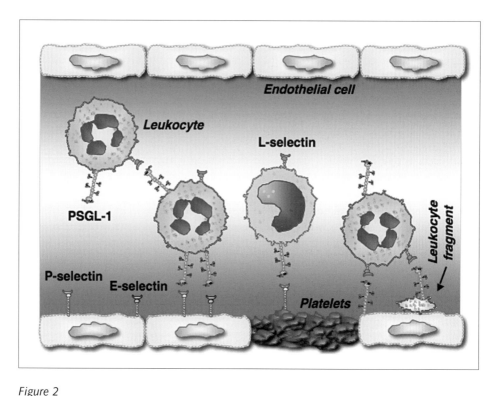

Figure 2

Multicellular interactions mediated by binding of PSGL-1 to selectins

Interactions of PSGL-1 with P-selectin promote leukocyte rolling on activated platelets and endothelial cells under flow. Interactions of PSGL-1 with L-selectin mediate leukocyte rolling on adherent leukocytes or leukocyte fragments, which amplifies leukocyte recruitment to vascular surfaces. At some sites of inflammation, L-selectin on leukocytes may interact with PSGL-1 expressed on endothelial cells. PSGL-1 mediates the initial tethering of flowing leukocytes to E-selectin on endothelial cells but is not required for subsequent rolling on E-selectin. For clarity, PSGL-1 is depicted as a monomer instead of a dimer. Not shown are PSGL-1–P-selectin interactions that recruit leukocyte microparticles to platelet thrombi or form platelet-leukocyte aggregates.

ligand on leukocytes. One of the more novel aspects of interactions of PSGL-1 with P-selectin is the potential to augment thrombosis by recruiting tissue factor-rich leukocyte microparticles to growing platelet thrombi [115, 116]. Mice with procoagulant phenotypes tend to have higher plasma levels of soluble P-selectin, which is proteolytically shed from the surfaces of activated platelets and endothelial cells [115, 117]. Generation of soluble P- and E-selectin in the circulation results from PSGL-1-dependent interactions of leukocytes with P-selectin, as PSGL-1-deficient mice have markedly reduced levels of soluble selectins in plasma [118]. How PSGL-1-dependent leukocyte adhesion causes proteolytic cleavage of P- and E-selectin is not known.

The endothelial cells of bone marrow constitutively express both P- and E-selectin, which contribute to the homing of circulating hematopoietic stem and progenitor cells to bone marrow, where they take up residence in specialized niches [119]. In addition to its role as the major P-selectin ligand, PSGL-1 on hematopoietic stem/progenitor cells serves as one of the E-selectin ligands that support homing [120]. Stem/progenitor cells from human cord blood may have impaired homing to bone marrow because of incomplete glycosylation of PSGL-1 [121]. Forced α1-3-fucosylation of the surface of these cells improves their ability to interact with both P- and E-selectin [122, 123] and enhances homing to bone marrows of irradiated NOD/SCID mice [122]. The endothelial cells of the thymus also constitutively express P-selectin, and interactions with PSGL-1 on bone marrow-derived thymic progenitors mediate homing of these cells to the thymus [124].

Interactions of PSGL-1 with other molecules

In vitro, the N terminus of PSGL-1 binds to some chemokines. Binding was shown to require tyrosine sulfation but not specific glycosylation of PSGL-1 [125]. It remains possible that glycosylation could positively or negatively modulate chemokine binding under some conditions. The ability of PSGL-1 to bind chemokines suggests new possibilities for how it could regulate leukocyte trafficking *in vivo*. For example, chemokines might transiently bind to PSGL-1 and then "transfer" to G protein-coupled chemokine receptors that activate leukocyte integrins. Binding of chemokines to PSGL-1 might explain why mAbs to the N terminus of PSGL-1 are more protective than a combination of mAbs to all three selectins in a murine model of Crohn's disease [17].

PSGL-1 has been reported to bind directly to von Willebrand factor and to support transient adhesion of transfected cells expressing PSGL-1 to immobilized von Willebrand factor under flow [126]. Although the physiological significance of this interaction has not been defined, it might provide another connection between the hemostatic and inflammatory responses to tissue injury.

In vitro, the proteoglycan versican binds to a membrane-proximal segment of PSGL-1 that does not require glycosylation [127]. The interaction was originally discovered in a yeast two-hybrid screen with versican as bait, and its biological relevance has not been demonstrated.

Pathogens use molecular mimicry of host molecules to facilitate infection and avoid host immune responses. Such adaptive mechanisms of pathogens have been extended to PSGL-1. The intracellular bacterium *Anaplasma phagocytophilum* adheres to and invades human neutrophils through interactions with PSGL-1 [128], but in a manner distinct from how P-selectin binds [129]. The bacteria appear to express distinct adhesins that bind cooperatively to a nonsulfated N-terminal peptide of human PSGL-1 and to the sLe^x determinant on PSGL-1 or on other glycoproteins [129]. A toxin from *Staphylococcus aureus* binds to PSGL-1 [130]. The binding is probably to an N-terminal region of PSGL-1, since the toxin inhibits adhesion of neutrophils to P-selectin. This suggests that secretion of the toxin reduces P-selectin-mediated inflammatory responses to the bacteria.

Conclusions

PSGL-1 has important interactions with each of the selectins in both health and disease. Its biological functions in cell adhesion and signaling are among the best characterized for a mucin, and it seems likely that additional functions will be revealed as its interactions with selectins and other molecules continue to be explored.

References

1   Vestweber D, Blanks JE (1999) Mechanisms that regulate the function of the selectins and their ligands. *Physiol Rev* 79: 181–213

2   McEver RP (2002) Selectins: lectins that initiate cell adhesion under flow. *Curr Opin Cell Biol* 14: 581–586

3   Somers WS, Tang J, Shaw GD, Camphausen RT (2000) Insights into the molecular basis of leukocyte tethering and rolling revealed by structures of P- and E-selectin bound to SLe(X) and PSGL-1. *Cell* 103: 467–479

4   Homeister JW, Thall AD, Petryniak B, Maly P, Rogers CE, Smith PL, Kelly RJ, Gersten KM, Askari SW, Cheng GY et al (2001) The α(1,3)fucosyltransferases FucT-IV and FucT-VII exert collaborative control over selectin-dependent leukocyte recruitment and lymphocyte homing. *Immunity* 15: 115–126

5   Smithson G, Rogers CE, Smith PL, Scheidegger EP, Petryniak B, Myers JT, Kim DSL, Homeister JW, Lowe JB (2001) Fuc-TVII is required for T helper 1 and T cytotoxic 1 lymphocyte selectin ligand expression and recruitment in inflammation, and together with Fuc-TIV regulates naive T cell trafficking to lymph nodes. *J Exp Med* 194: 601–614

6   Moore KL, Stults NL, Diaz S, Smith DL, Cummings RD, Varki A, McEver RP (1992) Identification of a specific glycoprotein ligand for P-selectin (CD62) on myeloid cells. *J Cell Biol* 118: 445–456

7   Norgard KE, Moore KL, Diaz S, Stults NL, Ushiyama S, McEver RP, Cummings RD, Varki A (1993) Characterization of a specific ligand for P-selectin on myeloid cells. A minor glycoprotein with sialylated O-linked oligosaccharides. *J Biol Chem* 268: 12764–12774

8   Sako D, Chang XJ, Barone KM, Vachino G, White HM, Shaw G, Veldman GM, Bean KM, Ahern TJ, Furie B et al (1993) Expression cloning of a functional glycoprotein ligand for P-selectin. *Cell* 75: 1179–1186

9   Veldman GM, Bean KM, Cumming DA, Eddy RL, Sait SNJ, Shows TB (1995) Genomic organization and chromosomal localization of the gene encoding human P-selectin glycoprotein ligand. *J Biol Chem* 270: 16470–16475

10 Li F, Erickson HP, James JA, Moore KL, Cummings RD, McEver RP (1996) Visualization of P-selectin glycoprotein ligand-1 as a highly extended molecule and mapping of protein epitopes for monoclonal antibodies. *J Biol Chem* 271: 6342–6348

11 Snapp KR, Craig R, Herron M, Nelson RD, Stoolman LM, Kansas GS (1998) Dimerization of P-selectin glycoprotein ligand-1 (PSGL-1) required for optimal recognition of P-selectin. *J Cell Biol* 142: 263–270

12 Epperson TK, Patel KD, McEver RP, Cummings RD (2000) Noncovalent association of P-selectin glycoprotein ligand-1 and minimal determinants for binding to P-selectin. *J Biol Chem* 275: 7839–7853

13 Yang J, Galipeau J, Kozak CA, Furie BC, Furie B (1996) Mouse P-selectin glycoprotein ligand-1: Molecular cloning, chromosomal localization, and expression of a functional P-selectin receptor. *Blood* 87: 4176–4186

14 Moore KL, Patel KD, Bruehl RE, Fugang L, Johnson DA, Lichenstein HS, Cummings RD, Bainton DF, McEver RP (1995) P-selectin glycoprotein ligand-1 mediates rolling of human neutrophils on P-selectin. *J Cell Biol* 128: 661–671

15 Laszik Z, Jansen PJ, Cummings RD, Tedder TF, McEver RP, Moore KL (1996) P-selectin glycoprotein ligand-1 is broadly expressed in cells of myeloid, lymphoid, and dendritic lineage and in some nonhematopoietic cells. *Blood* 88: 3010–3021

16 Frenette PS, Denis CV, Weiss L, Jurk K, Subbarao S, Kehrel B, Hartwig JH, Vestweber D, Wagner DD (2000) P-selectin glycoprotein ligand 1 (PSGL-1) is expressed on platelets and can mediate platelet-endothelial interactions *in vivo*. *J Exp Med* 191: 1413–1422

17 Rivera-Nieves J, Burcin TL, Olson TS, Morris MA, McDuffie M, Cominelli F, Ley K (2006) Critical role of endothelial P-selectin glycoprotein ligand 1 in chronic murine ileitis. *J Exp Med* 203: 907–917

18 da Costa Martins P, Garcia-Vallejo JJ, van Thienen JV, Fernandez-Borja M, van Gils J, Beckers C, Horrevoets AJ, Hordijk PL, Zwaginga JJ (2007) P-selectin glycoprotein ligand-1 Is expressed on endothelial cells and mediates monocyte adhesion to activated endothelium. *Arterioscler Thromb Vasc Biol* 27: 1023–1029

19 Wilkins PP, McEver RP, Cummings RD (1996) Structures of the O-glycans on P-selectin glycoprotein ligand-1 from HL–60 cells. *J Biol Chem* 271: 18732–18742

20 Kobzdej MMA, Leppänen A, Ramachandran V, Cummings RD, McEver RP (2002) Discordant expression of selectin ligands and sialyl Lewis x-related epitopes on murine myeloid cells. *Blood* 100: 485–494

21 Walcheck B, Moore KL, McEver RP, Kishimoto TK (1996) Neutrophil-neutrophil interactions under hydrodynamic shear stress involve L-selectin and PSGL-1: a mechanism

that amplifies initial leukocyte accumulation on P-selectin *in vitro*. *J Clin Invest* 98: 1081–1087

22 Tu LL, Chen AJ, Delahunty MD, Moore KL, Watson SR, McEver RP, Tedder TF (1996) L-selectin binds to P-selectin glycoprotein ligand-1 on leukocytes – Interactions between the lectin, epidermal growth factor, and consensus repeat domains of the selectins determine ligand binding specificity. *J Immunol* 157: 3995–4004

23 Spertini O, Cordey AS, Monai N, Giuffre L, Schapira M (1996) P-selectin glycoprotein ligand-1 (PSGL-1) is a ligand for L-selectin on neutrophils, monocytes and CD34$^+$ hematopoietic progenitor cells. *J Cell Biol* 135: 523–531

24 Guyer DA, Moore KL, Lynam E, Schammel CMG, Rogelj S, McEver RP, Sklar LA (1996) P-selectin glycoprotein ligand-1 (PSGL-1) is a ligand for L-selectin in neutrophil aggregation. *Blood* 88: 2415–2421

25 Borges E, Tietz W, Steegmaier M, Moll T, Hallmann R, Hamann A, Vestweber D (1997) P-selectin glycoprotein ligand-1 (PSGL-1) on T helper 1 but not on T helper 2 cells binds to P-selectin and supports migration into inflamed skin. *J Exp Med* 185: 573–578

26 Alon R, Fuhlbrigge RC, Finger EB, Springer TA (1996) Interactions through L-selectin between leukocytes and adherent leukocytes nucleate rolling adhesions on selectins and VCAM-1 in shear flow. *J Cell Biol* 135: 849–865

27 Wilkins PP, Moore KL, McEver RP, Cummings RD (1995) Tyrosine sulfation of P-selectin glycoprotein ligand-1 is required for high affinity binding to P-selectin. *J Biol Chem* 270: 22677–22680

28 Sako D, Comess KM, Barone KM, Camphausen RT, Cumming DA, Shaw GD (1995) A sulfated peptide segment at the amino terminus of PSGL-1 is critical for P-selectin binding. *Cell* 83: 323–331

29 Pouyani T, Seed B (1995) PSGL-1 recognition of P-selectin is controlled by a tyrosine sulfation consensus at the PSGL-1 amino terminus. *Cell* 83: 333–343

30 Li F, Wilkins PP, Crawley S, Weinstein J, Cummings RD, McEver RP (1996) Post-translational modifications of recombinant P-selectin glycoprotein ligand-1 required for binding to P- and E-selectin. *J Biol Chem* 271: 3255–3264

31 Liu WJ, Ramachandran V, Kang J, Kishimoto TK, Cummings RD, McEver RP (1998) Identification of N-terminal residues on P-selectin glycoprotein ligand-1 required for binding to P-selectin. *J Biol Chem* 273: 7078–7087

32 Ramachandran V, Nollert MU, Qiu H, Liu WJ, Cummings RD, Zhu C, McEver RP (1999) Tyrosine replacement in P-selectin glycoprotein ligand-1 affects distinct kinetic and mechanical properties of bonds with P- and L-selectin. *Proc Natl Acad Sci USA* 96: 13771–13776

33 Ellies LG, Tsuboi S, Petryniak B, Lowe JB, Fukuda M, Marth JD (1998) Core 2 oligosaccharide biosynthesis distinguishes between selectin ligands essential for leukocyte homing and inflammation. *Immunity* 9: 881–890

34 Ouyang YB, Crawley JT, Aston CE, Moore KL (2002) Reduced body weight and increased postimplantation fetal death in tyrosylprotein sulfotransferase–1-deficient mice. *J Biol Chem* 277: 23781–23787

35 Borghei A, Ouyang YB, Westmuckett AD, Marcello MR, Landel CP, Evans JP, Moore KL (2006) Targeted disruption of tyrosylprotein sulfotransferase-2, an enzyme that catalyzes post-translational protein tyrosine O-sulfation, causes male infertility. *J Biol Chem* 281: 9423–9431

36 Leppänen A, Mehta P, Ouyang YB, Ju T, Helin J, Moore KL, van Die I, Canfield WM, McEver RP, Cummings RD (1999) A novel glycosulfopeptide binds to P-selectin and inhibits leukocyte adhesion to P-selectin. *J Biol Chem* 274: 24838–24848

37 Leppänen A, White SP, Helin J, McEver RP, Cummings RD (2000) Binding of glyco-sulfopeptides to P-selectin requires stereospecific contributions of individual tyrosine sulfate and sugar residues. *J Biol Chem* 275: 39569–39578

38 Leppänen A, Yago T, Otto VI, McEver RP, Cummings RD (2003) Model glycosulfo-peptides from P-selectin glycoprotein ligand-1 require tyrosine sulfation and a core 2-branched O-glycan to bind to L-selectin. *J Biol Chem* 278: 26391–26400

39 Leppänen A, Penttilä L, Renkonen O, McEver RP, Cummings RD (2002) Glycosulfo-peptides with O-glycans containing sialylated and polyfucosylated polylactosamine bind with low affinity to P-selectin. *J Biol Chem* 277: 39749–39759

40 Mehta P, Cummings RD, McEver RP (1998) Affinity and kinetic analysis of P-selectin binding to P-selectin glycoprotein ligand-1. *J Biol Chem* 273: 32506–32513

41 Croce K, Freedman SJ, Furie BC, Furie B (1998) Interaction between soluble P-selectin and soluble P-selectin glycoprotein ligand 1: Equilibrium binding analysis. *Biochemistry* 37: 16472–16480

42 Poppe L, Brown GS, Philo JS, Nikrad PV, Shah BH (1997) Conformation of sLex tet-rasaccharide, free in solution and bound to E-, P-, and L-selectin. *J Am Chem Soc* 119: 1727–1736

43 Brandley BK, Kiso M, Abbas S, Nikrad P, Srivasatava O, Foxall C, Oda Y, Hasegawa A (1993) Structure-function studies on selectin carbohydrate ligands. Modifications to fucose, sialic acid and sulphate as a sialic acid replacement. *Glycobiology* 3: 633–639

44 Koenig A, Jain R, Vig R, Norgard-Sumnicht KE, Matta KL, Varki A (1997) Selectin inhibition: Synthesis and evaluation of novel sialylated, sulfated and fucosylated oligo-saccharides, including the major capping group of GlyCAM-1. *Glycobiology* 7: 79–93

45 Xia L, Ramachandran V, McDaniel JM, Nguyen KN, Cummings RD, McEver RP (2003) N-terminal residues in murine P-selectin glycoprotein ligand-1 required for bind-ing to murine P-selectin. *Blood* 101: 552–559

46 Moore KL, Eaton SF, Lyons DE, Lichenstein HS, Cummings RD, McEver RP (1994) The P-selectin glycoprotein ligand from human neutrophils displays sialylated, fucosylated, O-linked poly-N-acetyllactosamine. *J Biol Chem* 269: 23318–23327

47 Goetz DJ, Greif DM, Ding H, Camphausen RT, Howes S, Comess KM, Snapp KR, Kansas GS, Luscinskas FW (1997) Isolated P-selectin glycoprotein ligand-1 dynamic adhesion to P- and E-selectin. *J Cell Biol* 137: 509–519

48 Patel KD, Moore KL, Nollert MU, McEver RP (1995) Neutrophils use both shared and distinct mechanisms to adhere to selectins under static and flow conditions. *J Clin Invest* 96: 1887–1896

49 Vachino G, Chang XJ, Veldman GM, Kumar R, Sako D, Fouser LA, Berndt MC, Cumming DA (1995) P-selectin glycoprotein ligand-1 is the major counter-receptor for P-selectin on stimulated T cells and is widely distributed in non-functional form on many lymphocytic cells. *J Biol Chem* 270: 21966–21974

50 Ley K, Kansas GS (2004) Selectins in T-cell recruitment to non-lymphoid tissues and sites of inflammation. *Nat Rev Immunol* 4: 325–335

51 André P, Spertini O, Guia S, Rihet P, Dignat-George F, Brailly H, Sampol J, Anderson PJ, Vivier E (2000) Modification of P-selectin glycoprotein ligand-1 with a natural killer cell-restricted sulfated lactosamine creates an alternate ligand for L-selectin. *Proc Natl Acad Sci USA* 97: 3400–3405

52 Schakel K, Kannagi R, Kniep B, Goto Y, Mitsuoka C, Zwirner J, Soruri A, von Kietzell M, Rieber E (2002) 6-Sulfo LacNAc, a novel carbohydrate modification of PSGL-1, defines an inflammatory type of human dendritic cells. *Immunity* 17: 289–301

53 Rosen SD (2004) Ligands for L-selectin: homing, inflammation, and beyond. *Annu Rev Immunol* 22: 129–156

54 Lawrence MB, Springer TA (1991) Leukocytes roll on a selectin at physiologic flow rates: Distinction from and prerequisite for adhesion through integrins. *Cell* 65: 859–873

55 Ushiyama S, Laue TM, Moore KL, Erickson HP, McEver RP (1993) Structural and functional characterization of monomeric soluble P-selectin and comparison with membrane P-selectin. *J Biol Chem* 268: 15229–15237

56 Hattori R, Hamilton KK, Fugate RD, McEver RP, Sims PJ (1989) Stimulated secretion of endothelial von Willebrand factor is accompanied by rapid redistribution to the cell surface of the intracellular granule membrane protein GMP-140. *J Biol Chem* 264: 7768–7771

57 Yeo EL, Sheppard JAI, Feuerstein IA (1994) Role of P-selectin and leukocyte activation in polymorphonuclear cell adhesion to surface adherent activated platelets under physiologic shear conditions (an injury vessel wall model). *Blood* 83: 2498–2507

58 McEver RP, Martin MN (1984) A monoclonal antibody to a membrane glycoprotein binds only to activated platelets. *J Biol Chem* 259: 9799–9804

59 Hsu-Lin SC, Berman CL, Furie BC, August D, Furie B (1984) A platelet membrane protein expressed during platelet activation and secretion. Studies using a monoclonal antibody specific for thrombin-activated platelets. *J Biol Chem* 259: 9121–9126

60 Alon R, Hammer DA, Springer TA (1995) Lifetime of the P-selectin: carbohydrate bond and its response to tensile force in hydrodynamic flow. *Nature* 374: 539–542

61 Ramachandran V, Yago T, Epperson TK, Kobzdej MMA, Nollert MU, Cummings RD, Zhu C, McEver RP (2001) Dimerization of a selectin and its ligand stabilizes cell rolling and enhances tether strength in shear flow. *Proc Natl Acad Sci USA* 98: 10166–10171

62 Dembo M, Torney DC, Saxman K, Hammer D (1988) The reaction-limited kinetics of membrane-to-surface adhesion and detachment. *Proc R Soc Lond B Biol Sci* 234: 55–83

63   Marshall BT, Long M, Piper JW, Yago T, McEver RP, Zhu C (2003) Direct observation of catch bonds involving cell-adhesion molecules. *Nature* 423: 190–193

64   Finger EB, Puri KD, Alon R, Lawrence MB, Von Andrian UH, Springer TA (1996) Adhesion through L-selectin requires a threshold hydrodynamic shear. *Nature* 379: 266–269

65   Lawrence MB, Kansas GS, Kunkel EJ, Ley K (1997) Threshold levels of fluid shear promote leukocyte adhesion through selectins (CD62L, P, E). *J Cell Biol* 136: 717–727

66   Sarangapani KK, Yago T, Klopocki AG, Lawrence MB, Fieger CB, Rosen SD, McEver RP, Zhu C (2004) Low force decelerates L-selectin dissociation from P-selectin glycoprotein ligand-1 and endoglycan. *J Biol Chem* 279: 2291–2298

67   Yago T, Wu J, Wey CD, Klopocki AG, Zhu C, McEver RP (2004) Catch bonds govern adhesion through L-selectin at threshold shear. *J Cell Biol* 166: 913–923

68   Lou J, Yago T, Klopocki AG, Mehta P, Chen W, Zarnitsyna VI, Bovin NV, Zhu C, McEver RP (2006) Flow-enhanced adhesion regulated by a selectin interdomain hinge. *J Cell Biol* 174: 1107–1117

69   Phan UT, Waldron TT, Springer TA (2006) Remodeling of the lectin-EGF-like domain interface in P- and L-selectin increases adhesiveness and shear resistance under hydrodynamic force. *Nat Immunol* 7: 883–889

70   Yago T, Leppänen A, Qiu H, Marcus WD, Nollert MU, Zhu C, Cummings RD, McEver RP (2002) Distinct molecular and cellular contributions to stabilizing selectin-mediated rolling under flow. *J Cell Biol* 158: 787–799

71   Von Andrian UH, Hasslen SR, Nelson RD, Erlandsen SL, Butcher EC (1995) A central role for microvillous receptor presentation in leukocyte adhesion under flow. *Cell* 82: 989–999

72   Bruehl RE, Moore KL, Lorant DE, Borregaard N, Zimmerman GA, McEver RP, Bainton DF (1997) Leukocyte activation induces surface redistribution of P-selectin glycoprotein ligand-1. *J Leukoc Biol* 61: 489–499

73   Patel KD, Nollert MU, McEver RP (1995) P-selectin must extend a sufficient length from the plasma membrane to mediate rolling of neutrophils. *J Cell Biol* 131: 1893–1902

74   Huang J, Chen J, Chesla SE, Yago T, Mehta P, McEver RP, Zhu C, Long M (2004) Quantifying the effects of molecular orientation and length on two-dimensional receptor-ligand binding kinetics. *J Biol Chem* 279: 44915–22923

75   Shao JY, Ting-Beall HP, Hochmuth RM (1998) Static and dynamic lengths of neutrophil microvilli. *Proc Natl Acad Sci USA* 95: 6797–6802

76   Schmidtke DW, Diamond SL (2000) Direct observation of membrane tethers formed during neutrophil attachment to platelets or P-selectin under physiological flow. *J Cell Biol* 149: 719–729

77   Ramachandran V, Williams M, Yago T, Schmidtke DW, McEver RP (2004) Dynamic alterations of membrane tethers stabilize leukocyte rolling on P-selectin. *Proc Natl Acad Sci USA* 101: 13519–13524

78   Setiadi H, Sedgewick G, Erlandsen SL, McEver RP (1998) Interactions of the cytoplas-

mic domain of P-selectin with clathrin-coated pits enhance leukocyte adhesion under flow. *J Cell Biol* 142: 859–871

79 Setiadi H, McEver RP (2003) Signal-dependent distribution of cell surface P-selectin in clathrin-coated pits affects leukocyte rolling under flow. *J Cell Biol* 163: 1385–1395

80 Serrador JM, Urzainqui A, Alonso-Lebrero JL, Cabrero JR, Montoya MC, Vicente-Manzanares M, Yanez-Mo M, Sanchez-Madrid F (2002) A juxta-membrane amino acid sequence of P-selectin glycoprotein ligand-1 is involved in moesin binding and ezrin/radixin/moesin-directed targeting at the trailing edge of migrating lymphocytes. *Eur J Immunol* 32: 1560–1566

81 Snapp KR, Heitzig CE, Kansas GS (2002) Attachment of the PSGL-1 cytoplasmic domain to the actin cytoskeleton is essential for leukocyte rolling on P-selectin. *Blood* 99: 4494–4502

82 Marcus WD, McEver RP, Zhu C (2004) Forces required to initiate membrane tether extrusion from cell surface depend on cell type but not on the surface molecule. *Mech Chem Biosyst* 1: 245–251

83 Barkalow FJ, Barkalow KL, Mayadas TN (2000) Dimerization of P-selectin in platelets and endothelial cells. *Blood* 96: 3070–3077

84 Chen SQ, Springer TA (1999) An automatic braking system that stabilizes leukocyte rolling by an increase in selectin bond number with shear. *J Cell Biol* 144: 185–200

85 Lei X, Lawrence MB, Dong C (1999) Influence of cell deformation on leukocyte rolling adhesion in shear flow. *J Biomech Eng* 121: 636–643

86 Handa K, Jacobs F, Longenecker BM, Hakomori SI (2001) Association of MUC–1 and PSGL-1 with low-density microdomain in T-lymphocytes: a preliminary note. *Biochem Biophys Res Commun* 285: 788–794

87 Del Conde I, Shrimpton CN, Thiagarajan P, Lopez JA (2005) Tissue-factor-bearing microvesicles arise from lipid rafts and fuse with activated platelets to initiate coagulation. *Blood* 106: 1604–1611

88 Abbal C, Lambelet M, Bertaggia D, Gerbex C, Martinez M, Arcaro A, Schapira M, Spertini O (2006) Lipid raft adhesion receptors and Syk regulate selectin-dependent rolling under flow conditions. *Blood* 108: 3352–3359

89 Lorant DE, Patel KD, McIntyre TM, McEver RP, Prescott SM, Zimmerman GA (1991) Coexpression of GMP-140 and PAF by endothelium stimulated by histamine or thrombin: A juxtacrine system for adhesion and activation of neutrophils. *J Cell Biol* 115: 223–234

90 Lorant DE, Topham MK, Whatley RE, McEver RP, McIntyre TM, Prescott SM, Zimmerman GA (1993) Inflammatory roles of P-selectin. *J Clin Invest* 92: 559–570

91 Elstad MR, La Pine TR, Cowley FS, McEver RP, McIntyre TM, Prescott SM, Zimmerman GA (1995) P-selectin regulates platelet-activating factor synthesis and phagocytosis by monocytes. *J Immunol* 155: 2109–2122

92 Blanks JE, Moll T, Eytner R, Vestweber D (1998) Stimulation of P-selectin glycoprotein ligand-1 on mouse neutrophils activates β2-integrin mediated cell attachment to ICAM-1. *Eur J Immunol* 28: 433–443

93   Hidari KI-PJ, Weyrich AS, Zimmerman GA, McEver RP (1997) Engagement of P-selectin glycoprotein ligand-1 enhances tyrosine phosphorylation and activates mitogen-activated protein kinases in human neutrophils. *J Biol Chem* 272: 28750–28756

94   Celi A, Pellegrini G, Lorenzet R, De Blasi A, Ready N, Furie BC, Furie B (1994) P-selectin induces the expression of tissue factor on monocytes. *Proc Natl Acad Sci USA* 91: 8767–8771

95   Damle NK, Klussman K, Dietsch MR, Mohagheghpour N, Aruffo A (1992) GMP-140 (P-selectin/CD62) binds to chronically stimulated but not resting CD4⁺ T lymphocytes and regulates their production of proinflammatory cytokines. *Eur J Immunol* 22: 1789–1793

96   Haller H, Kunzendorf U, Sacherer K, Lindschau C, Walz G, Distler A, Luft FC (1997) T cell adhesion to P-selectin induces tyrosine phosphorylation of pp125 focal adhesion kinase and other substrates. *J Immunol* 158: 1061–1067

97   Evangelista V, Manarini S, Sideri R, Rotondo S, Martelli N, Piccoli A, Totani L, Piccardoni P, Vestweber D, de Gaetano G, Cerletti C (1999) Platelet/polymorphonuclear leukocyte interaction: P-selectin triggers protein-tyrosine phosphorylation-dependent CD11b/CD18 adhesion: Role of PSGL-1 as a signaling molecule. *Blood* 93: 876–885

98   Urzainqui A, Serrador JM, Viedma F, Yanez-Mo M, Rodriguez A, Corbi AL, Alonso-Lebrero JL, Luque A, Deckert M, Vazquez J, Sanchez-Madrid F (2002) ITAM-based interaction of ERM proteins with Syk mediates signaling by the leukocyte adhesion receptor PSGL-1. *Immunity* 17: 401–412

99   Mahoney TS, Weyrich AS, Dixon DA, McIntyre T, Prescott SM, Zimmerman GA (2001) Cell adhesion regulates gene expression at translational checkpoints in human myeloid leukocytes. *Proc Natl Acad Sci USA* 98: 10284–10289

100   Ma YQ, Plow EF, Geng JG (2004) P-selectin binding to P-selectin glycoprotein ligand-1 induces an intermediate state of αMβ2 activation and acts cooperatively with extracellular stimuli to support maximal adhesion of human neutrophils. *Blood* 104: 2549–2556

101   del Conde I, Nabi F, Tonda R, Thiagarajan P, Lopez JA, Kleiman NS (2005) Effect of P-selectin on phosphatidylserine exposure and surface-dependent thrombin generation on monocytes. *Arterioscler Thromb Vasc Biol* 25: 1065–1070

102   Atarashi K, Hirata T, Matsumoto M, Kanemitsu N, Miyasaka M (2005) Rolling of Th1 cells *via* P-selectin glycoprotein ligand-1 stimulates LFA-1-mediated cell binding to ICAM-1. *J Immunol* 174: 1424–1432

103   Woollard KJ, Kling D, Kulkarni S, Dart AM, Jackson S, Chin-Dusting J (2006) Raised plasma soluble P-selectin in peripheral arterial occlusive disease enhances leukocyte adhesion. *Circ Res* 98: 149–156

104   Dixon DA, Tolley ND, Bemis-Standoli K, Martinez ML, Weyrich AS, Morrow JD, Prescott SM, Zimmerman GA (2006) Expression of COX–2 in platelet-monocyte interactions occurs *via* combinatorial regulation involving adhesion and cytokine signaling. *J Clin Invest* 116: 2727–2738

105   Smith ML, Sperandio M, Galkina EV, Ley K (2004) Autoperfused mouse flow chamber

reveals synergistic neutrophil accumulation through P-selectin and E-selectin. *J Leukoc Biol* 76: 985–993

106 Turner M, Schweighoffer E, Colucci F, Di Santo JP, Tybulewicz VL (2000) Tyrosine kinase SYK: essential functions for immunoreceptor signalling. *Immunol Today* 21: 148–154

107 Piccardoni P, Sideri R, Manarini S, Piccoli A, Martelli N, De Gaetano G, Cerletti C, Evangelista V (2001) Platelet/polymorphonuclear leukocyte adhesion: a new role for SRC kinases in Mac-1 adhesive function triggered by P-selectin. *Blood* 98: 108–116

108 Norman KE, Moore KL, McEver RP, Ley K (1995) Leukocyte rolling *in vivo* is mediated by P-selectin glycoprotein ligand-1. *Blood* 86: 4417–4421

109 Borges E, Eytner R, Moll T, Steegmaier M, Campbell MA, Ley K, Mossman H, Vestweber D (1997) The P-selectin glycoprotein ligand-1 is important for recruitment of neutrophils into inflamed mouse peritoneum. *Blood* 90: 1934–1942

110 Yang J, Hirata T, Croce K, Merrill-Skoloff G, Tchernychev B, Williams E, Flaumenhaft R, Furie BC, Furie B (1999) Targeted gene disruption demonstrates that P-selectin glycoprotein ligand 1 (PSGL-1) is required for P-selectin-mediated but not E-selectin-mediated neutrophil rolling and migration. *J Exp Med* 190: 1769–1782

111 Xia L, Sperandio M, Yago T, McDaniel JM, Cummings RD, Pearson-White S, Ley K, McEver RP (2002) P-selectin glycoprotein ligand-1-deficient mice have impaired leukocyte tethering to E-selectin under flow. *J Clin Invest* 109: 939–950

112 Katayama Y, Hidalgo A, Chang J, Peired A, Frenette PS (2005) CD44 is a physiological E-selectin ligand on neutrophils. *J Exp Med* 201: 1183–1189

113 Sperandio M, Smith ML, Forlow SB, Olson TS, Xia L, McEver RP, Ley K (2003) P-selectin glycoprotein ligand-1 mediates L-selectin-dependent leukocyte rolling in venules. *J Exp Med* 197: 1355–1363

114 Hirata T, Merrill-Skoloff G, Aab M, Yang J, Furie BC, Furie B (2000) P-selectin glycoprotein ligand 1 (PSGL-1) is a physiological ligand for E-selectin in mediating T helper 1 lymphocyte migration. *J Exp Med* 192: 1669–1675

115 Hrachovinova I, Cambien B, Hafezi-Moghadam A, Kappelmayer J, Camphausen RT, Widom A, Xia L, Kazazian HH Jr., Schaub RG, McEver RP, Wagner DD (2003) Interaction of P-selectin and PSGL-1 generates microparticles that correct hemostasis in a mouse model of hemophilia A. *Nat Med* 9: 1020–1025

116 Falati S, Liu Q, Gross P, Merrill-Skoloff G, Chou J, Vandendries E, Celi A, Croce K, Furie BC, Furie B (2003) Accumulation of tissue factor into developing thrombi *in vivo* is dependent upon microparticle P-selectin glycoprotein ligand 1 and platelet P-selectin. *J Exp Med* 197: 1585–1598

117 André P, Hartwell D, Hrachovinová I, Saffaripour S, Wagner DD (2000) Pro-coagulant state resulting from high levels of soluble P-selectin in blood. *Proc Natl Acad Sci USA* 97: 13835–13840

118 Bodary PF, Homeister JW, Vargas FB, Wickenheiser KJ, Cudney SS, Bahrou KL, Ohman M, Rabbani AB, Eitzman DT (2007) Generation of soluble P- and E-selectins *in vivo*

is dependent on expression of P-selectin glycoprotein ligand-1. *J Thromb Haemost* 5: 599–603

119 Frenette PS, Subbarao S, Mazo IB, Von Andrian UH, Wagner DD (1998) Endothelial selectins and vascular cell adhesion molecule-1 promote hematopoietic progenitor homing to bone marrow. *Proc Natl Acad Sci USA* 95: 14423–14428

120 Katayama Y, Hidalgo A, Furie BC, Vestweber D, Furie B, Frenette PS (2003) PSGL-1 participates in E-selectin-mediated progenitor homing to bone marrow: evidence for cooperation between E-selectin ligands and α4 integrin. *Blood* 102: 2060–2067

121 Hidalgo A, Weiss LA, Frenette PS (2002) Functional selectin ligands mediating human CD34+ cell interactions with bone marrow endothelium are enhanced postnatally. *J Clin Invest* 110: 559–569

122 Xia L, McDaniel JM, Yago T, Doeden A, McEver RP (2004) Surface fucosylation of human cord blood cells augments binding to P-selectin and E-selectin and enhances engraftment in bone marrow. *Blood* 104: 3091–3096

123 Hidalgo A, Frenette PS (2005) Enforced fucosylation of neonatal CD34+ cells generates selectin ligands that enhance the initial interactions with microvessels but not homing to bone marrow. *Blood* 105: 567–575

124 Rossi FM, Corbel SY, Merzaban JS, Carlow DA, Gossens K, Duenas J, So L, Yi L, Ziltener HJ (2005) Recruitment of adult thymic progenitors is regulated by P-selectin and its ligand PSGL-1. *Nat Immunol* 6: 626–634

125 Hirata T, Furukawa Y, Yang BG, Hieshima K, Fukuda M, Kannagi R, Yoshie O, Miyasaka M (2004) Human P-selectin glycoprotein ligand-1 (PSGL-1) interacts with the skin-associated chemokine CCL27 *via* sulfated tyrosines at the PSGL-1 amino terminus. *J Biol Chem* 279: 51775–51782

126 Pendu R, Terraube V, Christophe OD, Gahmberg CG, de Groot PG, Lenting PJ, Denis CV (2006) P-selectin glycoprotein ligand 1 and β2-integrins cooperate in the adhesion of leukocytes to von Willebrand factor. *Blood* 108: 3746–3752

127 Zheng PS, Vais D, Lapierre D, Liang YY, Lee V, Yang BL, Yang BB (2004) PG-M/versican binds to P-selectin glycoprotein ligand-1 and mediates leukocyte aggregation. *J Cell Sci* 117: 5887–5895

128 Herron MJ, Nelson CM, Larson J, Snapp KR, Kansas GS, Goodman JL (2000) Intracellular parasitism by the human granulocytic ehrlichiosis bacterium through the P-selectin ligand, PSGL-1. *Science* 288: 1653–1656

129 Yago T, Leppanen A, Carlyon JA, Akkoyunlu M, Karmakar S, Fikrig E, Cummings RD, McEver RP (2003) Structurally distinct requirements for binding of P-selectin glycoprotein ligand-1 and sialyl Lewis x to *Anaplasma phagocytophilum* and P-selectin. *J Biol Chem* 278: 37987–37997

130 Bestebroer J, Poppelier MJ, Ulfman LH, Lenting PJ, Denis CV, van Kessel KP, van Strijp JA, de Haas CJ (2007) Staphylococcal superantigen-like 5 binds PSGL-1 and inhibits P-selectin-mediated neutrophil rolling. *Blood* 109: 2936–2943

L-selectin-mediated leukocyte adhesion and migration

Douglas A. Steeber, Hariharan Subramanian, Jamison J. Grailer, Rochelle M. Conway and Traci J. Storey

Department of Biological Sciences, Box 413, University of Wisconsin-Milwaukee, Milwaukee, WI 53201, USA

Introduction

L-selectin, which is expressed by the majority of leukocytes, plays a pivotal role in the generation of rapid and efficient immune responses by mediating the interaction of circulating leukocytes with the vascular endothelium. Specifically, L-selectin functions to mediate the attachment of leukocytes to the vascular wall at sites of inflammation and the attachment of lymphocytes to specialized post-capillary venules termed "high endothelial venules" (HEV) located within secondary lymphoid tissues. This initial adhesive event termed "capture" or "tethering" occurs through selective interaction of L-selectin with its appropriately displayed ligand(s). While this adhesive function has largely defined L-selectin, it is now clear that L-selectin functions in a highly complex manner during leukocyte recruitment. For example, L-selectin has been demonstrated to function cooperatively and synergistically with other adhesion molecules under *in vivo* inflammatory conditions to mediate optimal leukocyte recruitment [1, 2]. In addition, L-selectin function can be positively and negatively regulated following leukocyte activation, both of which significantly impact leukocyte recruitment. Specifically, L-selectin ligand binding activity is transiently increased following leukocyte activation [3] followed by its rapid endoproteolytic cleavage or "shedding" from the leukocyte surface [4]. Interestingly, the generation of this soluble form of L-selectin (sL-selectin) that retains functional activity may serve to dampen inflammatory responses by competing with cell-surface L-selectin for ligand binding [5, 6]. Furthermore, in addition to functioning as an adhesion molecule, L-selectin can also serve as a signaling molecule resulting in the activation of other classes of adhesion molecules, such as integrins [7, 8] and chemokine receptors [9, 10]. Thus, L-selectin performs multiple roles important for regulating the migration of leukocytes from the blood and into tissues. Continued interest in L-selectin function stems not only from its involvement in mediating leukocyte recruitment during normal immune responses, but also from its role in

Adhesion Molecules: Function and Inhibition, edited by Klaus Ley
© 2007 Birkhäuser Verlag Basel/Switzerland

regulating unwanted pathological leukocyte migration during conditions of chronic inflammation and autoimmune disease. A better understanding of L-selectin function and its role in pathological conditions may lead to the development of novel therapeutic agents.

Structure and expression

L-selectin (CD62L) belongs to the selectin family of adhesion molecules that also contains two other structurally related members, E- and P-selectin (CD62E and CD62P, respectively, see the chapter by D. C. Bullard). The selectins share a unique extracellular domain organization composed of an N-terminal calcium-dependent lectin domain, an epidermal growth factor (EGF)-like domain, and short consensus repeat (SCR) units homologous to domains found in complement regulatory proteins (Fig. 1). The selectin molecules are highly homologous in their lectin and EGF domains, demonstrating ~65% amino acid sequence identity, with lower sequence identity (~40%) found in the SCR domains. Human L-selectin contains two SCR domains, while E- and P-selectin contain six and nine SCR domains, respectively. Interestingly, L-selectin from all species described thus far contains two SCR domains, whereas the number of SCR domains in E- and P-selectin ranges from four to nine. By contrast to the extracellular domains, the cytoplasmic domains show little similarity among the selectins. The cytoplasmic tail of L-selectin is relatively short, containing just 17 amino acid residues, but is conserved across species. Despite its short length, the cytoplasmic domain of L-selectin has been shown to be important for optimal functioning of L-selectin and to be involved in a number of processes ranging from leukocyte rolling, regulation of ligand binding activity, and endoproteolytic cleavage to signal transduction and cytoskeletal association (a detailed discussion of the cytoplasmic domain is provided below). Although being closely related structurally, each of the selectins demonstrates a unique pattern of expression. Specifically, L-selectin is constitutively expressed by all classes of leukocytes, P-selectin is expressed by activated platelets and endothelial cells from intracellular stores, and E-selectin is expressed by activated endothelial cells following synthesis.

Human lymphocytes express a 74-kDa form of L-selectin, while neutrophils express a 90–110-kDa form of the molecule [11, 12]. These different isoforms of L-selectin are the result of changes in glycosylation rather than differences in the protein core [13]. Unique to L-selectin is the presence of a membrane proximal region containing an endoproteolytic cleavage site that is recognized by endogenous membrane-bound proteases following cell activation (Fig. 1) [4, 14]. Tumor necrosis factor-α converting enzyme (TACE)/ADAM17 has been shown to be responsible for a majority of the activation-induced cleavage of L-selectin [15, 16]. However, TACE-independent cleavage mechanisms also exist since significant L-selectin cleav-

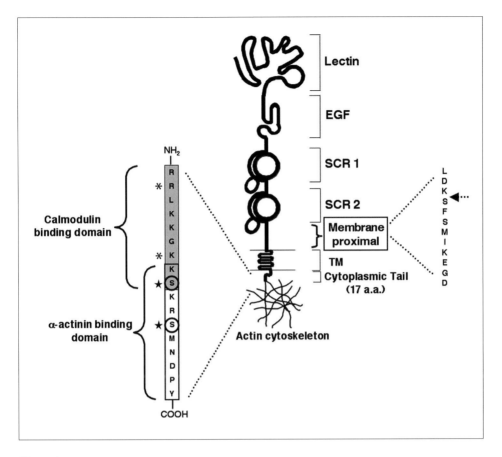

Figure 1
Structure of human L-selectin
The five different domains: lectin, epidermal growth factor (EGF)-like, short consensus repeats (SCR 1 and SCR 2), transmembrane (TM), and cytoplasmic are shown. The lectin domain is responsible for the majority of ligand recognition. The relatively short (17 amino acids – a.a.) cytoplasmic domain binds to the actin cytoskeleton and is important for mediating signal transduction following ligation of L-selectin. The cytoplasmic tail and the membrane proximal region are expanded to show the individual amino acids. The serine residues of the cytoplasmic tail, Ser364 and Ser367 (circles) undergo rapid protein kinase C-mediated phosphorylation (★) after chemoattractant-induced activation of leukocytes. The cytoplasmic domain also binds to calmodulin (CaM), α-actinin, and the ezrin-radixin-moesin (ERM) family of proteins. The amino acids involved in CaM binding (shaded box) and α-actinin (open box) are highlighted. Arg357 and Lys362 in the cytoplasmic domain (shown by ∗) are important for binding of moesin. L-selectin is endoproteolytically cleaved in the membrane proximal region (dashed arrow) by cell surface proteases.

age has been reported to occur following activation in TACE-deficient transfectants [17]. TACE activity is also involved in regulating constitutive cleavage of L-selectin from the cell surface. Specifically, in radiation-chimeric mice reconstituted with TACE-deficient fetal liver cells, significantly increased expression levels of L-selectin were found on neutrophils relative to wild-type controls [16]. Cleavage of L-selectin results in the generation of a large soluble extracellular fragment, sL-selectin, and a small ~6-kDa cell-associated fragment containing the transmembrane and cytoplasmic domains [14]. Consistent with the above-noted size differences in cell-associated L-selectin, two major isoforms of sL-selectin are found within human serum: a lymphocyte-derived ~62-kDa form and a neutrophil-derived 75–100-kDa form [5]. Both lymphocyte- and neutrophil-derived forms of sL-selectin have also been described in mouse [6] and rat [18] serum. Once cleaved, sL-selectin is stable with a $t_{1/2}$ of ~20 h, a finding that helps explain the presence of relatively high concentrations (1–2 µg/ml) in human and mouse serum [5, 6, 19]. Importantly, sL-selectin retains functional activity as demonstrated by its ability to inhibit L-selectin-dependent leukocyte attachment to activated endothelium during *in vitro* [5] and *in vivo* [20] adhesion assays, and by inhibition of L-selectin-dependent *in vivo* lymphocyte migration [6]. In fact, near physiological levels of sL-selectin inhibit lymphocyte migration to peripheral lymph nodes (PLN) by ~30%. Thus, the generation of sL-selectin following leukocyte activation may function to dampen immune responses by decreasing leukocyte migration. Furthermore, the presence of sL-selectin in the serum may establish a threshold level for L-selectin ligand expression that is required for leukocyte migration.

In addition to the generation of a functionally active soluble receptor, L-selectin cleavage has been proposed to be involved in the regulation of leukocyte/endothelial cell interactions. This idea is supported in part by the finding that leukocytes show decreased L-selectin expression following interaction with, or migration through, the endothelium [21–23]. Studies using hydroxamic acid-based protease inhibitors that block L-selectin cleavage through inhibition of a *cis*-acting metalloproteinase [24] have generated supporting data. Specifically, blocking L-selectin cleavage resulted in decreased leukocyte rolling velocities along with increased leukocyte recruitment during *in vitro* and *in vivo* assays [25–27]. By contrast, other studies found no effect of inhibiting L-selectin cleavage on leukocyte rolling, adhesion, or transmigration across the endothelium [28]. However, it should be noted that differences between adhesion molecule expression in these studies, such as the presence of E-selectin, may have contributed to these contradictory results. The role of L-selectin cleavage in leukocyte recruitment has been further investigated using a genetically targeted mouse expressing a mutant form of L-selectin, denoted L(E), that is resistant to endoproteolytic cleavage [29]. Inhibition of L-selectin cleavage in these mice resulted in an ~twofold increase in receptor expression on all leukocytes, demonstrating that spontaneous cleavage of L-selectin is necessary for maintenance of homeostatic expression levels. Furthermore, this increase in L-selectin expression resulted in increased neutrophil migration into the inflamed peritoneum. In this same study, another line of mice was created by crossing L(E) and L-selectin-deficient mice, denoted L(E)^SAME, that expressed near wild-type levels of L-selectin on leukocytes. Results using these mice showed that cleavage of L-selectin was not required for regulating normal neutrophil rolling interactions or for migration of lymphocytes to PLN. Similarly, studies using a transgenic line of mice that express a cleavage-resistant form of L-selectin (LΔP) on T cells found these cells had equivalent rolling interactions and migration to PLN as wild-type cells [30]. Given these results, it is possible that studies using protease inhibitors may acutely increase the expression level of L-selectin on leukocytes, which could thereby influence subsequent leukocyte/endothelial interactions.

L-selectin ligands

Characterization

The ability of the selectins to bind ligands is conferred primarily through the lectin domain, although the EGF and SCR domains also contribute [31–37]. SCR domains can facilitate ligand binding by extending the lectin domain out away from the cell surface [38, 39]. While it would seem to be functionally unfavorable for L-selectin to have few SCR domains, and thus a relatively short length, this may be overcome by L-selectin being preferentially localized to the tips of microvilli [40, 41]. In fact, a chimeric L-selectin/CD44 receptor targeted to the cell body was found to function poorly in capturing transfectants under flow conditions [42]. Therefore, cooperative interaction among the lectin, EGF, and SCR domains is necessary for optimal ligand binding. Consistent with the function of the lectin domain being the dominant ligand binding site for the selectins, their ligands must have carbohydrate-based epitopes for recognition. The high degree of homology among the lectin domains of the selectin molecules results in overlapping ligand specificity. For example, all of the selectins recognize the tetrasaccharide sialyl Lewis x (sLex), albeit with relatively low affinity, and thus sLex can be considered the "prototype" selectin ligand [43]. For L-selectin specifically, the basic functional ligand recognition unit is 6-sulfo sLex [44].

In vivo, L-selectin ligands are expressed on a variety of sites including HEV of lymph nodes [45], leukocytes [46], myelin [47], endometrium during implantation [48], and tumor cells [49]. High avidity binding of ligands to L-selectin *in vivo* is dependent on the correct alignment of protein scaffolds displaying O-linked carbohydrate side chains that are fucosylated, sialylated, and sulfated [50]. To date, nine such protein scaffolds have been identified: glycosylation-dependent cell adhesion molecule-1 (GlyCAM-1), CD34, podocalyxin, nepmucin, Sgp200, endomucin, endoglycan, mucosal addressin cell adhesion molecule-1 (MAdCAM-1), and P-selectin glycoprotein ligand-1 (PSGL-1) (Tab. 1). With the exception of endoglycan, MAdCAM-1, and PSGL-1, these ligands are collectively termed "peripheral node addressins" (PNAd), and can be detected using the MECA-79 mAb [51]. It is important to note that the MECA-79 mAb does not bind directly to the L-selectin binding site [52]. However, despite the identification of numerous ligands that can be recognized by L-selectin, defining any of these molecules as the physiological ligand for L-selectin remains difficult. For example, genetically targeted mice deficient in expression of GlyCAM-1, CD34 or both demonstrate normal L-selectin-dependent leukocyte adhesion and migration [53, 54]. Therefore, either these molecules do not function as physiological L-selectin ligands *in vivo* or there is a high degree of redundancy among ligands.

Biosynthesis

Most L-selectin ligand protein scaffolds are expressed throughout the body, but it is only after post-translational modification by specific enzymes that they become physiologically active ligands. These proteins contain firm, extended, serine/threonine-rich mucin-like domains that are decorated with O-linked oligosaccharides (reviewed in [55]). The oligosaccharide extensions are categorized as core 1, core 2, or biantennary extensions. Core 1 extensions are considerably shorter than core 2, but both are capped by a 6-sulfo sLex group. The sLex moiety is a tetrasaccharide consisting of galactose, N-acetylglucosamine, fucose, and sialic acid. As such, the oligosaccharide must be glycosylated, fucosylated and sialylated. Several glycosyltransferases have been implicated in the generation of functional L-selectin ligands including α1,3-fucosyltransferase VII (Fuc-T VII) [56], Fuc-T IV [57], α2,3-sialyltransferase-IV (ST3Gal-IV) [58], ST3Gal-VI, and ST3Gal-III (reviewed in [55]). In addition, oligosaccharide branches must be sulfated for high affinity L-selectin binding. Two sulfotransferases, GlcNAc6ST-1 (GST-2) and GlcNAc6ST-2 (LSST), have been described that are required for MECA-79 epitope generation and high avidity L-selectin-Ig fusion protein binding [59–62]. The result of these modifications is addition of a 6-sulfo sLex capping group. Therefore, generation of functional L-selectin ligands requires modification of core protein scaffolds with O-linked oligosaccharides that are fucosylated, sialylated, and sulfated.

Expression

High endothelial venules
L-selectin ligands, in their physiologically active role, are displayed primarily by HEV in secondary lymphoid tissues. HEV exhibit a plump, cuboidal shape, and express the vascular addressins and chemokines necessary for supporting efficient

Table 1 - L-selectin ligands

Name (alternate)	Size (kDa)	Expression	Comments	Refs.
GlyCAM-1 (Sgp50)	50	HEV of PLN and MLN, lung, mammary endothelium, breast milk (non-functional)	Constitutively expressed, secreted protein	[45, 270, 271]
Sgp200	200	PLN HEV, lung, mammary endothelium	Membrane-bound and secreted form	[77, 272]
CD34 (Sgp90)	90	Vascular endothelium, hematopoietic stem cells, tumor cells	CD34 family, constitutively expressed type I membrane protein, non-reactive forms in non-HEV, non-inflamed vessels	[45, 273–275]
Podocalyxin (Podocalyxin-like protein, PCLP)	160	PLN HEV, some mucosal HEV, vascular endothelium, kidney glomeruli, platelets, hematopoietic stem cells	CD34 family, can be pro- or anti-adhesive, regulated by post-translational modification, type I membrane protein, non-reactive forms in non-HEV, non-inflamed vessels	[66, 276, 277]
Endoglycan	165	Vascular endothelium, smooth muscle, leukocytes, hematopoietic stem cells	CD34 family, not dependent on carbohydrate 6-O-sulfotransferases for L-selectin binding, selectin binding similarity to PSGL-1	[102, 103]
Nepmucin	75, 95	HEV of PLN and MLN, dendritic cells, some myeloid cells, spleen, heart	Can mediate lymphocyte binding through Ig domain independent of LFA-1 or VLA-4, preferentially located with MAdCAM-1	[65]
MAdCAM-1	58–66	HEV of PP and MLN, intestinal lamina propria vessels, PLN HEV of young mice	Primarily mediates $\alpha_4\beta_7$ integrin binding through Ig domain, L-selectin binding regulated by tissue-specific glycosylation	[67, 168]
Endomucin	90–100	HEV of PLN and MLN, vascular endothelium, embryonic cells	Non-reactive forms in non-HEV, non-inflamed vessels	[64, 278–280]
PSGL-1	220	Hematopoietic stem cells, some myeloid cells, leukocytes	Homodimer, binds L-, P-, and E-selectin through overlapping regions, must also be modified with tyrosine O-sulfation for proper function	[32, 46, 94, 281]

GlyCAM-1, glycosylation dependent cell adhesion molecule-1; PLN, peripheral lymph node; HEV, high endothelial venules; MLN, mesenteric lymph node; LFA-1, lymphocyte function associated antigen-1; VLA-4, very late antigen-4; MAdCAM-1, mucosal addressin cell adhesion molecule-1; PP, Peyer's patches; Ig, immunoglobulin; PSGL-1, P-selectin glycoprotein ligand-1.

lymphocyte migration to these tissues (reviewed in [63]). The specific L-selectin ligands located on HEV in PLN that have been identified to date are GlyCAM-1, CD34, Sgp200, podocalyxin, nepmucin, and endomucin [44, 64–66]. MAdCAM-1 expressed in high levels on mucosally located HEV also serves as a L-selectin ligand [67].

The expression of the HEV phenotype in post-capillary venules of lymph nodes is regulated by factors entering in the afferent lymph. Interruption of the afferent lymph results in conversion of the HEV to a flat-walled morphology and loss of ability to support lymphocyte adhesion [68, 69]. In addition, loss of afferent lymph results in a lack of luminal MECA-79 mAb staining, a large reduction in sulfate incorporation, loss of Fuc-T VII mRNA expression, and a profound down-regulation of GlyCAM-1 mRNA levels [70–72]. Therefore, homeostatic control of HEV and expression of L-selectin ligands are determined by factors in the afferent lymph that are yet to be elucidated.

Within days of antigen stimulation, the vasculature of the PLN, including the HEV, undergoes rapid growth [73, 74]. This increase in HEV contributes to an increase in the number of lymphocytes migrating into the PLN [75]. Recently, it has been shown that injection of antigen-primed CD11c+ dendritic cells is sufficient to produce these increases in PLN vasculature and MECA-79+ HEV [76]. Antigen stimulation also has effects on L-selectin ligand expression. Importantly, GlyCAM-1 and soluble Sgp200 mRNA levels were down-regulated by day 4 following antigen challenge, whereas CD34 and membrane-bound Sgp200 levels remained unchanged [77]. Antigen stimulation may also influence the enzymes that produce L-selectin ligands. Specifically, in inflamed lymph nodes, Fuc-T VII staining was reduced by days 2–3 following challenge, but returned to normal levels by day 4 [78]. Thus, during an immune response, dramatic changes occur in the HEV that have the potential to significantly affect L-selectin-dependent lymphocyte migration.

Inflamed endothelium

L-selectin ligands can be induced on endothelial cells by treatment with pro-inflammatory cytokines. Specifically, tumor necrosis factor (TNF)-α greatly increased L-selectin-dependent adhesion on bovine [79] and porcine [80] aortic endothelium, as well as on human coronary endothelium [81] and human umbilical vein endothelial cells (HUVEC, [82]). In addition, TNF-α increased LSST mRNA levels in HUVEC nearly threefold [59]. L-selectin ligands can also be found in chronically inflamed tissues. Specifically, blood vessels that are phenotypically and functionally similar to HEV, in that they express PNAd or MAdCAM-1 and support lymphocyte migration, have been reported in several chronic inflammatory diseases (including multiple sclerosis [83], Grave's disease and Hashimoto's thyroiditis [84], rheumatoid arthritis [85], diabetes [86], and asthma [87] as well as during kidney allograft rejection [88]). Recently, PSGL-1 was found to be expressed on the venular endothelium of mesenteric lymph nodes (MLN) and small intestine, and to be involved in mediating leukocyte recruitment in a chronic murine ileitis model [89]. In some cases, the presence of specific chemokines alone is sufficient for induction of PNAd and/or MAdCAM-1 expression [90, 91]. Since the protein scaffolds necessary for L-selectin ligand generation are widely expressed, but ligand activity requires appropriate post-translational modification, induction of the enzymes that modify these proteins may be a general mechanism for regulating the presence of L-selectin ligands on non-HEV vessels. In support of this idea, the normally HEV-restricted sulfotransferase LSST was found colocalized with MECA-79 mAb staining in the synovium of rheumatoid arthritic tissue [92] and inflamed pulmonary vessels during asthma [87]. Furthermore, ectopic lymphotoxin production was able to drive the up-regulation of LSST and PNAd expression in the pancreas [93]. Therefore, vascular L-selectin ligands can be induced by inflammatory cytokines in extralymphoid tissues where they serve to promote leukocyte recruitment.

Leukocytes

In addition to endothelium, leukocytes also express L-selectin ligands. The dominant L-selectin ligand expressed on leukocytes is PSGL-1 [46], which is broadly expressed on most classes of leukocytes [94]. PSGL-1 was originally identified as a specific ligand for P-selectin [95, 96] and was subsequently confirmed to be the dominant physiological ligand for P-selectin [97]. In addition to the post-translational modifications creating a core 2 branched O-glycan capped with 6-sulfo sLex, it is essential that PSGL-1 be sulfated on three distinct tyrosines (Tyr46, Tyr48, and Tyr51); however, this is more important for P-selectin- than for L-selectin-dependent adhesion [98, 99]. Interestingly, the affinity of PSGL-1 for L-selectin is 20-fold stronger than that for GlyCAM-1 to L-selectin [100, 101]. This finding suggests two distinct types of L-selectin ligands; one robustly expressed but with lower affinity in HEV, and another with high affinity but lower expression levels. It is clear that there are also PSGL-1-independent L-selectin ligands expressed on leukocytes. One such molecule, endoglycan, is a CD34-family sialomucin that bears significant homology to PSGL-1 [102] and is expressed on leukocyte subpopulations and vascular endothelium [103].

The presence of L-selectin ligands on leukocytes allows for L-selectin-dependent adhesive interactions to occur between leukocytes [104]. This process of leukocytes interacting with already adhered leukocytes on the vascular endothelium has been termed "secondary tethering" and can significantly contribute to leukocyte recruitment [105]. This event has been shown to be completely dependent on L-selectin, with PSGL-1 on leukocytes being the dominant L-selectin ligand involved [106, 107]. Secondary tethering is of particular importance under conditions in which the primary interactions between leukocytes and the vascular wall appear to be L-selectin-independent [106].

Adhesive and signaling interactions

General leukocyte-endothelial cell interactions

Leukocyte interactions with the vascular endothelium encompass at least three sequential steps: transient contact with the endothelium called tethering or capture, followed by rolling, and then strong adhesion or arrest on the endothelium, ultimately resulting in transendothelial migration (Fig. 2) [108]. Free-flowing leukocytes are captured onto the vascular wall by the selectins where they are converted into rolling cells *via* the constant formation and dissociation of receptor-ligand interactions. Subsequent cellular activation by interaction of appropriately displayed chemokines on the endothelium with leukocyte chemokine receptors leads to increased binding avidity of integrins [109, 110]. Firm arrest of rolling leukocytes is then mediated by integrins such as lymphocyte function associated antigen-1 (LFA-1) and very late antigen-4 (VLA-4) binding to the immunoglobulin superfamily members, intercellular adhesion molecule-1 (ICAM-1) and vascular cell adhesion molecule-1 (VCAM-1), respectively. Numerous studies have now established that this process is mediated by the overlapping and synergistic interactions of multiple families of adhesion molecules [111–113]. L- and P-selectin are efficient at mediating the initial tethering of free-flowing leukocytes and supporting subsequent fast rolling interactions. By contrast, E-selectin does not mediate leukocyte tethering but does support rolling at very slow velocities and thereby facilitates the transition from rolling to firm arrest [114, 115]. The role of L-selectin in mediating leukocyte/endothelial cell interactions was subsequently confirmed by studies of L-selectin-deficient mice [116]. Mice deficient in L-selectin initially have normal levels of trauma-induced leukocyte rolling, but show a marked decrease in rolling at later time points. The initial leukocyte rolling is mediated by P-selectin in these mice [112]. By contrast, P-selectin-deficient mice show reduced trauma-induced leukocyte rolling at early time points, which begins to recover at later time points [117]. Consistent with these observations in the single adhesion molecule-deficient mice, leukocytes from mice lacking both L- and P-selectin demonstrate an absence of trauma-induced leukocyte rolling [118]. These findings demonstrate both unique and overlapping roles for the selectins in mediating leukocyte recruitment during inflammation.

The mechano-kinetic properties of leukocyte capture and rolling mediated by L-selectin remains an active area of research. Selectin-ligand interactions are dynamic and involve the constant formation (K_{on}) and dissociation (K_{off}) of bonds depending on the shear forces exerted on the leukocyte due to the flow of blood. Forces above a specific threshold are required for the formation of functional tethers between L-selectin and its ligand [119, 120]. Importantly, L-selectin mediates rolling at faster velocities as compared to the other selectins implying that the K_{on} and K_{off} for L-selectin are higher than that for P- and E-selectin. The frequency of L-selectin interaction with its ligand is enhanced under increasing flow rates resulting in formation

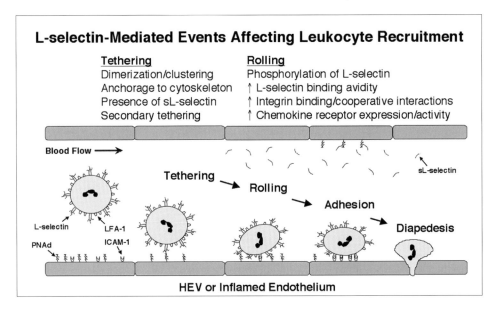

L-selectin-Mediated Events Affecting Leukocyte Recruitment

Tethering
Dimerization/clustering
Anchorage to cytoskeleton
Presence of sL-selectin
Secondary tethering

Rolling
Phosphorylation of L-selectin
↑ L-selectin binding avidity
↑ Integrin binding/cooperative interactions
↑ Chemokine receptor expression/activity

Blood Flow →

Tethering

Rolling

Adhesion

Diapedesis

sL-selectin

L-selectin   LFA-1
PNAd   ICAM-1

HEV or Inflamed Endothelium

Figure 2
L-selectin-mediated leukocyte recruitment
*Expression of constitutive peripheral node addressin (PNAd) on the surface of high endothe-
lial venules (HEV) or inducible vascular ligands for L-selectin at sites of inflammation mediate the initial tethering or capture of free-flowing leukocytes from the blood. Alternatively,
leukocytes can interact with already adhered leukocytes in a process termed "secondary
tethering" that is dependent on binding between L-selectin and PSGL-1. L-selectin is preferentially located on microvilli of leukocytes thereby facilitating interactions with ligand.
By contrast, β_2 integrins such as lymphocyte function associated antigen-1 (LFA-1), are
localized to the cell body. Following initial tethering, L-selectin supports leukocyte rolling at
high velocities but can function synergistically with LFA-1 to stabilize and decrease rolling
velocities. During rolling, leukocytes become activated by endothelial-displayed chemokines,
which up-regulate the binding of integrins to ligands such as intercellular adhesion molecule-
1 (ICAM-1). Firm adhesion to the endothelium, primarily mediated through high-affinity
binding of the integrins to their ligands, precedes transendothelial migration (diapedesis).
The presence of relatively high concentrations of soluble L-selectin (sL-selectin) can decrease
L-selectin-mediated leukocyte adhesion and therefore may play a role in dampening inflammatory responses. Important events affecting L-selectin function during tethering and rolling
are listed.*

of "catch bonds" [121, 122]. The formation of many such catch bonds decreases the
flow rate of the leukocyte on the cell surface. Low levels of force allow for enhancement in the half-life of the L-selectin/ligand interaction and stabilization of catch

bonds. At higher forces, the catch bonds convert to slip bonds [122]. Recently, Lou et al. [123] reported that the hinge domain between the lectin and EGF domains of L-selectin can exist in both open and closed conformations. Moreover, mutations in the hinge domain that increase flexibility decrease the shear threshold for rolling mediated by L-selectin. Thus, conformational changes can occur in the structure of L-selectin subsequent to ligand binding and these changes may affect L-selectin-mediated tethering and rolling of leukocytes.

Interestingly, the α_4 integrins ($\alpha_4\beta_7$ and $\alpha_4\beta_1$) can also mediate the initial capture and rolling of certain leukocyte populations [124–126]. Recently, Nandi et al. [127] extended these findings by describing the formation of a biomolecular complex between CD44 and VLA-4 as a prerequisite for VLA-4-mediated firm adhesion. Although L-selectin has not been definitively shown to be associated with other adhesion molecules, the localization of L-selectin on the tips of microvilli along with other adhesion molecules involved in the capture and rolling of leukocytes (e.g., β_7 and β_1 integrins, and PSGL-1) suggests that these molecules can function synergistically to enhance the binding avidity of leukocytes to the endothelium [40, 128]. In addition, L-selectin and ICAM-1 have been shown in both in vitro and in vivo assays to function cooperatively to mediate optimal leukocyte/endothelial cell interactions [1, 2, 129]. Specifically, L-selectin-mediated leukocyte rolling on a transfected endothelial cell line expressing L-selectin ligand was enhanced by ~70% and leukocyte rolling velocities were decreased by ~30% in the presence of ICAM-1 [129]. Additionally, in mice deficient in both L-selectin and ICAM-1, the number of rolling leukocytes was reduced by >50% and the rolling velocities were increased as compared to L-selectin-deficient mice after TNF-α stimulation [1]. Therefore, functional synergism between different classes of adhesion molecules supports leukocyte/endothelial cell interactions.

L-selectin as a signal transducing molecule

L-selectin has been reported to generate transmembrane signals following ligand engagement that can enhance the adhesive interactions of leukocytes to the vascular endothelium. Antibody cross-linking studies using human neutrophils or L-selectin-expressing cell lines have shown that L-selectin ligation can induce rapid increases in intracellular calcium concentrations [130, 131], superoxide generation [130], protein tyrosine phosphorylation [132], and F-actin polymerization [133]. In addition, L-selectin ligation leads to activation of mitogen-activated protein kinase (MAP kinase, [132]), Jun N-terminal kinase [134], Ras [135], and nuclear factor of activated T lymphocytes (NFAT, [136]). Sulfatide, which has been reported to bind L-selectin, has also been suggested to trigger an increase in intracellular free calcium levels and enhance expression of TNF-α and interleukin-8 (IL-8) in human neutrophils [131]. However, sulfatides can also bind sulfatide receptors and activate leukocytes in an L-selectin-independent manner [137]. Therefore, experiments using sulfatides to study effects of L-selectin activation need to be interpreted with caution. In addition, when using antibody-mediated ligation, differences in strength of binding, number of receptors bound, duration of engagement, and the possibility of activation through Fc receptors must be considered.

Simon et al. [138] reported that activation of L-selectin through antibody cross-linking enhanced the adhesive properties of neutrophils to LPS-treated HUVECs and that this adhesive interaction required the β_2 integrin, Mac-1. Cross-linking of L-selectin on lymphocytes with the anti-L-selectin mAb MEL-14 induces homotypic lymphocyte adhesion by a lymphocyte LFA-1-independent mechanism [139]. Furthermore, activation of L-selectin using the LAM1-116 mAb that binds to the lectin domain of L-selectin enhances the expression of β_1 and β_2 integrin activation epitopes and results in the rapid homotypic adhesion of leukocytes [7]. This L-selectin-induced adhesion required energy metabolism, an intact cytoskeleton, and kinase function. Importantly, the binding of many other L-selectin mAbs to regions of the molecule other than the lectin domain did not trigger activation. In addition, activation of L-selectin through GlyCAM-1 binding results in increased integrin-mediated adhesion [8, 140]. Taken together, these findings suggest an important role for L-selectin-induced activation of leukocyte integrins in regulating adhesive interactions with the vascular endothelium (Fig. 2).

Other reports suggest a role for L-selectin-mediated signaling in influencing leukocyte chemotaxis. Specifically, in addition to reductions in numbers of leukocytes recruited in an antigen challenge model, leukocytes in L-selectin-deficient mice migrated a shorter distance away from the vessel wall than did wild-type cells [141]. In a subsequent study, leukocytes in L-selectin-deficient mice showed a significant reduction in emigration to platelet-activating factor and keratinocyte-derived cytokine [142]. A severe impairment in leukocyte chemotaxis to some but not all chemokines was also observed in L(E)^{SAME} mice expressing cleavage-resistant L-selectin [29]. Furthermore, L-selectin ligation experiments with mAb along with cytokine activation synergistically enhance neutrophil chemotaxis during *in vitro* assays [9]. All these findings demonstrate a role for L-selectin in mediating efficient leukocyte chemotaxis during inflammation. However, the mechanism(s) underlying the role of L-selectin in influencing leukocyte chemotaxis remains unclear. Several possibilities can be proposed to explain these effects including up-regulation of chemokine receptor expression and/or close physical proximity of L-selectin and the chemokine receptor allowing for "cross talk" to occur between these two molecules. In fact, Ding et al. [10] reported that activation through L-selectin enhanced stromal cell-derived factor-1α (SDF-1α)-mediated adhesion and transendothelial migration of leukocytes by up-regulating the functional expression of its receptor CXCR4. Studies from our lab indicate that activation through L-selectin by LAM1-116 mAb ligation increased the overall chemotactic ability of lymphocytes to secondary lymphoid tissue chemokine (SLC) during *in vitro* chemotaxis assays [143].

Specifically, L-selectin activation significantly enhanced the chemotaxis of CD8[+] T cells by 30–40% and CD4[+] T cells by 25–30% to SLC. Importantly, in this case L-selectin ligation-induced enhanced chemotaxis was not due to up-regulation of CCR7 expression, the chemokine receptor for SLC. Therefore, these results suggest that signals mediated following ligation of L-selectin lead to sensitization of CCR7. It remains to be determined whether there is any physical association between L-selectin and CCR7 that could account for these results as has been reported for β_1 integrin and the IL-3 receptor [144].

The activation of intracellular signaling pathways can affect L-selectin function, most likely through interactions with the cytoplasmic domain. The first evidence for this was the demonstration that the ligand binding activity of L-selectin was significantly increased following leukocyte activation [3]. Haribabu et al. [145] extended these results by showing that chemokine activation of leukocytes resulted in rapid protein kinase C (PKC)-dependent phosphorylation of serine residues within the cytoplasmic domain of L-selectin (Fig. 1). Furthermore, pharmacological inhibitors of PKC eliminated the activation-induced increase in L-selectin ligand binding avidity, suggesting that phosphorylation of L-selectin may be a physiologically important mechanism for regulating L-selectin-mediated adhesion. Specific isoforms of PKC responsible for phosphorylation of L-selectin have now been identified [146]. Although the mechanism by which phosphorylation of L-selectin could influence ligand-binding activity is unknown, indirect evidence suggests that dimerization of the receptor may be involved. Specifically, enforced dimerization of L-selectin using the coumermycin/GyrB system enhances L-selectin-dependent leukocyte adhesion with a marked reduction in leukocyte rolling velocities on endothelial cell monolayers [147]. Furthermore, mAb-induced L-selectin dimerization enhances leukocyte capture on purified L-selectin ligands [148].

Consistent with a role for the cytoplasmic domain in regulating L-selectin function is the finding that cells expressing a mutant form of L-selectin lacking the C-terminal 11 amino acids of the cytoplasmic tail (LΔcyto) have reduced tethering and rolling during *in vitro* and *in vivo* assays [149, 150]. Furthermore, a 15-amino acid truncation of the cytoplasmic domain (L358stop) results in an even more severe impairment of L-selectin function [150]. The cytoplasmic tail of L-selectin can dynamically associate with a number of proteins including α-actinin [151], members of the ezrin-radixin-moesin (ERM) family of proteins [152], and calmodulin (CaM, Fig. 1) [153]. Interestingly, truncation of the cytoplasmic domain of L-selectin prevents association with α-actinin but not localization to microvilli [151]. Other studies have reported L-selectin association with the actin cytoskeleton [154–156]. The results of these studies suggest that linkage of L-selectin to the cytoskeleton is required for stabilization and optimal functioning of L-selectin.

The cytoplasmic domain of L-selectin can also regulate receptor function by playing a role in endoproteolytic cleavage (reviewed in [157]). Specifically, binding of CaM to the cytoplasmic tail of L-selectin is thought to negatively regulate endoproteolytic cleavage by inducing a conformational change in the extracellular domain that renders it resistant to proteolysis [153]. Following leukocyte activation, dissociation of CaM from the cytoplasmic domain of L-selectin exposes the cleavage site. Interestingly, the LΔcyto mutant lacks part of the proposed CaM binding site but is not more sensitive to cleavage [4]. Possible explanations for this finding are that the partial deletion is not sufficient to fully disrupt the association of CaM with L-selectin or that deletion of the cytoplasmic domain may result in conformational changes to the extracellular membrane proximal region of L-selectin and thereby alter its reactivity with the protease(s). Further studies will be needed to fully resolve these contradictory reports.

Lymphocyte migration

It has been known for nearly half a century since the landmark experiments of Gowans and Knight that lymphocytes migrate continuously between lymphoid tissues *via* blood and lymph fluid [158]. This event is of paramount importance to the generation of rapid and efficient immune responses due to the fact that there are a limited number of lymphocytes specific for any particular antigen. The highly selective nature of this migration occurring only at the HEV within lymphoid tissues suggested the involvement of specific adhesion molecule/ligand interactions [159].

Tissue-specific migration

HEV are found in secondary lymphoid tissues such as PLN, MLN, Peyer's patches, appendix, tonsils, and nasal associated lymphoid tissue (NALT), but are not found in the spleen or primary lymphoid tissues. L-selectin was first identified as a peripheral lymphoid tissue homing receptor by demonstration that an anti-L-selectin mAb (MEL-14) blocked lymphocyte adherence to HEV of PLN and MLN, but not mucosal lymphoid tissues [160]. Studies using L-selectin-deficient mice provided further insight into the role that this molecule plays in lymphocyte migration. Specifically, PLN from L-selectin-deficient mice are reduced by 70–90% in cellularity and lymphocytes from these mice are unable to attach to HEV of PLN during *in vitro* binding assays [116, 161]. Furthermore, lymphocytes from L-selectin-deficient mice are unable to migrate into either resting or antigen-activated PLN of normal mice during *in vivo* migration assays. Migration of L-selectin-deficient lymphocytes into MLN and unexpectedly into Peyer's patches was severely decreased in short-term migration assays, but partially recovered during long-term experiments. Interestingly, lymphocytes unable to enter peripheral lymphoid tissues accumulated in the spleen, resulting in a proportionate increase in spleen cellularity [116, 161].

The incomplete inhibition of lymphocyte migration to the MLN and Peyer's patches is due to the presence of a second homing receptor, the $\alpha_4\beta_7$ integrin [162]. Subsequent generation of mice deficient in β_7 integrin expression confirmed the role of $\alpha_4\beta_7$ integrins in mediating lymphocyte migration to mucosal sites. Specifically, β_7 integrin-deficient mice have dramatically hypocellular Peyer's patches and significantly decreased numbers of lamina propria and intraepithelial lymphocytes [163]. Lymphocytes from these mice show severely decreased migration into Peyer's patches during *in vivo* migration experiments while migration into the MLN is much less affected [163, 164]. Furthermore, intravital microscopy of exteriorized Peyer's patches from β_7 integrin-deficient mice revealed that although normal numbers of lymphocytes can be observed rolling along the endothelium, the number of cells firmly adhering to the endothelium is decreased dramatically [163, 165]. The use of blocking antibodies in these mice showed rolling to be almost entirely mediated by L-selectin.

Generation of mice lacking expression of both L-selectin and β_7 integrins confirmed that these two homing receptors regulate virtually all lymphocyte migration across HEV in lymphoid tissues [164, 166]. Specifically, lymphocytes from L-selectin/β_7 integrin-deficient mice are unable to migrate into PLN, MLN, or Peyer's patches. When compared to β_7 integrin-deficient mice, L-selectin/β_7 integrin-deficient mice have even fewer resident cells in Peyer's patches, further demonstrating a role for L-selectin in migration to Peyer's patches [164]. However, under inflammatory conditions, P-selectin can support low levels of lymphocyte rolling on HEV of Peyer's patches in the absence of L-selectin and β_7 integrins [165].

While most naïve lymphocytes express both L-selectin and β_7 integrins, the counter receptors or "vascular addressins" for these receptors are not expressed by all HEV. As discussed above, HEV located in the PLN express a variety of L-selectin ligands (PNAd) that all express the MECA-79 epitope [45, 51]. By contrast, mucosal tissues such as Peyer's patches express MAdCAM-1, which is the dominant physiological ligand for the $\alpha_4\beta_7$ integrin [167, 168]. Interestingly, HEV located in the MLN and NALT express both PNAd and MAdCAM-1 and thus can support both L-selectin- and β_7 integrin-mediated pathways of lymphocyte migration [51, 169].

L-selectin-dependent leukocyte migration has clearly been shown to be required for generation of peripheral as well as mucosal immune responses. For example, L-selectin-deficient mice exhibit a significant decrease in neutrophil, monocyte, and lymphocyte migration to the inflamed peritoneum and reduced contact and delayed-type hypersensitivity reactions, and delayed rejection of allogeneic skin grafts [170–172]. Furthermore, these mice have significantly delayed humoral immune responses when immunized subcutaneously with antigen, but display a normal immune response when immunized intraperitoneally [173, 174]. In addition, these mice exhibit a >30-fold decrease in vaginal IgA production and a significant reduction in antigen-specific antibody-forming cells in the nasal passages following intranasal immunizations, suggesting that L-selectin plays a dominant role in lymphocyte migration to the NALT [169, 175]. Furthermore, infection studies involving the oral immunization of L-selectin-deficient mice with a *Salmonella* vaccine vector expressing the colonization factor Ag 1 from enterotoxigenic *Escherichia coli* have shown that these mice have greatly decreased mucosal IgA responses [176]. L-selectin also mediates the migration of lymphocytes to non-lymphoid tissues. Specifically, migration of L-selectin-deficient T and B cells is reduced by 50–60% to the aortic wall [177]. Therefore, L-selectin regulates the migration of lymphocytes to both peripheral and mucosal lymphoid tissues.

Subset-specific migration

Studies performed in sheep showing different migration patterns among CD4[+] T helper (Th) cells, CD8[+] T cytotoxic cells, and B cells, including naïve and memory populations, have suggested that in addition to organ-specific mechanisms, subset-specific lymphocyte migration patterns exist [178–180]. However, other studies have found no selective difference in lymphocyte subset migration to the PLN when analyzed at the level of the HEV [181]. Additional work with mice demonstrated a propensity for T cells to migrate to the PLN and B cells to the Peyer's patches and spleen [182–184]. This can be explained, at least in part, by the finding that although most lymphocytes express both L-selectin and $\alpha_4\beta_7$ integrins, T cells have twofold higher levels of L-selectin expression than B cells, while the opposite is found for $\alpha_4\beta_7$ integrin expression [185, 186]. Importantly, reducing the expression level of L-selectin on lymphocytes by 50% results in a nearly 70% decrease in their migration to the PLN [185]. Binding assays demonstrate that CD4[+] T cells preferentially bind to HEV of PLN, and are more capable of entering these tissues, whereas CD8[+] cells prefer HEV from gut-associated lymphoid tissue [187, 188]. These binding properties of T and B cells are obvious regardless of their tissue source or whether they bind to resting or inflamed HEV [182]. Furthermore, CD4[+] T cells are capable of recirculating more efficiently than CD8[+] cells during *in vivo* experiments [179, 189]. Therefore, differences in expression levels of adhesion molecules by T and B cells contribute significantly to differences in their observed migration routes.

Activated lymphocytes

L-selectin expression on naïve lymphocytes is relatively uniform. However, activation of lymphocytes induces rapid changes in adhesion molecule expression, including the up-regulation of β_1 and β_2 integrins and CD44, and the down-regulation of L-selectin [190–194]. Specifically, L-selectin is down-regulated from the T cell surface within 60 min after stimulation due to accelerated proteolytic cleavage [190].

L-selectin is then re-expressed at three- to fourfold above resting levels by 48 h following activation as a result of decreased mRNA turnover and subsequent increase in protein synthesis. L-selectin expression is finally lost following transcriptional down-regulation 2–3 days after stimulation, with a total loss of expression by days 6–7. These findings have been confirmed using lymphocytes from L(E)^{SAME} mice expressing a cleavage-resistant form of L-selectin. In these studies, wild-type T cells lost ~80% of L-selectin expression by 2 h following activation through the T cell receptor, whereas L(E)^{SAME} T cells had no decrease in expression level [29]. Furthermore, by 5 days following activation, T cells from both wild-type and L(E)^{SAME} mice had the same low level of L-selectin expression demonstrating that this late loss of L-selectin was not due to cleavage from the cell surface. These studies also suggest that the major function of L-selectin cleavage in lymphocytes is to redirect activated cells away from the PLN. Specifically, activation through the T cell receptor led to a fourfold reduction in migration of wild-type cells to the PLN but only a twofold reduction in L(E)^{SAME} cells [29]. Similarly, another study using the L-selectin cleavage-resistant LΔP transgenic mice, reported that activated T cells expressing levels of L-selectin similar to unactivated cells migrated less efficiently to PLN than did unactivated cells [30]. These findings suggest that other factors in addition to L-selectin expression levels affect the entry of activated lymphocytes into lymph nodes.

Memory and effector lymphocytes

Upon activation, CD4$^+$ T cells differentiate into subclasses of Th cells based on the type of cytokines present. Reports have indicated that cells exhibiting a Th1 type cytokine profile express L-selectin, whereas Th2 cells do not [195, 196]. Furthermore, maintained expression of L-selectin on Th1 cells was shown to be dependent on the presence of IL-12 [197]. Currently, more is known about the use of E- and P-selectin by Th cells than for L-selectin. Specifically, PSGL-1 on Th1 cells, and to a lesser extent Th2 cells, mediates adhesion with P- and E-selectin expressed at sites of inflammation [198–202]. This difference in adhesive capability is in part regulated by the opposite effects of IL-12 and IL-4 on Fuc-T VII expression, which plays an important role in regulating the binding ability of selectin ligands [203–205]. Also of importance is the finding that E-selectin may facilitate cutaneous lymphocyte antigen (CLA)-mediated adhesion of Th2 cells [206]. The adhesive mechanisms used by Th cells are likely to become more complex as other Th subsets, such as Th17, are described [207, 208]. Adhesive mechanisms utilized by Th17 cells are not yet known.

While L-selectin has been shown to be down-regulated following antigen activation, there are numerous conflicting reports concerning L-selectin expression on memory cells. Specifically, memory CD4$^+$ T cells have been shown to be L-selectin$^-$ in some reports [193, 209], while other studies have shown expression of L-selectin on significant proportions of memory CD4+ T cells [11, 161, 210]. Selectin expression on memory lymphocytes may be regulated in part by the anatomical site in which the cell was activated [211, 212], and may contribute to lymphocyte homing. The varied expression of L-selectin between naïve and memory lymphocytes is consistent with observed differences in migratory capabilities of these cells. Specifically, naïve lymphocytes are very efficient in migrating directly from the blood to secondary lymphoid organs, while memory cells exhibit an enhanced ability to enter extralymphoid sites such as the gut and skin [178, 213–215]. The ambiguity in studies of naïve *versus* memory cells may exist because of the complexity of memory cell subsets. For example, two populations of such memory T cells, central memory (T_{CM}) and effector memory (T_{EM}), have received much attention. T_{CM} cells have been shown to express both L-selectin and the chemokine receptor CCR7, whereas T_{EM} cells lack CCR7 and express L-selectin at lower and more heterogeneous levels [216, 217]. Not unexpectedly, these two subsets of memory T cells display different migratory patterns. Specifically, T_{EM} cells preferentially migrate to inflamed, extralymphoid sites, whereas T_{CM} primarily migrate through the PLN [217]. The CCR7+L-selectin+ phenotype of T_{CM} cells is comparable to that of naïve T cells, and thus promotes a similar migration pattern. However, T_{CM} cells also express inflammatory chemokine receptors that enable them to enter sites of inflammation. However, the true *in vivo* situation is likely not this straightforward as demonstrated by the recent finding of a CCR7+L-selectin− subset of memory T cells, indicating a more heterogeneous population than first described [218]. Further complicating the issue is the finding that T_{EM} cells are able to transition and proliferate as T_{CM} cells, and re-express L-selectin [219]. Studies of the production of memory cells indicate that more L-selectin+ T_{CM} cells are produced upon weak antigenic challenge (increased ratio of cells to antigen) than are produced with strong stimulation. Also, the ratio of L-selectin+ T_{CM} cells to L-selectin− T_{EM} cells increases with time (reviewed in [220]). Thus, subsets of memory and effector T cells express L-selectin differently, which contributes to their circulation patterns *in vivo*.

Regulatory T cells

Regulatory T cells (T_{reg}) represented by a CD25+CD4+ phenotype and expression of the transcription factor Foxp3, are necessary for a wide array of immune responses including immune tolerance, inflammatory responses, maintenance of graft tolerance, and T cell homeostasis [221–226]. Previous work done on T_{reg} cells has shown that L-selectin is expressed by the majority of CD25+ T_{reg} cells and is involved in their migration to the PLN [227–229]. Recently, through the use of L-selectin-deficient as well as cleavage-resistant L(E)^{SAME} mice, it has been shown that L-selectin is essential for T_{reg} cell migration and normal tissue distribution [230]. Specifically, numbers of T_{reg} cells from PLN of L-selectin-deficient mice are reduced by 90%, while spleen T_{reg} cell numbers are increased correspondingly. Interestingly, T_{reg} cells were found to turn over cell-surface L-selectin at a faster rate than $CD25^-CD4^+$ T cells. However, T_{reg} cells maintained physiologically appropriate L-selectin expression by having a twofold higher level of L-selectin mRNA. As a result, T_{reg} cells from $L(E)^{SAME}$ mice expressed 30–40% more cell-surface L-selectin, which resulted in a twofold increase in T_{reg} cell migration to the PLN compared to wild-type cells. Therefore, like naïve CD4+ T cells, T_{reg} cells also require L-selectin for migration to peripheral lymphoid tissues, but use a unique mechanism for regulating appropriate L-selectin expression levels.

Role of L-selectin in disease

Aberrant leukocyte recruitment, either in location or duration, is a major hallmark of inflammatory disease. As such, identification of the molecular pathways involved in mediating this recruitment has received a great deal of attention. An accumulating body of evidence, from both human patients and animal models, shows L-selectin to be a key regulator of leukocyte migration in a number of disease conditions. Therefore, selective targeting of L-selectin function may be of therapeutic benefit for these patients.

The reperfusion of ischemic tissue often results in severe cell and tissue damage that can lead to loss of organ function. Ischemia-reperfusion injury can result from a number of disease conditions including myocardial infarction, stroke, shock, trauma, and hypotension. Although the mechanisms mediating tissue damage are not fully understood, a significant component involves the rapid influx of leukocytes, primarily neutrophils. A number of reports indicate that L-selectin, in addition to P- and E-selectin, is involved in mediating neutrophil interactions with ischemic vessels. For example, blocking L-selectin function with mAb treatment reduces damage to skeletal muscle in rat models of ischemia-reperfusion injury [231, 232]. Furthermore, L-selectin was shown to be important in the generation of hepatic injury using a murine model of normothermic hepatic ischemia [233]. In these studies, L-selectin-deficient mice showed reduced neutrophil adhesion, absence of microcirculatory failure, reduced tissue damage, and increased survival following partial hepatic ischemia compared to wild-type mice. Thus, L-selectin-mediated neutrophil recruitment is an important factor in the development of ischemic injury.

While the above findings demonstrate a role for L-selectin in recruitment of neutrophils to the ischemic liver, whether L-selectin also contributes to T cell recruitment during hepatic inflammation was unclear. Several studies using the Concanavalin A (Con A)-induced model of human T cell hepatitis [234] reported critical roles for P- and E-selectin, VCAM-1, ICAM-1, and LFA-1 in lymphocyte recruitment [235–237]. A recent study by Kawasuji et al. [238] using this model in L-selectin-deficient mice showed a prominent role for L-selectin in development of liver injury following

Con A treatment. Specifically, L-selectin-deficient mice had reductions in levels of plasma transaminase, numbers of infiltrating CD4+ T cells, TNF-α mRNA levels, and liver necrosis. Therefore, L-selectin is involved in mediating T cell recruitment and subsequent injury to the liver under inflammatory conditions.

Airway inflammation involving intense influx of leukocytes is a prominent factor in the development of chronic airway hyperresponsiveness found in asthma. Leukocyte/endothelial interactions in the lung vascular bed have been considered to be regulated differently from that of the systemic circulation due to the presence of a highly interconnected vascular bed in which the diameter of the vessels are often smaller than that of the leukocytes [239]. Because of this, it was thought that specialized adhesion molecules such as the selectins were not required for leukocyte recruitment to the lung. However, multiple studies have demonstrated a role for each of the selectins in this process. Specifically, L-selectin has been shown to support leukocyte rolling on inflamed and non-inflamed lung vessels [240–242]. Furthermore, Nishio et al. [242] found that *in vivo* administration of fucoidin, an L-selectin competitor, reduced the number of leukocytes rolling on pulmonary arterioles in a hyperoxia-induced rat lung model. Additionally, an important role for L-selectin in mediating the migration of leukocytes during pulmonary inflammation in the murine ovalbumin (OVA)-induced asthma model has been described [243]. In these studies, L-selectin-deficient mice displayed a significant reduction in the number of CD3+ cells present in the broncheoalveolar lavage following challenge compared to control mice despite demonstrating equivalent levels of sensitization. In a sheep model of allergic bronchoconstriction, treatment with an anti-L-selectin mAb as well as low-molecular weight selectin antagonists were shown to be therapeutic [244]. Specifically, mAb blockade of L-selectin resulted in significant reductions in severity of early and late airway responses and a lack of postchallenge airway hyperresponsiveness. Similar effects including reductions in histamine release and neutrophil numbers in bronchoalveolar lavage were found using the selectin inhibitors in challenged animals. Subsequent studies in this model showed that blockade of L-selectin ligands using the MECA-79 mAb dramatically reduced the accumulation of neutrophils, macrophages, lymphocytes, and eosinophils in the bronchoalveolar fluid of allergic sheep, and ameliorated both the late-phase airway response and airway hyperresponsiveness induced by airway allergen challenge [87]. Other studies by Hamaguchi et al. [245] using a bleomycin-induced lung injury model demonstrated that L-selectin was important for the trafficking of leukocytes during the development of pulmonary fibrosis. Specifically, L-selectin-deficient mice had reduced infiltration of neutrophils and lymphocytes and reduced collagen deposition in the lung following bleomycin treatment compared to normal mice. Therefore, L-selectin-mediated leukocyte migration has been shown to be critically involved in the generation of pulmonary inflammatory responses.

Type I diabetes is characterized by the destruction of insulin-producing β cells of the pancreas as a result of T cell recruitment into the pancreatic tissue. L-selectin ligands, specifically PNAd and MAdCAM-1, are expressed and functional in the pancreas of nonobese diabetic (NOD) mice, a murine model for type I diabetes [86, 246]. Because of the presence of L-selectin ligands, it was suggested that L-selectin-mediated interactions are involved in lymphocyte trafficking to the pancreas. In fact, administration of anti-L-selectin mAb to NOD mice can protect against the development of insulitis and diabetes [247, 248]. By contrast, L-selectin-deficient NOD mice exhibit similar islet infiltrates and similar disease progression to L-selectin+ NOD mice [249, 250]. These results, along with the demonstration that other adhesion molecule pairs are likely involved [247], suggest that L-selectin function is not an absolute requirement for disease in NOD mice. It is also possible that L-selectin-mediated T cell migration to pancreatic lymph nodes is necessary for disease development in this model [250]. Another interesting finding is that L-selectin+ T_{reg} cells are involved in controlling pathogenesis in a murine transgenic model of diabetes [251]. Therefore, further studies will be needed before a definitive role for L-selectin in the development of diabetes can be established.

In addition to cell-surface adhesion molecules, the soluble forms of these molecules have been receiving an increasing amount of attention. While soluble adhesion molecules have been used successfully as markers of inflammation or disease activity, their role in physiological processes must also be considered (reviewed in [252]). Specifically, significantly increased levels of sL-selectin have been reported to be associated with a number of different disease conditions including chronic myeloid and lymphocytic leukemia [253–255], sepsis [19, 256], HIV infection [19], atopic dermatitis [257], psoriasis [258], and lupus [259]. As discussed above, since sL-selectin retains functional activity, these increased levels may have important physiological effects on leukocyte migration in these patients. In fact, higher levels of sL-selectin in acute myeloid leukemia patients at the time of diagnosis correlated with decreased probability of achieving complete remission, shorter event-free survival, and shorter overall survival [260]. Interestingly, in severe trauma patients, decreased levels of sL-selectin correlate with an increased risk of progression to acute respiratory distress syndrome [261]. In these patients, low levels of sL-selectin may indicate the presence of newly expressed L-selectin ligands in large vascular beds such as the lung. This loss of L-selectin buffering activity in the plasma may thereby allow increased leukocyte infiltration to occur at remote sites. Thus, membrane-bound and sL-selectin contribute significantly to leukocyte recruitment in inflammatory disease with sL-selectin levels having prognostic value.

L-selectin-targeted therapies

The demonstrated role of L-selectin in the development of pathology in numerous autoimmune and inflammatory disorders makes it an attractive therapeutic target. However, despite a wealth of successes using blockade of L-selectin function in animal models of disease, few of these successes have translated to humans. Recently, pan-selectin antagonists have received attention for their ability to reduce leukocyte infiltration to diseased tissues by competing with selectins for ligand binding. Specifically, bimosiamose, developed by Kogan et al., a low molecular weight nonoligosaccharide selectin inhibitor, prevents P-, E-, and L-selectin-mediated adhesion *in vitro* [262, 263]. Additional *in vivo* studies suggest that bimosiamose does not inhibit leukocyte rolling but functions primarily by blocking E-selectin-mediated adhesion [264]. Recent limited clinical trials using bimosiamose have shown some promise. Specifically, patients with mild allergic asthma treated with inhaled bimosiamose had a 50% reduction in late asthmatic reactions following allergen challenge [265]. Similarly, psoriatic patients treated with bimosiamose, demonstrated reductions in epidermal thickness and lymphocyte infiltration [266]. In addition to pan-selectin antagonists, specific blockade of L-selectin function using humanized anti-L-selectin mAbs (HuDREG-55 and HuDREG-200) is being explored [267]. Specifically, the HuDREG-55 mAb (aselizumab) was found to significantly decrease mortality and increase survival time in a baboon model of hemorrhagic-traumatic shock [268]. However, no significant benefit of aselizumab treatment was found in a Phase II clinical trial of patients suffering from sustained trauma due to blunt or penetrating injury [269]. Another possible therapeutic strategy is the targeting of selectin ligands. Support for this comes from studies by Rosen et al. [87], demonstrating that blocking L-selectin ligands using MECA-79 mAb treatment results in decreases in late-phase airway responses and airway hyperresponsiveness following airway allergen challenge in a sheep model of asthma. Therefore, despite some promising initial results, whether or not selectin-based therapies will provide efficacious treatment for human disease remains to be determined.

Conclusions and future challenges

Leukocyte recruitment is a complex process involving cooperative interactions between several classes of adhesion molecules. Binding of L-selectin to its ligand on the endothelium is crucial for initial leukocyte capture from the blood. L-selectin functions synergistically with integrins and chemokine receptors to mediate rolling and firm adhesion of leukocytes, followed by transendothelial migration into peripheral and mucosal tissues. Early adhesive events are controlled by expression of complex carbohydrates expressing 6-sulfo sLex on the surface of the endothelium, increased avidity of L-selectin for these ligands and endoproteolytic release of L-selectin from the cell surface. L-selectin ligand expression is controlled by enzymes that sulfate, fucosylate, and sialylate oligosaccharide branches. Importantly, L-selectin has been shown to play an important role in diseases such as ischemia/reperfusion injury, pulmonary inflammation, and diabetes, and as such it remains an attractive therapeutic target.

Despite over 20 years of research, much about the regulation and function of L-selectin remains unknown. For example, while many different ligands that can be recognized by L-selectin have been described, definitive evidence demonstrating any of these molecules to be the physiological ligand for L-selectin is lacking. Furthermore, the PSGL-1-independent L-selectin ligands expressed by leukocytes remain to be identified. Important to L-selectin function is the generation of transmembrane signals through L-selectin following ligation. These signals, described following binding of ligands and L-selectin-specific mAbs, result in increased leukocyte effector functions such as enhancement in leukocyte binding activity and subsequent chemotaxis. However, it is unclear how such a short cytoplasmic tail, lacking known binding domains for signaling molecules, mediates such a variety of functions. Much of the differences observed in the migration of lymphocyte subsets can be explained by differential expression patterns of L-selectin. An important question remaining is why L-selectin is expressed on some populations of memory and effector cells and not on others. Interestingly, the L-selectin expressing subset is often the one possessing the majority of the biological activity described to that cell population. In addition, L-selectin surface expression is actively regulated by proteases. However, despite identification of TACE as one such enzyme, other enzymes having similar activity remain to be described, as does the regulation of these enzymes themselves. Perhaps the largest hurdle remaining is the current inability to demonstrate therapeutic efficacy of targeting L-selectin in human disease. Diseases including rheumatoid arthritis, asthma, lupus, psoriasis, multiple sclerosis, inflammatory bowel diseases, and many other inflammatory disorders could all potentially benefit from L-selectin-targeted therapies. However, issues such as low affinity of ligand mimetics and redundancy among adhesion molecules will continue to be a challenge. New approaches such as targeting selectin ligand biosynthesis remain to be proven. It is possible that selectin-based therapies will not be beneficial alone but will have to be used in combination with other treatments. With increasing knowledge of the regulation and function of L-selectin, it is expected that progress will be made in the development of therapeutic approaches to disease treatment.

Acknowledgements

We thank the National Institutes of Health (grant 05-SC-NIH-1040) and the UWM Chancellor's Office for their financial support, Drs. Thomas Tedder and Guglielmo Venturi for providing valuable information included in this chapter, and Andrew Karalewitz for help in the preparation of the manuscript.

References

1   Steeber DA, Campbell MA, Basit A, Ley K, Tedder TF (1998) Optimal selectin-medi-

ated rolling of leukocytes during inflammation *in vivo* requires intercellular adhesion molecule-1 expression. *Proc Natl Acad Sci USA* 95: 7562–7567

2   Steeber DA, Tang MLK, Green NE, Zhang X-Q, Sloane JE, Tedder TF (1999) Leukocyte entry into sites of inflammation requires overlapping interactions between the L-selectin and intercellular adhesion molecule-1 pathways. *J Immunol* 163: 2176–2186

3   Spertini O, Kansas GS, Munro JM, Griffin JD, Tedder TF (1991) Regulation of leukocyte migration by activation of the leukocyte adhesion molecule-1 (LAM-1) selectin. *Nature* 349: 691–694

4   Chen A, Engel P, Tedder TF (1995) Structural requirements regulate endoproteolytic release of the L-selectin (CD62L) adhesion receptor from the cell surface of leukocytes. *J Exp Med* 182: 519–530

5   Schleiffenbaum BE, Spertini O, Tedder TF (1992) Soluble L-selectin is present in human plasma at high levels and retains functional activity. *J Cell Biol* 119: 229–238

6   Tu L, Poe JC, Kadono T, Venturi GM, Bullard DC, Tedder TF, Steeber DA (2002) A functional role for circulating mouse L-selectin in regulating leukocyte/endothelial cell interactions *in vivo*. *J Immunol* 169: 2034–2043

7   Steeber DA, Engel P, Miller AS, Sheetz MP, Tedder TF (1997) Ligation of L-selectin through conserved regions within the lectin domain activates signal transduction pathways and integrin function in human, mouse, and rat leukocytes. *J Immunol* 159: 952–963

8   Giblin PA, Hwang ST, Katsumoto TR, Rosen SD (1997) Ligation of L-selectin on T lymphocytes activates β_1 integrins and promotes adhesion to fibronectin. *J Immunol* 159: 3498–3507

9   Tsang YTM, Neelamegham S, Hu Y, Berg EL, Burns AR, Smith CW, Simon SI (1997) Synergy between L-selectin signaling and chemotactic activation during neutrophil adhesion and transmigration. *J Immunol* 159: 4566–4577

10   Ding Z, Issekutz TB, Downey GP, Waddell TK (2003) L-selectin enhances functional expression of surface CXCR4 in lymphocytes: implication for cellular activation during adhesion and migration. *Blood* 101: 4245–4252

11   Tedder TF, Matsuyama T, Rothstein DM, Schlossman SF, Morimoto C (1990) Human antigen-specific memory T cells express the homing receptor necessary for lymphocyte recirculation. *Eur J Immunol* 20: 1351–1355

12   Griffin JD, Spertini O, Ernst TJ, Belvin MP, Levine HB, Kanakura Y, Tedder TF (1990) GM-CSF and other cytokines regulate surface expression of the leukocyte adhesion molecule-1 on human neutrophils, monocytes, and their precursors. *J Immunol* 145: 576–584

13   Ord DC, Ernst TJ, Zhou LJ, Rambaldi A, Spertini O, Griffin JD, Tedder TF (1990) Structure of the gene encoding the human leukocyte adhesion molecule-1 (TQ1, Leu-8) of lymphocytes and neutrophils. *J Biol Chem* 265: 7760–7767

14   Kahn J, Ingraham RH, Shirley F, Magaki GI, Kishimoto TK (1994) Membrane proximal cleavage of L-selectin: identification of the cleavage site and a 6-kD transmembrane peptide fragment of L-selectin. *J Cell Biol* 125: 461–470

15 Peschon JJ, Slack JL, Reddy P, Stocking KL, Sunnarborg SW, Lee DC, Russell WE, Cast-
    ner BJ, Johnson RS, Fitzner JN et al (1998) An essential role for ectodomain shedding
    in mammalian development. *Science* 282: 1281–1284

16 Li Y, Brazzell J, Herrera A, Walcheck B (2006) ADAM17 deficiency by mature neutro-
    phils has differential effects on L-selectin shedding. *Blood* 108: 2275–2279

17 Walcheck B, Alexander SR, St. Hill CA, Matala E (2003) ADAM-17-independent shed-
    ding of L-selectin. *J Leukoc Biol* 74: 389–394

18 Tamatani T, Kitamura F, Kuida K, Shirao M, Mochizuki M, Suematsu M, Schmid-Schö-
    bein G, Watanabe K, Tsurufuji S, Miyasaka M (1993) Characterization of rat LECAM-1
    (L-selectin) by the use of monoclonal antibodies and evidence for the presence of soluble
    LECAM-1 in rat sera. *Eur J Immunol* 23: 2181–2188

19 Spertini O, Schleiffenbaum B, White-Owen C, Ruiz Jr P, Tedder TF (1992) ELISA
    for quantitation of L-selectin shed from leukocytes *in vivo. J Immunol Methods* 156:
    115–123

20 Ferri LE, Swartz D, Christou NV (2001) Soluble L-selectin at levels present in septic
    patients diminishes leukocyte-endothelial cell interactions in mice *in vivo*: a mechanism
    for decreased leukocyte delivery to remote sites in sepsis. *Crit Care Med* 29: 117–122

21 Faveeuw C, Preece G, Ager A (2001) Transendothelial migration of lymphocytes across
    high endothelial venules into lymph nodes is affected by metalloproteinases. *Blood* 98:
    688–695

22 Jutila MA, Rott L, Berg EL, Butcher EC (1989) Function and regulation of the neutro-
    phil MEL-14 antigen *in vivo*: comparison with LFA-1 and MAC-1. *J Immunol* 143:
    3318–3324

23 Smith CW, Kishimoto TK, Abbass O, Hughes B, Rothlein R, McIntire LV, Butcher E,
    Anderson DC (1991) Chemotactic factors regulate lectin adhesion molecule 1 (LECAM-
    1)-dependent neutrophil adhesion to cytokine-stimulated endothelial cells *in vitro. J Clin
    Invest* 87: 609–618

24 Preece G, Murphy G, Ager A (1996) Metalloproteinase-mediated regulation of L-selec-
    tin levels on leucocytes. *J Biol Chem* 271: 11634–11640

25 Walcheck B, Kahn J, Fisher JM, Wang BB, Fisk RS, Payan DG, Feehan C, Betageri R,
    Darlak K, Spatola AF et al (1996) Neutrophil rolling altered by inhibition of L-selectin
    shedding *in vitro. Nature* 380: 720–723

26 Hafezi-Moghadam A, Ley K (1999) Relevance of L-selectin shedding for leukocyte roll-
    ing *in vivo. J Exp Med* 189: 939–947

27 Hafezi-Moghadam A, Thomas KL, Prorock AJ, Huo Y, Ley K (2001) L-selectin shed-
    ding regulates leukocyte recruitment. *J Exp Med* 193: 863–872

28 Allport JR, Ding HT, Ager A, Steeber DA, Tedder TF, Luscinskas FW (1997) L-selectin
    shedding does not regulate human neutrophil attachment, rolling or transmigration
    across human vascular endothelium *in vitro. J Immunol* 158: 4365–4372

29 Venturi GM, Tu L, Kadono T, Khan AI, Fujimoto Y, Oshel P, Bock CB, Miller AS,
    Albrecht RM, Kubes P et al (2003) Leukocyte migration is regulated by L-selectin endo-
    proteolytic release. *Immunity* 19: 713–724

30 Galkina E, Tanousis K, Preece G, Tolaini M, Kioussis D, Florey O, Haskard DO, Tedder TF, Ager A (2003) L-selectin shedding does not regulate constitutive T cell trafficking but controls the migration pathways of antigen-activated T lymphocytes. *J Exp Med* 198: 1323–1335

31 Spertini O, Kansas GS, Reimann KA, Mackay CR, Tedder TF (1991) Functional and evolutionary conservation of distinct epitopes on the leukocyte adhesion molecule-1 (LAM-1) that regulate leukocyte migration. *J Immunol* 147: 942–949

32 Tu L, Chen A, Delahunty MD, Moore KL, Watson S, McEver RP, Tedder TF (1996) L-selectin binds to P-selectin glycoprotein ligand-1 on leukocytes. Interactions between the lectin, EGF and consensus repeat domains of the selectins determine ligand binding specificity. *J Immunol* 156: 3995–4004

33 Siegelman MH, Cheng IC, Weissman IL, Wakeland EK (1990) The mouse lymph node homing receptor is identical with the lymphocyte cell surface marker Ly-22: Role of the EGF domain in endothelial binding. *Cell* 61: 611–622

34 Kansas GS, Spertini O, Stoolman LM, Tedder TF (1991) Molecular mapping of functional domains of the leukocyte receptor for endothelium, LAM-1. *J Cell Biol* 114: 351–358

35 Kansas GS, Saunders KB, Ley K, Zakrzewicz A, Gibson RM, Furie BC, Furie B, Tedder TF (1994) A role for the epidermal growth factor-like domain of P-selectin in ligand recognition and cell adhesion. *J Cell Biol* 124: 609–618

36 Gibson RM, Kansas GS, Tedder TF, Furie B, Furie BC (1995) Lectin and epidermal growth factor domains of P-selectin at physiological density are the recognition unit for leukocyte binding. *Blood* 85: 151–158

37 Pigott R, Needham LA, Edwards RM, Walker C, Power C (1991) Structural and functional studies of the endothelial activation antigen endothelial leukocyte adhesion molecule-1 using a panel of monoclonal antibodies. *J Immunol* 147: 130–135

38 Patel KD, Nollert MU, McEver RP (1995) P-selectin must extend a sufficient length from the plasma membrane to mediate rolling of neutrophils. *J Cell Biol* 131: 1893–1902

39 Li SH, Burns DK, Rumberger JM, Presky DH, Wilkinson VL, Anosterio Jr M, Wolitzky BA, Norton CR, Familletti PC, Kim KJ et al (1994) Consensus repeat domains of E-selectin enhance ligand binding. *J Biol Chem* 269: 4431–4437

40 Erlandsen SL, Hasslen SR, Nelson RD (1993) Detection and spatial distribution of the beta 2 integrin (Mac-1) and L-selectin (LECAM-1) adherence receptors on human neutrophils by high-resolution field emission SEM. *J Histochem Cytochem* 41: 327–333

41 Bruehl RE, Springer TA, Bainton DF (1996) Quantitation of L-selectin distribution on human leukocyte microvilli by immunogold labeling and electron microscopy. *J Histochem Cytochem* 44: 835–844

42 von Andrian UH, Hasslen SR, Nelson SL, Erlandsen SL, Butcher EC (1995) A central role for microvillus receptor presentation in leukocyte adhesion under flow. *Cell* 82: 989–999

43 Lowe JB (2002) Glycosylation in the control of selectin counter-receptor structure and function. *Immunol Rev* 186: 19–36

44   Rosen SD (2004) Ligands for L-selectin: homing, inflammation, and beyond. *Annu Rev Immunol* 22: 129–156

45   Imai Y, Singer MS, Fennie C, Lasky LA, Rosen SD (1991) Identification of a carbo-hydrate-based endothelial ligand for a lymphocyte homing receptor. *J Cell Biol* 113: 1213–1221

46   Spertini O, Cordey A-S, Monai N, Giuffrè L, Schapira M (1996) P-selectin glycoprotein ligand-1 is a ligand for L-selectin on neutrophils, monocytes, and CD34⁺ hematopoietic progenitor cells. *J Cell Biol* 135: 523–531

47   Huang K, Geoffroy JS, Singer MS, Rosen SD (1991) A lymphocyte homing receptor (L-selectin) mediates the *in vitro* attachment of lymphocytes to myelinated tracts of the central nervous system. *J Clin Invest* 88: 1778–1783

48   Genbacev OD, Prakobphol A, Foulk RA, Krtolica AR, Ilic D, Singer MS, Yang ZQ, Kiessling LL, Rosen SD, Fisher SJ (2003) Trophoblast L-selectin-mediated adhesion at the maternal-fetal interface. *Science* 299: 405–408

49   Borsig L, Wong R, Hynes RO, Varki NM, Varki A (2002) Synergistic effects of L- and P-selectin in facilitating tumor metastasis can involve non-mucin ligands and implicate leukocytes as enhancers of metastasis. *Proc Natl Acad Sci USA* 99: 2193–2198

50   Rosen SD, Bertozzi CR (1994) The selectins and their ligands. *Curr Opin Cell Biol* 6: 663–673

51   Streeter PR, Rouse BTN, Butcher EC (1988) Immunohistologic and functional charac-terization of a vascular addressin involved in lymphocyte homing into peripheral lymph nodes. *J Cell Biol* 107: 1853–1862

52   Maly P, Thall AD, Petryniak B, Rogers CE, Smith PL, Marks RM, Kelly RJ, Gersten KM, Cheng G, Saunders TL et al (1996) The α(1,3) fucosyltransferase Fuc-TVII con-trols leukocyte trafficking through an essential role in L-, E-, and P-selectin ligand bio-synthesis. *Cell* 86: 643–653

53   Suzuki A, Andrew DP, Gonzalo J, Fukumoto M, Spellberg J, Hashiyama M, Takimoto H, Gerwin N, Webb I, Molineux G et al (1996) CD34-deficient mice have reduced eosinophil accumulation after allergen exposure and show a novel crossreactive 90-kD protein. *Blood* 87: 3550–3562

54   Kansas GS (1996) Selectins and their ligands: current concepts and controversies. *Blood* 88: 3259–3287

55   Sperandio M (2006) Selectins and glycosyltransferases in leukocyte rolling *in vivo*. *FEBS J* 273: 4377–4389

56   Kimura N, Mitsuoka C, Kanamori A, Hiraiwa N, Uchimura K, Muramatsu T, Tamatani T, Kansas GS, Kannagi R (1999) Reconstitution of functional L-selectin ligands on a cultured human endothelial cell line by cotransfection of α1-->3 fusosyltransferase VII and newly cloned GlcNAcβ:6-sulfotransferase cDNA. *Proc Natl Acad Sci USA* 96: 4530–4535

57   M'Rini C, Cheng G, Schweitzer C, Cavanagh LL, Palframan RT, Mempel TR, Warnock RA, Lowe JB, Quackenbush EJ, Von Andrian UH (2003) A novel endothelial L-selectin

ligand activity in lymph node medulla that is regulated by α(1,3)-fucosyltransferase-IV. *J Exp Med* 198: 1301–1312

58   Sperandio M, Frommhold D, Babushkina I, Ellies LG, Olson TS, Smith ML, Fritzsching B, Pauly E, Smith DF, Nobiling R et al (2006) Alpha2,3-Sialyltransferase-IV is essential for L-selectin ligand function in inflammation. *Eur J Immunol* 36: 3207–3215

59   Li X, Tu L, Murphy PG, Kadono T, Steeber DA, Tedder TF (2001) CHST1 and CHST2 sulfotransferase expression by vascular endothelial cells regulates shear-resistant leukocyte rolling *via* L-selectin. *J Leukoc Biol* 69: 565–574

60   Li X, Tedder TF (1999) CHST1 and CHST2 sulfotransferases expressed by human vascular endothelial cells: cDNA cloning, expression, and chromosomal localization. *Genomics* 55: 345–347

61   Uchimura K, Gauguet JM, Singer MS, Tsay D, Kannagi R, Muramatsu T, von Andrian UH, Rosen SD (2005) A major class of L-selectin ligands is eliminated in mice deficient in two sulfotransferases expressed in high endothelial venules. *Nat Immunol* 6: 1105–1113

62   Kawashima H, Petryniak B, Hiraoka N, Mitoma J, Huckaby V, Nakayama J, Uchimura K, Kadomatsu K, Muramatsu T, Lowe JB et al (2005) N-Acetylglucosamine-6-O-sulfotransferases 1 and 2 cooperatively control lymphocyte homing through L-selectin ligand biosynthesis in high endothelial venules. *Nat Immunol* 6: 1096–1104

63   Miyasaka M, Tanaka T (2004) Lymphocyte trafficking across high endothelial venules: dogmas and enigmas. *Nat Rev Immunol* 4: 360–370

64   Kanda H, Tanaka T, Matsumoto M, Umemoto E, Ebisuno Y, Kinoshita M, Noda M, Kannagi R, Hirata T, Murai T et al (2004) Endomucin, a sialomucin expressed in high endothelial venules, supports L-selectin-mediated rolling. *Int Immunol* 16: 1265–1274

65   Umemoto E, Tanaka T, Kanda H, Jin S, Tohya K, Otani K, Matsutani T, Matsumoto M, Ebisuno Y, Jang MH et al (2006) Nepmucin, a novel HEV sialomucin, mediates L-selectin-dependent lymphocyte rolling and promotes lymphocyte adhesion under flow. *J Exp Med* 203: 1603–1614

66   Sassetti C, Tangemann K, Singer MS, Kershaw DB, Rosen SD (1998) Identification of podocalyxin-like protein as an HEV ligand for L-selectin: parallels to CD34. *J Exp Med* 187: 1965–1975

67   Berg EL, McEvoy LM, Berlin C, Bargatze RF, Butcher EC (1993) L-selectin-mediated lymphocyte rolling on MAdCAM-1. *Nature* 366: 695–698

68   Drayson MT, Ford WL (1984) Afferent lymph and lymph borne cells: their influence on lymph node function. *Immunobiology* 168: 362–379

69   Hendriks HR, Duijvestijn AM, Kraal G (1987) Rapid decrease in lymphocyte adherence to high endothelial venules in lymph nodes deprived of afferent lymphatic vessels. *Eur J Immunol* 17: 1691–1695

70   Mebius RE, Streeter PR, Breve J, Duijvestijn AM, Kraal G (1991) The influence of afferent lymphatic vessel interruption on vascular addressin expression. *J Cell Biol* 115: 85–95

71   Mebius RE, Dowbenko D, Williams A, Fennie C, Lasky LA, Watson SR (1993) Expres-

sion of GlyCAM-1, an endothelial ligand for L-selectin, is affected by afferent lymphatic flow. *J Immunol* 151: 6769–6776

72 Lacorre DA, Baekkevold ES, Garrido I, Brandtzaeg P, Haraldsen G, Amalric F, Girard JP (2004) Plasticity of endothelial cells: rapid dedifferentiation of freshly isolated high endothelial venule endothelial cells outside the lymphoid tissue microenvironment. *Blood* 103: 4164–4172

73 Mebius RE, Breve J, Duijvestijn AM, Kraal G (1990) The function of high endothelial venules in mouse lymph nodes stimulated by oxazolone. *Immunology* 71: 423–427

74 Steeber DA, Erickson CM, Hodde KC, Albrecht RM (1987) Vascular changes in popliteal lymph nodes due to antigen challenge in normal and lethally irradiated mice. *Scanning Microsc* 1: 831–839

75 Kittas C, Henry L (1981) BCG-induced changes in the post capillary venules of the guinea pig lymph node. *Lymphology* 14: 24–28

76 Webster B, Ekland EH, Agle LM, Chyou S, Ruggieri R, Lu TT (2006) Regulation of lymph node vascular growth by dendritic cells. *J Exp Med* 203: 1903–1913

77 Hoke D, Mebius RE, Dybdal N, Dowbenko D, Gribling P, Kyle C, Baumhueter S, Watson SR (1995) Selective modulation of the expression of L-selectin ligands by an immune response. *Curr Biol* 5: 670–678

78 Swarte VV, Joziasse DH, Van den Eijnden DH, Petryniak B, Lowe JB, Kraal G, Mebius RE (1998) Regulation of fucosyltransferase-VII expression in peripheral lymph node high endothelial venules. *Eur J Immunol* 28: 3040–3047

79 Giuffrè L, Cordey A-S, Monai N, Tardy Y, Schapira M, Spertini O (1997) Monocyte adhesion to activated aortic endothelium: role of L-selectin and heparan sulfate proteoglycans. *J Cell Biol* 136: 945–956

80 Robinson LA, Tu LL, Steeber DA, Preis O, Platt JL, Tedder TF (1998) The role of adhesion molecules in human leukocyte attachment to porcine vascular endothelium: implications for xenotransplantation. *J Immunol* 161: 6931–6938

81 Zakrzewicz A, Grafe M, Terbeek D, Bongrazio M, Auch-Schwelk W, Walzog B, Graf K, Fleck E, Ley K, Gaehtgens P (1997) L-selectin-dependent leukocyte adhesion to microvascular but not to macrovascular endothelial cells of the human coronary system. *Blood* 89: 3228–3235

82 Spertini O, Luscinskas FW, Kansas GS, Munro JM, Griffin JD, Gimbrone Jr MA, Tedder TF (1991) Leukocyte adhesion molecule-1 (LAM-1, L-selectin) interacts with an inducible endothelial cell ligand to support leukocyte adhesion. *J Immunol* 147: 2565–2573

83 Cannella B, Cross AH, Raine CS (1990) Upregulation and coexpression of adhesion molecules correlate with relapsing autoimmune demyelination in the central nervous system. *J Exp Med* 172: 1521–1524

84 Kabel PJ, Voorbij HAM, de Haan-Meulman M, Pals ST, Drexhage HA (1989) High endothelial venules present in lymphoid cell accumulations in thyroids affected by autoimmune disease: a study in men and BB rats of functional activity and development. *J Clin Endocrinol Metab* 68: 744–751

85 van Dinther-Janssen ACHM, Pals ST, Scheper R, Breedveld F, Meijer CJLM (1990)

Dendritic cells and high endothelial venules in the rheumatoid synovial membrane. *J Rheumatol* 17: 11–17

86 Hanninen A, Taylor C, Streeter PR, Stark LS, Sarte JM, Shizuru JA, Simell O, Michie SA (1993) Vascular addressins are induced on islet vessels during insulitis in nonobese diabetic mice and are involved in lymphoid binding to islet endothelium. *J Clin Invest* 92: 2509–2515

87 Rosen SD, Tsay D, Hemmerich S, Abraham WM (2005) Therapeutic targeting of endothelial ligands for L-selectin (PNAd) in a sheep model of asthma. *Am J Pathol* 166: 935–944

88 Kirveskari J, Paavonen T, Hayry P, Renkonen R (2000) De novo induction of endothelial L-selectin ligands during kidney allograft rejection. *J Am Soc Nephrol* 11: 2358–2365

89 Rivera-Nieves J, Burcin TL, Olson TS, Morris MA, McDuffie M, Cominelli F, Ley K (2006) Critical role of endothelial P-selectin glycoprotein ligand 1 in chronic murine ileitis. *J Exp Med* 203: 907–917

90 Lee MS, Sarvetnick N (1994) Induction of vascular addressins and adhesion molecules in the pancreas of IFN-gamma transgenic mice. *J Immunol* 152: 4597–4603

91 Sikorski EE, Hallmann R, Berg EL, Butcher EC (1993) The Peyer's patch high endothelial receptor for lymphocytes, the mucosal addressin, is induced on a murine endothelial cell line by tumor necrosis factor-α and IL-1. *J Immunol* 151: 5239–5250

92 Pablos JL, Santiago B, Tsay D, Singer MS, Palao G, Galindo M, Rosen SD (2005) A HEV-restricted sulfotransferase is expressed in rheumatoid arthritis synovium and is induced by lymphotoxin-α/β and TNF-α in cultured endothelial cells. *BMC Immunol* 6: 6–15

93 Drayton DL, Yang X, Lee J, Lesslauer W, Ruddle NH (2003) Ectopic LTαβ directs lymphoid organ neogenesis with concomitant expression of peripheral node addressin and a HEV-restricted sulfotransferase. *J Exp Med* 197: 1153–1163

94 Laszik Z, Jansen PJ, Cummings RD, Tedder TF, McEver RP, Moore KL (1996) P-selectin glycoprotein ligand-1 is broadly expressed in cells of myeloid, lymphoid, and dendritic lineage and in some non-hematopoietic cells. *Blood* 88: 3010–3021

95 Sako D, Chang X-J, Barone KM, Vachino G, White HM, Shaw G, Veldman KM, Bean KM, Ahern TJ, Furie B et al (1993) Expression cloning of a functional glycoprotein ligand for P-selectin. *Cell* 75: 1179–1186

96 Moore KL, Stults NL, Diaz S, Smith DF, Cummings RD, Varki A, McEver RP (1992) Identification of a specific glycoprotein ligand for P-selectin (CD62) on myeloid cells. *J Cell Biol* 118: 445–456

97 Yang J, Hirata T, Croce K, Merrill-Skoloff G, Tchernychev B, Williams E, Flaumenhaft R, Furie BC, Furie B (1999) Targeted gene disruption demonstrates that P-selectin glycoprotein ligand-1 (PSGL-1) is required for P-selectin-mediated but not E-selectin-mediated neutrophil rolling and migration. *J Exp Med* 190: 1769–1782

98 Kanamori A, Kojima N, Uchimura K, Muramatsu T, Tamatani T, Berndt MC, Kansas GS, Kannagi R (2002) Distinct sulfation requirements of selectins disclosed using cells that support rolling mediated by all three selectins under shear flow. L-selectin prefers

carbohydrate 6-sulfation to tyrosine sulfation, whereas P-selectin does not. *J Biol Chem* 277: 32578–32586

99 Bernimoulin MP, Zeng XL, Abbal C, Giraud S, Martinez M, Michielin O, Schapira M, Spertini O (2003) Molecular basis of leukocyte rolling on PSGL-1. Predominant role of core-2 O-glycans and of tyrosine sulfate residue 51. *J Biol Chem* 278: 37–47

100 Nicholson MW, Barclay AN, Singer MS, Rosen SD, van der Merwe PA (1998) Affinity and kinetic analysis of L-selectin (CD62L) binding to glycosylation-dependent cell-adhesion molecule-1. *J Biol Chem* 273: 763–770

101 Leppanen A, Yago T, Otto VI, McEver RP, Cummings RD (2003) Model glycosulfopeptides from PSGL-1 require tyrosine sulfation and a core-2 branched O-glycan to bind to L-selectin. *J Biol Chem* 278: 26391–26400

102 Fieger CB, Sassetti CM, Rosen SD (2003) Endoglycan, a member of the CD34 family, functions as an L-selectin ligand through modifications with tyrosine sulfation and sialyl Lewis X. *J Biol Chem* 278: 27390–27398

103 Sassetti C, Van Zante M, Rosen SD (2000) Identification of endoglycan, a member of the CD34/podocalyxin family of sialomucins. *J Biol Chem* 275: 9001–9010

104 Bargatze RF, Kurk S, Butcher EC, Jutila MA (1994) Neutrophils roll on adherent neutrophils bound to cytokine-induced endothelial cells *via* L-selectin on the rolling cells. *J Exp Med* 180: 1785–1792

105 Alon R, Fuhlbrigge RC, Finger EB, Springer TA (1996) Interactions through L-selectin between leukocytes and adherent leukocytes nucleate rolling adhesions on selectins and VCAM-1 in shear flow. *J Cell Biol* 135: 849–865

106 Eriksson EE, Xie X, Werr J, Thoren P, Lindbom L (2001) Importance of primary capture and L-selectin-dependent secondary capture in leukocyte accumulation in inflammation and atherosclerosis *in vivo*. *J Exp Med* 194: 205–218

107 Sperandio M, Smith ML, Forlow SB, Olson TS, Xia L, McEver RP, Ley K (2003) P-selectin glycoprotein ligand-1 mediates L-selectin-dependent leukocyte rolling in venules. *J Exp Med* 197: 1355–1363

108 Butcher EC (1991) Leukocyte-endothelial cell recognition: three (or more) steps to specificity and diversity. *Cell* 67: 1033–1036

109 Constantin G, Majeed M, Giagulli C, Piccio L, Kim JY, Butcher EC, Laudanna C (2000) Chemokines trigger immediate $\beta2$ integrin affinity and mobility changes: differential regulation and roles in lymphocyte arrest under flow. *Immunity* 13: 759–769

110 Chan JR, Hyduk SJ, Cybulsky MI (2001) Chemoattractants induce a rapid and transient upregulation of monocyte $\alpha4$ integrin affinity for vascular cell adhesion molecule-1 which mediate arrest: an early step in the process of emigration. *J Exp Med* 193: 1149–1158

111 Steeber DA, Tedder TF (2000) Adhesion molecule cascades direct lymphocyte recirculation and leukocyte migration during inflammation. *Immunol Res* 22: 299–317

112 Ley KE, Bullard D, Arbones ML, Bosse R, Vestweber D, Tedder TF, Beaudet AL (1995) Sequential contribution of L- and P-selectin to leukocyte rolling *in vivo*. *J Exp Med* 181: 669–675

113 Tedder TF, Steeber DA, Chen A, Engel P (1995) The selectins: vascular adhesion molecules. *FASEB J* 9: 866–873

114 Kunkel EJ, Ley K (1996) Distinct phenotype of E-selectin-deficient mice. E-selectin is required for slow leukocyte rolling *in vivo*. *Circ Res* 79: 1196–1204

115 Ley K, Allietta M, Bullard DC, Morgan SJ (1998) The importance of E-selectin for firm leukocyte adhesion *in vivo*. *Circ Res* 83: 287–294

116 Arbones ML, Ord DC, Ley K, Radich H, Maynard-Curry C, Capon DJ, Tedder TF (1994) Lymphocyte homing and leukocyte rolling and migration are impaired in L-selectin-deficient mice. *Immunity* 1: 247–260

117 Mayadas TN, Johnson RC, Rayburn H, Hynes RO, Wagner DD (1993) Leukocyte rolling and extravasation are severely compromised in P-selectin-deficient mice. *Cell* 74: 541–554

118 Jung U, Ley K (1999) Mice lacking two and all three selectins demonstrate overlapping and distinct functions for each selectin. *J Immunol* 162: 6755–6762

119 Finger EB, Puri KD, Alon R, Lawrence MB, von Andrian UH, Springer TA (1996) Adhesion through L-selectin requires a threshold hydrodynamic shear. *Nature* 379: 266–269

120 Alon R, Chen S, Puri KD, Finger EB, Springer TA (1997) The kinetics of L-selectin tethers and the mechanics of selectin-mediated rolling. *J Cell Biol* 138: 1169–1180

121 Marshall BT, Long M, Piper JW, Yago T, McEver RP, Zhu C (2003) Direct observation of catch bonds involving cell adhesion molecules. *Nature* 423: 190–193

122 Sarangapani KK, Yago T, Klopocki AG, Lawrence MB, Fieger CB, Rosen SD, McEver RP, Zhu C (2004) Low force decelerates L-selectin dissociation from P-selectin glycoprotein ligand-1 and endoglycan. *J Biol Chem* 279: 2291–2298

123 Lou J, Yago T, Klopocki AG, Mehta P, Chen W, Zarnitsyna VI, Bovin NV, Zhu C, McEver RP (2006) Flow-enhanced adhesion regulated by a selectin interdomain hinge. *J Cell Biol* 174: 1107–1117

124 Alon R, Kassner PD, Carr MC, Finger EB, Hemler ME, Springer TA (1995) The integrin VLA-4 supports tethering and rolling in flow on VCAM-1. *J Cell Biol* 128: 1243–1253

125 Berlin C, Bargatze RF, Campbell JJ, von Andrian UH, Szabo MC, Hasslen SR, Nelson RD, Berg EL, Erlandsen SL, Butcher EC (1995) α4 integrins mediate lymphocyte attachment and rolling under physiologic flow. *Cell* 80: 413–422

126 Bargatze RF, Jutila MA, Butcher EC (1995) Distinct roles of L-selectin and integrins α4β7 and LFA-1 in lymphocyte homing to Peyer's patch-HEV in situ: the multistep hypothesis confirmed and refined. *Immunity* 3: 99–108

127 Nandi A, Estess P, Siegelman M (2004) Bimolecular complex between rolling and firm adhesion receptors required for cell arrest: CD44 association with VLA-4 in T cell extravasation. *Immunity* 20: 455–465

128 Abitorabi MA, Pachynski RK, Ferrando RE, Tidswell M, Erle DJ (1997) Presentation of integrins on leukocyte microvilli: a role for the extracellular domain in determining membrane localization. *J Cell Biol* 139: 563–571

129 Kadono T, Venturi GM, Steeber DA, Tedder TF (2002) Leukocyte rolling velocities and migration are optimized by cooperative L-selectin and intercellular adhesion molecule-1 functions. *J Immunol* 169: 4542–4550

130 Waddell TK, Fialkow L, Chan CK, Kishimoto TK, Downey GP (1994) Potentiation of the oxidative burst of human neutrophils. A signaling role for L-selectin. *J Biol Chem* 269: 18485–18491

131 Laudanna C, Constantin G, Baron P, Scarpini E, Scarlato G, Cabrini G, Dechecchi C, Rossi F, Cassatella MA, Berton G (1994) Sulfatides trigger increase of cytosolic free calcium and enhanced expression of tumor necrosis factor-α and interleukin-8 mRNA in human neutrophils. Evidence for a role of L-selectin as a signaling molecule. *J Biol Chem* 269: 4021–4026

132 Waddell TK, Fialkow L, Chan CK, Kishimoto TK, Downey GP (1995) Signaling functions of L-selectin. Enhancement of tyrosine phosphorylation and activation of MAP kinase. *J Biol Chem* 270: 15403–15411

133 Chen C, Ba X, Xu T, Cui L, Hao S, Zeng X (2006) c-Abl is involved in F-actin assembly triggered by L-selectin crosslinking. *J Biochem* 140: 229–235

134 Brenner B, Weinmann S, Grassme H, Lang F, Linderkamp O, Gulbins E (1997) L-selectin activates JNK *via* src-like tyrosine kinases and the small G-protein Rac. *Immunology* 92: 214–219

135 Brenner B, Gulbins E, Schlottmann K, Koppenhoefer U, Busch GL, Walzog B, Steinhausen M, Coggeshall KM, Linderkamp O, Lang F (1996) L-selectin activates the Ras pathway *via* the tyrosine kinase p56[lck]. *Proc Natl Acad Sci USA* 93: 15376–15381

136 Brenner B, Kadel S, Grigorovich S, Linderkamp O (2002) Mechanisms of L-selectin-induced activation of nuclear factor of activated T lymphocytes (NFAT). *Biochem Biophys Res Commun* 291: 237–244

137 Ding Z, Kawashima H, Miyasaka M (2000) Sulfatide binding and activation of leukocytes through an L-selectin-independent pathway. *J Leukoc Biol* 68: 65–72

138 Simon SI, Burns AR, Taylor AD, Gopalan PK, Lynam EB, Sklar LA, Smith CW (1995) L-selectin (CD62L) cross-linking signals neutrophil adhesive functions *via* the Mac-1 (CD11b/CD18) β_2-integrin. *J Immunol* 155: 1502–1514

139 Strauch UG, Holzmann B (1993) Triggering of L-selectin (gp90[MEL-14]) induces homotypic lymphocyte adhesion by a mechanism independent of LFA-1. *Int Immunol* 5: 393–398

140 Hwang ST, Singer MS, Giblin PA, Yednock TA, Bacon KB, Simon SI, Rosen SD (1996) GlyCAM-1, a physiologic ligand for L-selectin, activates β2 integrins on naive peripheral lymphocytes. *J Exp Med* 184: 1343–1348

141 Kanwar S, Steeber DA, Tedder TF, Hickey MJ, Kubes P (1999) Overlapping roles for L-selectin and P-selectin in antigen-induced immune responses in the microvasculature. *J Immunol* 162: 2709–2716

142 Hickey MJ, Forster M, Mitchell D, Kaur J, De Caigny C, Kubes P (2000) L-selectin facilitates emigration and extravascular locomotion of leukocytes during acute inflammatory responses *in vivo*. *J Immunol* 165: 7164–7170

143 Subramanian H, Kodera M, Conway RM, Steeber DA (2006) Signaling through L-selectin enhances T cell chemotaxis to secondary lymphoid tissue chemokine (SLC). *J Immunol* 176: S37–38

144 Defilippi P, Rosso A, Dentelli P, Calvi C, Garbarino G, Tarone G, Pegoraro L, Brizzi MF (2005) β1 Integrin and IL-3R coordinately regulate STAT5 activation and anchorage-dependent proliferation. *J Cell Biol* 168: 1099–1108

145 Haribabu B, Steeber DA, Ali H, Richardson RM, Snyderman R, Tedder TF (1997) Chemoattractant receptor-induced phosphorylation of L-selectin. *J Biol Chem* 272: 13961–13965

146 Kilian K, Dernedde J, Mueller E-C, Bahr I, Tauber R (2004) The interaction of protein kinase C isozymes α, ι, θ with the cytoplasmic domain of L-selectin is modulated by phosphorylation of the receptor. *J Biol Chem* 279: 34472–34480

147 Li X, Steeber DA, Tang MLK, Farrar MA, Perlmutter RM, Tedder TF (1998) Regulation of L-selectin-mediated rolling through receptor dimerization. *J Exp Med* 188: 1385–1390

148 Dwir O, Steeber DA, Schwarz US, Camphausen RT, Kansas GS, McEver RP, Tedder TF, Alon R (2002) L-selectin dimerization enhances tether formation to properly clustered ligand: evidence that shear stress increases local availability of selectin ligands at adhesive contacts. *J Biol Chem* 277: 21130–21139

149 Kansas GS, Ley K, Munro JM, Tedder TF (1993) Regulation of leukocyte rolling and adhesion to HEV through the cytoplasmic domain of L-selectin. *J Exp Med* 177: 833–838

150 Dwir O, Kansas GS, Alon R (2001) Cytoplasmic anchorage of L-selectin controls leukocyte capture and rolling by increasing the mechanical stability of the selectin tether. *J Cell Biol* 155: 145–156

151 Pavalko FM, Walker DM, Graham L, Goheen M, Doerschuk CM, Kansas GS (1995) The cytoplasmic domain of L-selectin interacts with cytoskeletal proteins *via* alpha-actinin: receptor positioning in microvilli does not require interaction with alpha-actinin. *J Cell Biol* 129: 1155–1164

152 Ivetic A, Deka J, Ridley A, Ager A (2002) The cytoplasmic tail of L-selectin interacts with members of the ezrin-radixin-moesin (ERM) family of proteins: cell activation-dependent binding of moesin but not ezrin. *J Biol Chem* 277: 2321–2329

153 Kahn J, Walcheck B, Migaki GI, Jutila MA, Kishimoto TK (1998) Calmodulin regulates L-selectin adhesion molecule expression and function through a protease-dependent mechanism. *Cell* 92: 809–818

154 Leid JG, Steeber DA, Tedder TF, Jutila MA (2001) Antibody binding to a conformation-dependent epitope induces L-selectin association with the detergent-resistant cytoskeleton. *J Immunol* 166: 4899–4907

155 Evans SS, Schleider DM, Bowman LA, Francis ML, Kansas GS, Black JD (1999) Dynamic association of L-selectin with the lymphocyte cytoskeletal matrix. *J Immunol* 162: 3615–3624

156 Mattila PE, Green CE, Schaff U, Simon SI, Walcheck B (2005) Cytoskeletal interactions regulate inducible L-selectin clustering. *Am J Physiol Cell Physiol* 289: C323-C332

157 Smalley DM, Ley K (2005) L-selectin: mechanisms and physiological significance of ectodomain cleavage. *J Cell Mol Med* 9: 255–266

158 Gowans JL, Knight EJ (1964) The route of recirculation of lymphocytes in the rat. *Proc R Soc Lond B Biol Sci* 159: 257–282

159 Stamper Jr HB, Woodruff JJ (1976) Lymphocyte homing into lymph nodes: *in vitro* demonstration of the selective affinity of recirculating lymphocytes for high-endothelial venules. *J Exp Med* 144: 828–833

160 Gallatin WM, Weissman IL, Butcher EC (1983) A cell-surface molecule involved in organ-specific homing of lymphocytes. *Nature* 304: 30–34

161 Steeber DA, Green NE, Sato S, Tedder TF (1996) Lymphocyte migration in L-selectin-deficient mice: altered subset migration and aging of the immune system. *J Immunol* 157: 1096–1106

162 Holzmann B, McIntyre BW, Weissman IL (1989) Identification of a murine Peyer's patch-specific lymphocyte homing receptor as an integrin molecule with an α chain homologous to human VLA-4α. *Cell* 56: 37–46

163 Wagner N, Lohler J, Kunkel EJ, Ley K, Leung E, Krissansen G, Rajewsky K, Müller W (1996) Critical role for β7 integrins in formation of the gut-associated lymphoid tissue. *Nature* 382: 366–370

164 Steeber DA, Tang MLK, Zhang X-Q, Müller W, Wagner N, Tedder TF (1998) Efficient lymphocyte migration across high endothelial venules of mouse Peyer's patches requires overlapping expression of L-selectin and β7 integrin. *J Immunol* 161: 6638–6647

165 Kunkel EJ, Ramos CL, Steeber DA, Müller W, Wagner N, Tedder TF, Ley K (1998) The roles of L-selectin, β7 integrins and P-selectin in leukocyte rolling and adhesion in high endothelial venules of Peyer's patches. *J Immunol* 161: 2449–2456

166 Wagner N, Löhler J, Tedder TF, Rajewsky K, Müller W, Steeber DA (1998) L-selectin and β7 integrin synergistically mediate lymphocyte migration to mesenteric lymph nodes. *Eur J Immunol* 28: 3832–3839

167 Nakache M, Berg E, Streeter P, Butcher E (1989) The mucosal vascular addressin is a tissue-specific endothelial cell adhesion molecule for circulating lymphocytes. *Nature* 337: 179–181

168 Streeter PR, Berg EL, Rouse BN, Bargatze RF, Butcher EC (1988) A tissue-specific endothelial cell molecule involved in lymphocyte homing. *Nature* 331: 41–46

169 Csencsits KL, Jutila MA, Pascual DW (1999) Nasal-associated lymphoid tissue: phenotypic and functional evidence for the primary role of peripheral node addressin in naive lymphocyte adhesion to high endothelial venules in a mucosal site. *J Immunol* 163: 1382–1389

170 Tedder TF, Steeber DA, Pizcueta P (1995) L-selectin deficient mice have impaired leukocyte recruitment into inflammatory sites. *J Exp Med* 181: 2259–2264

171 Catalina MD, Carroll MC, Arizpe H, Takashima A, Estess P, Siegelman MH (1996)

The route of antigen entry determines the requirement for L-selectin during immune responses. *J Exp Med* 184: 2341–2351

172 Tang MLK, Hale LP, Steeber DA, Tedder TF (1997) L-selectin is involved in lymphocyte migration to sites of inflammation in the skin: delayed rejection of allografts in L-selectin-deficient mice. *J Immunol* 158: 5191–5199

173 Xu J, Grewal IS, Geba GP, Flavell RA (1996) Impaired primary T cell responses in L-selectin-deficient mice. *J Exp Med* 183: 589–598

174 Steeber DA, Green NE, Sato S, Tedder TF (1996) Humoral immune responses in L-selectin-deficient mice. *J Immunol* 157: 4899–4907

175 Csencsits KL, Walters N, Pascual DW (2001) Dichotomy of homing receptor dependence by mucosal effector B cells: α_E *versus* L-selectin. *J Immunol* 167: 2441–2445

176 Pascual DW, White MD, Larson TL, Walters N (2001) Impaired mucosal immunity in L-selectin-deficient mice orally immunized with a *Salmonella* vaccine vector. *J Immunol* 167: 407–415

177 Galkina E, Kadl A, Sanders J, Varughese D, Sarembock IJ, Ley K (2006) Lymphocyte recruitment into the aortic wall before and during development of atherosclerosis is partially L-selectin dependent. *J Exp Med* 203: 1273–1282

178 Mackay CR, Kimpton WG, Brandon MR, Cahill RNP (1988) Lymphocyte subsets show marked differences in their distribution between blood and the afferent and efferent lymph of peripheral lymph nodes. *J Exp Med* 167: 1755–1765

179 Abernethy NJ, Hay JB, Kimpton WG, Washington EA, Cahill RNP (1990) Nonrandom recirculation of small, CD4+ and CD8+ T lymphocytes in sheep: evidence for lymphocyte subset-specific endothelial cell recognition. *Int Immunol* 2: 231–238

180 Kimpton WG, Washington EA, Cahill RNP (1989) Recirculation of lymphocyte subsets (CD5+, CD4+, CD8+, SBU-T19+, and B cells) through gut and peripheral lymph nodes. *Immunology* 66: 69–75

181 Westermann J, Nagahori Y, Walter S, Heerwagen C, Miyasaka M, Pabst R (1994) B and T lymphocyte subsets enter peripheral lymph nodes and Peyer's patches without preference *in vivo*: no correlation occurs between their localization in different types of high endothelial venules and the expression of CD44, VLA-4, LFA-1, ICAM-1, CD2 or L-selectin. *Eur J Immunol* 24: 2312–2316

182 Stevens SK, Weissman IL, Butcher EC (1982) Differences in the migration of B and T lymphocytes: organ-selective localization *in vivo* and the role of lymphocyte-endothelial cell recognition. *J Immunol* 128: 844–851

183 Sprent J, Basten A (1973) Circulating T and B lymphocytes of the mouse. II. Lifespan. *Cell Immunol* 7: 40–59

184 Kraal G, Weissman IL, Butcher EC (1983) Differences in *in vivo* distribution and homing of T cell subsets to mucosal vs nonmucosal lymphoid organs. *J Immunol* 130: 1097–1102

185 Tang MLK, Steeber DA, Zhang X-Q, Tedder TF (1998) Intrinsic differences in L-selectin expression levels affect T and B lymphocyte subset-specific recirculation pathways. *J Immunol* 160: 5113–5121

186 Andrew DP, Rott LS, Kilshaw PJ, Butcher EC (1996) Distribution of α4β7 and αEβ7 integrins on thymocytes, intestinal epithelial lymphocytes and peripheral lymphocytes. *Eur J Immunol* 26: 897–905

187 Witherden DA, Kimpton WG, Washington EA, Cahill RNP (1990) Non-random migration of CD4⁺, CD8⁺ and γ/δ⁺T19⁺ lymphocytes through peripheral lymph nodes. *Immunology* 70: 235–240

188 Washington EA, Kimpton WG, Cahill RNP (1988) CD4⁺ lymphocytes are extracted from blood by peripheral lymph nodes at different rates than other T cell subsets and B cells. *Eur J Immunol* 18: 2093–2096

189 Reynolds JD, Chin W, Shmoorkoff J (1988) T and B cells have similar recirculation kinetics in sheep. *Eur J Immunol* 18: 835–840

190 Chao CC, Jensen R, Dailey MO (1997) Mechanisms of L-selectin regulation by activated T cells. *J Immunol* 159: 1686–1694

191 Tedder TF, Penta AC, Levine HB, Freedman AS (1990) Expression of the human leukocyte adhesion molecule, LAM1. Identity with the TQ1 and Leu-8 differentiation antigens. *J Immunol* 144: 532–540

192 Jung TM, Gallatin WM, Weissman IL, Dailey MO (1988) Down-regulation of homing receptors after T cell activation. *J Immunol* 141: 4110–4117

193 Bradley LM, Atkins GG, Swain SS (1992) Long-term memory CD4⁺ T cells from spleen lack MEL-14, the lymph node homing receptor. *J Immunol* 148: 324–331

194 Mobley JL, Dailey MO (1992) Regulation of adhesion molecule expression by CD8 T cells *in vivo*. I. Differential regulation of gp90^MEL14 (LECAM-1), Pgp-1, LFA-1, and VLA-4α during the differentiation of CTL induced by allografts. *J Immunol* 148: 2348–2356

195 Kanegane H, Kasahare Y, Niida Y, Yachie A, Sugii S, Takatsu K, Taniguchi N, Miyawaki T (1996) Expression of L-selectin (CD62L) discriminates Th1- and Th2-like cytokine-producing memory CD4⁺ T cells. *Immunology* 87: 186–190

196 Matsuzaki S, Shinozaki K, Kobayashi N, Agematsu K (2005) Polarization of Th1/Th2 in human CD4 T cells separated by CD62L: analysis by transcription factors. *Allergy* 60: 780–787

197 van Wely CA, Beverley PCL, Brett SJ, Britten CJ, Tite JP (1999) Expression of L-selectin on Th1 cells is regulated by IL-12. *J Immunol* 163: 1214–1221

198 Austrup F, Vestweber D, Borges E, Lohning M, Brauer R, Herz U, Renz H, Hallmann R, Scheffold A, Radbruch A et al (1996) P- and E-selectin mediate recruitment of T-helper-1 but not T-helper-2 cells into inflamed tissues. *Nature* 358: 81–83

199 Borges E, Pendl G, Eythner R, Steegmaier M, Zöllner O, Vestweber D (1997) The binding of T cell-expressed P-selectin glycoprotein ligand-1 to E- and P-selectin is differentially regulated. *J Biol Chem* 272: 28786–28792

200 Bonder CS, Norman MU, MacRae T, Mangan PR, Weaver CT, Bullard DC, McCafferty D-M, Kubes P (2005) P-selectin can support both Th1 and Th2 lymphocyte rolling in the intestinal microvasculature. *Am J Pathol* 167: 1647–1660

201 Mangan PR, O'Quinn DB, Harrington L, Bonder CS, Kubes P, Kucik DF, Bullard DC,

Weaver CT (2005) Both Th1 and Th2 cells require P-selectin glycoprotein ligand-1 for optimal rolling on inflamed endothelium. *Am J Pathol* 167: 1661–1675

202 Tietz W, Allemand Y, Borges E, von Laer D, Hallmann R, Vestweber D, Hamann A (1998) CD4⁺ T cells migrate into inflamed skin only if they express ligands for E- and P-selectin. *J Immunol* 161: 963–970

203 Wagers AJ, Waters CM, Stoolman LM, Kansas GS (1998) Interleukin 12 and interleukin 4 control T cell adhesion to endothelial selectins through opposite effects on α1,3-fucosyltransferase VII gene expression. *J Exp Med* 188: 2225–2231

204 Lim Y, Henault L, Wagers AJ, Kansas GS, Luscinskas FW, Lichtman AH (1999) Expression of functional selectin ligands on Th cells is differentially regulated by IL-12 and IL-4. *J Immunol* 162: 3193–3201

205 Blander JM, Visintin I, Janeway Jr. CA, Medzhitov R (1999) α(1,3)-Fucosyltransferase VII and α(2,3)-sialyltransferase IV are up-regulated in activated CD4 T cells and maintained after their differentiation into Th1 and migration into inflammatory sites. *J Immunol* 163: 3746–3752

206 Biedermann T, Schwarzler C, Lawmetschewandtner G, Thoma G, Carballido-Perrig N, Kund J, de Vries JE, Rot A, Carballido JM (2002) Targeting CLA/E-selectin interactions prevents CCR4-mediated recruitment of human Th2 memory cells to human skin *in vivo*. *Eur J Immunol* 32: 3171–3180

207 Bettelli E, Carrier Y, Gao W, Korn T, Strom TB, Oukka M, Weiner HL, Kuchroo VK (2006) Reciprocal developmental pathways for the generation of pathogenic effector Th17 and regulatory T cells. *Nature* 441: 235–238

208 Mangan PR, Harrington LE, O'Quinn DB, Helms WS, Bullard DC, Elson CO, Hatton RD, Wahl SM, Schoeb TR, Weaver CT (2006) Transforming growth factor-β induces development of the Th17 lineage. *Nature* 441: 231–234

209 Swain SL, Bradley LM, Croft M, Tonkonogy S, Atkins G, Weinberg AD, Duncan DD, Hedrick SM, Dutton RW, Huston G (1991) Helper T cell subsets: phenotype, function, and the role of lymphokines in regulating their development. *Immunol Rev* 123: 115–144

210 Tsuji T, Nibu R, Iwai K, Kanegane H, Yachie A, Seki H, Miyawaki T, Taniguchi N (1994) Efficient induction of immunoglobulin production in neonatal naive B cells by memory CD4⁺ T cell subset expressing homing receptor L-selectin. *J Immunol* 152: 4417–4424

211 Picker LJ (1993) Regulation of tissue-selective T-lymphocyte homing receptors during the virgin to memory/effector cell transition in human secondary lymphoid tissues. *Am Rev Respir Dis* 148: S47–54

212 Campbell DJ, Butcher EC (2002) Rapid acquisition of tissue-specific homing phenotypes by CD4⁺ T cells activated in cutaneous or mucosal lymphoid tissues. *J Exp Med* 195: 135–141

213 Mackay CR, Marston WL, Dudler L, Spertini O, Tedder TF, Hein WR (1992) Tissue-specific migration pathways by phenotypically distinct subpopulations of memory T cells. *Eur J Immunol* 22: 887–895

214 Mackay CR, Marston W, Dudler L (1992) Altered patterns of T cell migration through lymph nodes and skin following antigen challenge. *Eur J Immunol* 22: 2205–2210

215 Mackay CR, Marston WL, Dudler L (1990) Naive and memory T cells show distinct pathways of lymphocyte recirculation. *J Exp Med* 171: 801–817

216 Sallusto F, Lenig D, Forster R, Lipp M, Lanzavecchia A (1999) Two subsets of memory T lymphocytes with distinct homing potentials and effector functions. *Nature* 401: 708–712

217 Masopust D, Vezys V, Marzo AL, Lefrancois L (2001) Preferential localization of effector memory cells in nonlymphoid tissue. *Science* 291: 2413–2417

218 Unsoeld H, Pircher H (2005) Complex memory T-cell phenotypes revealed by coexpression of CD62L and CCR7. *J Virol* 79: 4510–4513

219 Marzo AL, Klonowski KD, Le Bon A, Borrow P, Tough DF, Lefrancois L (2005) Initial T cell frequency dictates memory CD8[+] T cell lineage commitment. *Nat Immunol* 6: 793–799

220 Lefrancois L (2006) Development, trafficking, and function of memory T-cell subsets. *Immunol Rev* 211: 93–103

221 Sakaguchi S, Sakaguchi N, Asano M, Itoh M, Toda M (1995) Immunologic self-tolerance maintained by activated T cell expressing IL-2 receptor α-chains (CD25): breakdown of a single mechanism of self-tolerance causes various autoimmune diseases. *J Immunol* 155: 1151–1164

222 Sakaguchi S, Sakaguchi N, Shimizu J, Yamazaki S, Sakihama T, Itoh M, Kuniyasu Y, Nomura T, Toda M, Takahashi T (2001) Immunologic tolerance maintained by CD25[+]CD4[+] regulatory T cells: their common role in controlling autoimmunity, tumor immunity, and transplantation tolerance. *Immunol Rev* 182: 18–32

223 Shevach EM (2002) CD4[+]CD25[+] suppressor T cells: more questions than answers. *Nat Rev Immunol* 2: 389–400

224 Maloy KJ, Powrie F (2001) Regulatory T cells in the control of immune pathology. *Nat Immunol* 2: 816–822

225 Lee IV MK, Moore DJ, Jarrett BP, Lian MM, Deng S, Huang X, Markmann JW, Chiaccio M, Barker CF, Caton AJ et al (2004) Promotion of allograft survival by CD4[+]CD25[+] regulatory T cells: evidence for *in vivo* inhibition of effector cell proliferation. *J Immunol* 172: 6539–6544

226 Annacker O, Pimenta-Araujo R, Burlen-Defranoux O, Barbosa TC, Cumano A, Bandeira A (2001) CD25[+]CD4[+] T cells regulate the expansion of peripheral CD4 T cells through the production of IL-10. *J Immunol* 166: 3008–3018

227 Huehn J, Siegmund K, Lehmann JC, Siewart C, Haubold U, Feuerer M, Debes GF, Lauber J, Frey O, Przybylski GK (2004) Developmental stage, phenotype, and migration distinguish naive- and effector/memory-like CD4[+] regulatory T cells. *J Exp Med* 199: 303–313

228 Ochando JC, Yopp AC, Yang Y, Garin A, Li Y, Boros P, Llodra J, Ding Y, Lira SA, Kreiger NR et al (2005) Lymph node occupancy is required for the peripheral development of alloantigen-specific Foxp3[+] regulatory T cells. *J Immunol* 174: 6993–7005

229 Kim CH (2006) Migration and function of FoxP3⁺ regulatory T cells in the hematolym-phoid system. *Exp Hematol* 34: 1033–1040

230 Venturi GM, Conway RM, Steeber DA, Tedder TF (2007) CD25⁺CD4⁺ regulatory T cell migration requires L-selectin expression: L-selectin transcriptional regulation balances constitutive receptor turnover. *J Immunol* 178: 291–300

231 Seekamp A, Till GO, Mulligan MS, Paulson JC, Anderson DC, Miyasaka M, Ward PA (1994) Role of selectins in local and remote tissue injury following ischemia and reperfu-sion. *Am J Pathol* 144: 592–598

232 Yan ZQ, Bolognesi MP, Steeber DA, Tedder TF, Chen LE, Seaber AV, Urbaniak JR (2000) Blockade of L-selectin attenuates reperfusion injury in a rat model. *J Reconstr Microsurg* 16: 227–233

233 Yadav SS, Howell DN, Gao W, Steeber DA, Harland RC, Clavien P-A (1998) L-selectin and ICAM-1 mediate reperfusion injury and neutrophil adhesion in the warm ischemic mouse liver. *Am J Physiol* 275: G1341-G1352

234 Tiegs G, Hentschel J, Wendel A (1992) A T cell-dependent experimental liver injury in mice inducible by concanavalin A. *J Clin Invest* 90: 196–203

235 Morikawa H, Hachiya K, Mizuhara H, Fujiwara H, Nishiguchi S, Shiomi S, Kuroki T, Kaneda K (2000) Sublobular veins as the main site of lymphocyte adhesion/transmigra-tion and adhesion molecule expression in the porto-sinusoidal-hepatic venous system during concanavalin A-induced hepatitis in mice. *Hepatology* 31: 83–94

236 Massaguer A, Perez-Del-Pulgar S, Engel P, Serratosa J, Bosch J, Pizcueta P (2002) Con-canavalin-A-induced liver injury is severely impaired in mice deficient in P-selectin. *J Leukoc Biol* 72: 262–270

237 Watanabe Y, Morita M, Akaike T (1996) Concanavalin A induces perforin-mediated but not Fas-mediated hepatic injury. *Hepatology* 24: 702–710

238 Kawasuji A, Hasegawa M, Horikawa M, Fujita T, Matsushita Y, Matsushita T, Fuji-moto M, Steeber DA, Tedder TF, Takehara K et al (2006) L-selectin and intercellular adhesion molecule-1 regulate the development of concanavalin A-induced liver injury. *J Leukoc Biol* 79: 696–705

239 Doerschuk CM, Beyers N, Coxson HO, Wiggs B, Hogg JC (1993) Comparison of neu-trophil and capillary diameters and their relation to neutrophil sequestration in the lung. *J Appl Physiol* 74: 3040–3045

240 Gebb SA, Graham JA, Hanger CC, Godbey PS, Capen RL, Doerschuk CM (1995) Sites of leukocyte sequestration in the pulmonary microcirculation. *J Appl Physiol* 79: 493–497

241 Yamaguchi K, Nishio K, Sato N, Tsumura H, Ichihara A, Kudo H (1997) Leukocyte kinetics in pulmonary microcirculation: observations using real-time confocal lumines-cence microscopy coupled with high-speed video analysis. *Lab Invest* 76: 809–822

242 Nishio K, Suzuki Y, Aoki T, Suzuki K, Miyata A, Sato N (1998) Differential contribu-tion of various adhesion molecules to leukocyte kinetics in pulmonary microvessels of hyperoxia-exposed rat lungs. *Am J Respir Crit Care Med* 157: 599–609

243 Fiscus LC, Van Herpen J, Steeber DA, Tedder TF, Tang MLK (2001) L-selectin is

required for the development of airway hyperresponsiveness but not airway inflammation in a murine model of asthma. *J Allergy Clin Immunol* 107: 1019–1024

244 Abraham WM, Ahmed A, Sabater JR, Lauredo IT, Botvinnikova Y, Bjercke RJ, Hu X, Revelle M, Kogan TP, Scott IL et al (1999) Selectin blockade prevents antigen-induced late bronchial responses and airway hyperresponsiveness in allergic sheep. *Am J Respir Crit Care Med* 159: 1205–1214

245 Hamaguchi Y, Nishizawa Y, Yasui M, Hasegawa M, Kaburagi Y, Komura K, Nagaoka T, Saito E, Shimada Y, Takehara K et al (2002) Intercellular adhesion molecule-1 and L-selectin regulate bleomycin-induced lung fibrosis. *Am J Pathol* 161: 1607–1618

246 Faveeuw C, Gagnerault MC, Lepault F (1994) Expression of homing and adhesion molecules in infiltrated islets of Langerhans and salivary glands of nonobese diabetic mice. *J Immunol* 152: 5969–5978

247 Yang X-D, Karin N, Tisch R, Steinman L, McDevitt HO (1993) Inhibition of insulitis and prevention of diabetes in nonobese diabetic mice by blocking L-selectin and very late antigen 4 adhesion receptors. *Proc Natl Acad Sci USA* 90: 10494–10498

248 Lepault F, Gagnerault MC, Faveeuw C, Bazin H, Boitard C (1995) Lack of L-selectin expression by cells transferring diabetes in NOD mice: insights into the mechanisms involved in diabetes prevention by Mel-14 antibody treatment. *Eur J Immunol* 25: 1502–1507

249 Friedline RH, Wong CP, Steeber DA, Tedder TF, Tisch R (2002) L-selectin is not required for T cell-mediated autoimmune diabetes. *J Immunol* 168: 2659–2666

250 Mora C, Grewal IS, Wong SF, Flavell RA (2004) Role of L-selectin in the development of autoimmune diabetes in non-obese diabetic mice. *Int Immunol* 16: 257–264

251 You S, Slehoffer G, Barriot S, Bach J, Chatenoud L (2004) Unique role of CD4+CD62L+ regulatory T cells in the control of autoimmune diabetes in T cell receptor transgenic mice. *Proc Natl Acad Sci USA* 101: 14580–14585

252 Gearing AJ, Newman W (1994) Circulating adhesion molecules in disease. *Immunol Today* 14: 506–512

253 Spertini O, Callegari P, Cordey AS, Hauert J, Joggi J, von Fliedner V, Schapira M (1994) High levels of the shed form of L-selectin (sL-selectin) are present in patients with acute leukemia and inhibit blast cell adhesion to activated endothelium. *Blood* 84: 1249–1256

254 Stucki A, Cordey AS, Monai N, de Flaugergues JC, Schapira M, Spertini O (1995) Cleaved L-selectin concentrations in meningeal leukaemia. *Lancet* 345: 286–289

255 Zetterberg E, Richter J (1993) Correlation between serum level of soluble L-selectin and leukocyte count in chronic myeloid and lymphocytic leukemia and during bone marrow transplantation. *Eur J Haematol* 51: 113–119

256 McGill SN, Ahmed NA, Hu F, Michel RP, Christou NV (1996) Shedding of L-selectin as a mechanism for reduced polymorphonuclear neutrophil exudation in patients with the systemic inflammatory response syndrome. *Arch Surg* 131: 1141–1147

257 Shimada Y, Sato S, Hasegawa M, Tedder TF, Takehara K (1999) Elevated serum L-selectin levels and abnormal regulation of L-selectin expression on leukocytes in atopic

dermatitis: soluble L-selectin levels indicate disease severity. *J Allergy Clin Immunol* 104: 163–168

258 Inaoki M, Sato S, Shimada Y, Kawara S, Steeber DA, Tedder TF, Takehara K (2000) Decreased expression levels of L-selectin on subsets of leukocytes and increased serum L-selectin in severe psoriasis. *Clin Exp Immunol* 122: 484–492

259 Font J, Pizcueta P, Ramos-Casals M, Cervera R, Garcia-Carrasco M, Navarro M, Ingelmo M, Engel P (2000) Increased serum levels of soluble L-selectin (CD62L) in patients with active systemic lupus erythematosus (SLE). *Clin Exp Immunol* 119: 169–174

260 Extermann M, Bacchi M, Monai N, Fopp M, Fey M, Tichelli A, Schapira M, Spertini O (1998) Relationship between cleaved L-selectin levels and the outcome of acute myeloid leukemia. *Blood* 92: 3115–3122

261 Donnelly SC, Haslett C, Dransfield I, Robertson CE, Carter DC, Ross JA, Grant IS, Tedder TF (1994) Role of selectins in development of adult respiratory distress syndrome. *Lancet* 344: 215–219

262 Kogan TP, Dupre B, Bui H, McAbee KL, Kassir JM, Scott IL, Hu X, Vanderslice P, Beck PJ, Dixon RA (1998) Novel synthetic inhibitors of selectin-mediated cell adhesion: synthesis of 1,6-bis[3-(3-carboxymethylphenyl)-4-(2-α-D-mannopyranosyloxy)phenyl]-hexane (TBC-1269). *J Med Chem* 41: 1099–1111

263 Davenpeck KL, Berens KL, Dixon RAF, Dupre B, Bochner BS (2000) Inhibition of adhesion of human neutrophils and eosinophils to P-selectin by the sialyl Lewis(x) antagonist TBC–1269. Preferential activity against neutrophil adhesion *in vitro*. *J Allergy Clin Immunol* 105: 769–775

264 Hicks AER, Abbitt KB, Dodd P, Ridger VC, Hellewell PG, Norman KE (2005) The anti-inflammatory effects of a selectin ligand mimetic, TBC-1269, are not a result of competitive inhibition of leukocyte rolling *in vivo*. *J Leukoc Biol* 77: 59–66

265 Beeh KM, Beier J, Meyer M, Buhl R, Zahlten R, Wolff G (2006) Bimosiamose, an inhaled small-molecule pan-selectin antagonist, attenuates late asthmatic reactions following allergen challenge in mild asthmatics: a randomized, double-blind, placebo-controlled clinical cross-over-trial. *Pulm Pharmacol Ther* 19: 233–241

266 Friedrich M, Bock D, Philipp S, Ludwig N, Sabat R, Wolfk K, Schroeter-Maas S, Aydt E, Kang S, Dam TN et al (2006) Pan-selectin antagonism improves psoriasis manifestation in mice and man. *Arch Dermatol Res* 297: 345–351

267 Co MS, Landolfi NF, Nagy JO, Tan JH, Vexler V, Vasquez M, Roark L, Yuan S, Hinton PR, Melrose J et al (1999) Properties and pharmacokinetics of two humanized antibodies specific for L-selectin. *Immunotechnology* 4: 253–266

268 Schlag G, Redl HR, Till GO, Davies J, Martin U, Dumont L (1999) Anti-L-selectin antibody treatment of hemorrhagic-traumatic shock in baboons. *Crit Care Med* 27: 1900–1907

269 Seekamp A, van Griensven M, Dhondt E, Diefenbeck M, Demeyer I, Vundelinckx G, Haas N, Schaechinger U, Wolowicka L, Rammelt S et al (2004) The effect of anti-L-selectin (aselizumab) in multiple traumatized patients-results of a phase II clinical trial. *Crit Care Med* 32: 2021–2028

270 Lasky LA, Singer MS, Dowbenko D, Imai Y, Henzel WJ, Grimley C, Fennie C, Gillett N, Watson SR, Rosen SD (1992) An endothelial ligand for L-selectin is a novel mucin-like molecule. *Cell* 69: 927–938

271 Brustein M, Kraal G, Mebius RE, Watson SR (1992) Identification of a soluble form of a ligand for the lymphocyte homing receptor. *J Exp Med* 176: 1415–1419

272 Hemmerich S, Butcher EC, Rosen SD (1994) Sulfation-dependent recognition of HEV-ligands by L-selectin and MECA 79, an adhesion-blocking mAb. *J Exp Med* 180: 2219–2226

273 Baumhueter S, Singer MS, Henzel W, Hemmerich S, Renz M, Rosen SD, Lasky LA (1993) Binding of L-selectin to the vascular sialomucin CD34. *Science* 262: 436–438

274 Jalkanen S, Bargatze RF, de los Toyos J, Butcher EC (1987) Lymphocyte recognition of high endothelium: antibodies to distinct epitopes of an 85–95-kD glycoprotein antigen differentially inhibit lymphocyte binding to lymph node, mucosal, or synovial endothelial cells. *J Cell Biol* 105: 983–990

275 Greaves MF, Brown J, Molgaard HV, Spurr NK, Robertson D, Delia D, Sutherland DR (1992) Molecular features of CD34: A hematopoietic progenitor cell-associated molecule. *Leukemia* 6: 31–36

276 Kershaw DB, Beck SG, Wharram BL, Wiggins JE, Goyal M, Thomas PE, Wiggins RC (1997) Molecular cloning and characterization of human podocalyxin-like protein. *J Biol Chem* 272: 15708–15714

277 Kershaw DB, Thomas PE, Wharram BL, Goyal M, Wiggins JE, Whiteside CI, Wiggins RC (1995) Molecular cloning, expression, and characterization of podocalyxin-like protein 1 from rabbit as a transmembrane protein of glomerular podocytes and vascular endothelium. *J Biol Chem* 270: 29439–29446

278 Morgan SM, Samulowitz U, Darley L, Simmons DL, Vestweber D (1999) Biochemical characterization and molecular cloning of a novel endothelial-specific sialomucin. *Blood* 93: 165–175

279 Samulowitz U, Kuhn A, Brachtendorf G, Nawroth R, Braun A, Bankfalvi A, Bocker W, Vestweber D (2002) Human endomucin: Distribution pattern, expression on high endothelial venules, and decoration with the MECA-79 epitope. *Am J Pathol* 160: 1669–1681

280 Brachtendorf G, Kuhn A, Samulowitz U, Knorr R, Gustafsson E, Potocnik A, Fassler R, Vestweber D (2001) Early expression of endomucin on endothelium of the mouse embryo and on putative hematopoietic clusters in the dorsal aorta. *Dev Dyn* 222: 410–419

281 Asa D, Raycroft L, Ma L, Aeed PA, Kaytes PS, Elhammer AP, Geng JG (1995) The P-selectin glycoprotein ligand functions as a common human leukocyte ligand for P- and E-selectins. *J Biol Chem* 270: 11662–11672

P- and E-selectin

Daniel C. Bullard

Department of Genetics, Kaul Building 640A, 720 South 20th Street, University of Alabama at Birmingham, Birmingham, AL 35294, USA

Initial characterization and cloning

P- and E-selectin, commonly referred to as the "endothelial" selectins, were initially described and characterized in the mid-late 1980s. P-selectin was first identified in 1984, using antibodies raised against activated platelets [1, 2]. In these initial publications, P-selectin was described as a protein of molecular weight of approximately 140 000, which was not found on resting platelets, but showed up-regulated expression following activation with thrombin or other mediators. In a subsequent investigation, Stenberg et al. [3] referred to this protein as "granule membrane protein-140" (GMP-140) due to its localization in the α granules of unstimulated platelets. In this study, as well as another report by Berman et al. [4], P-selectin was shown to translocate from the α granules in platelets to the plasma membrane following activation. In this latter publication, P-selectin was termed "platelet activation-dependent granule-external membrane protein" (PADGEM; see Table 1 for a listing of the other common abbreviations and designations for P- and E-selectin). In 1989, this adhesion molecule was shown to be expressed on the surface of cultured human endothelial cells following stimulation with histamine, thrombin, C5b-9, and other activators, and stored in resting cells in the Weibel-Palade bodies [5–8]. During this same year, a cDNA for P-selectin was cloned, and sequence analysis suggested a cysteine-rich protein similar to that reported for a new endothelial-expressed protein termed ELAM-1 (endothelial leukocyte adhesion molecule-1; see below) [9, 10]. The structure of both selectins was predicted to consist of an N-terminal calcium-dependent lectin binding domain, an epidermal growth factor (EGF)–like domain, a series of complement control protein module (CCP)/short consensus repeats (SCR) domains often referred to as CR repeats and found in a number of complement proteins, a transmembrane domain, and a short cytoplasmic tail [9]. In further studies, P-selectin was shown to function as an adhesion molecule for platelet interactions with neutrophils and monocytes [11, 12], as well as a mediator of neutrophil adhesion to activated endothelial cells [13, 14]. Antibodies against P-selectin were also shown to inhibit platelet/leukocyte interactions, leading to reduced fibrin deposition in an arteriovenous shunt model in baboons [15].

Table 1 - Common names and abbreviations for P- and E-selectin

P-selectin	GMP-140	Granule membrane protein-140
	PADGEM	Platelet activation dependent granule-external membrane protein
	LECAM-3	Leukocyte endothelial cell adhesion molecule-3
	CD62P antigen	--------
	Selp	Selectin P*
E-selectin	ELAM-1	Endothelial leukocyte adhesion molecule-1
	LECAM-2	Leukocyte endothelial cell adhesion molecule-2
	CD62E antigen	---------
	Sele	Selectin E*

Name used by genomic databases

The mechanism for this phenomenon was not understood until later, when studies in mice showed that platelet-expressed P-selectin was involved in recruiting tissue factor-containing microparticles into thrombi through interactions with P-selectin glycoprotein ligand 1 (PSGL-1) [16].

In 1987, Bevilacqua et al. [17] described a protein termed ELAM-1 expressed on human umbilical vein endothelial cells (HUVEC) following stimulation with various inflammatory mediators including IL-1β. In this initial publication, E-selectin was identified by two different monoclonal antibodies, which immunoprecipitated on a polypeptide with approximate molecular weight of 115 000 on activated, but not resting endothelial cells. Further studies showed that E-selectin protein was initially detected on IL-1β-stimulated HUVEC after 30 min, reached maximal expression at 4 h, and declined thereafter to basal levels by 16–24 h. In addition, one of the antibodies was shown to significantly inhibit adhesion of HL-60 cells, a promyelo-monocytic cell line, and neutrophils to cytokine-activated HUVEC. In subsequent publications, E-selectin was also shown to mediate adhesion of T cells, NK cells, eosinophils, monocytes, and a human carcinoma cell line [18–22].

Leukocyte emigration in response to an inflammatory stimulus occurs primarily in postcapillary venules, and proceeds through a series of steps including capture and rolling, firm adhesion, and transendothelial migration [23, 24]. One of the first demonstrations that P-selectin was involved in mediating leukocyte rolling occurred in 1991. Lawrence and Springer [25] showed using *in vitro* flow chambers that neutrophils rolled at physiological shear rates on lipid bilayers containing purified P-selectin, but not those containing purified ICAM-1. Subsequent studies showed that P-selectin could mediate leukocyte rolling *in vivo*, and that E-selectin was also involved [26–30]. The importance of P- and E-selectin rolling for the process of leukocyte emigration during inflammatory responses has been shown in many different studies, and some of these are discussed in other sections of this chapter.

Structure and domain function

All three selectins show a similar genomic and protein structure and share high homology at both the nucleotide and amino acid level in their extracellular domains [31, 32]. As described briefly above, the extracellular portion of these proteins consists of three basic units, the lectin binding domain, EGF-like domain, and the CR repeats, and a number of studies of the selectins have focused on defining the three-dimensional structure, as well as the contributions of these different domains in ligand binding. The N-terminal lectin binding domain, also known as the carbo-hydrate-recognition domain (CRD), is homologous to the Ca^{2+}-dependent or C-type lectins (CTL) [33]. The requirement for Ca^{2+} was demonstrated early on by Geng et al. [13], who showed that P-selectin-mediated adhesion of neutrophils and HL-60 cells only in the presence of extracellular Ca^{2+}. Mutational analyses or competitive peptide binding assays were first used to demonstrate the absolute requirement of the lectin domain, as well as specific amino acid residues present in this region, for binding of both P- and E-selectin to leukocytes or carbohydrates such as tetrasac-charide sialyl Lewis^x (sLex) [34–38]. Investigations using chimeric proteins, where different domains were exchanged between P- and L-selectin showed that exchange of exons encoding for the lectin region resulted in a change in binding specificities to that of the parent molecule, thus further showing the importance of this domain for adhesive interactions [39, 40].

The EGF-like domain has also been shown to contribute to ligand binding [32]. Revelle et al. [41] found that mutagenic substitution of single amino acid residues in the EGF-like domain of both P-and E-selectin significantly altered the abilities of these proteins to bind to sLex, heparin, or sulfatide. In addition, peptides specific for a region of the EGF-like domain of P-selectin were also shown to inhibit monocyte adhesion to activated endothelial cells [42]. Finally, domain deletion studies showed that both the lectin and EGF regions were required for E-selectin binding to its ligands on U937 cells [38]. The crystal structure of the E-selectin lectin and EGF-like domains was first reported by Graves et al. [43]. In this analysis, the lectin domain showed a similar folding pattern to that of the rat mannose-binding protein, while the EGF domain exhibited the same general folding pattern, as well as a similar arrangement of disulfide bonds to that of other proteins with EGF-like domains. Furthermore, the crystal structure also showed limited interactions between the lec-tin and EGF-like domains and a single bound Ca^{2+} within the lectin region. In latter studies, the crystal structure of the P- and E-selectin lectin and EGF-like domains bound with sLex or PSGL-1 was reported [44]. These findings are summarized and discussed in Chapter 1 of this book.

73

Although all three selectin proteins are structurally similar, they do differ in the number of CR repeats. In humans, P-selectin contains nine CR domains, while E-selectin and L-selectin have six and two, respectively. In addition, species differences have been documented in the number of these domains for each individual selectin gene [31] (for the latest comparisons, see genome databases such as *Ensembl*). The specific roles of the CR repeats are still not entirely clear, although studies suggest that they are involved in promoting adhesion to their ligands, as well as extending the lectin and EGF-like domains away from the cell surface to facilitate interactions [45–47]. Unlike the extracellular membrane regions of the selectins, the cytoplasmic tails of the selectins do not share significant homology [32]. However, this domain does show high conservation between species for each of the individual selectin genes [32]. Deletion studies have further shown that loss of the cytoplasmic domain in either P- or E-selectin, unlike L-selectin, does not prevent expression of these mutant proteins and leukocyte rolling and adhesion mediated through these receptors can still occur [48, 49]. Mutations in the cytoplasmic domain of P-selectin, however, can inhibit the localization of this protein in Weibel-Palade bodies, internalization, recycling, degradation, and rolling efficiency [50–53]. Also, several studies have identified specific amino acid residues in the intracellular portion of these molecules that are important for intracellular signaling in platelets or endothelial cells [54, 55]. More work is necessary, however, to identify the specific functions and key regions of the intracellular domains of P- and E-selectin that promote intracellular signaling events, as well as identify the physiological changes that occur in cells in response to activation of these signaling pathways.

The genomic organization of the selectin genes is similar with separate exons encoding for each of the individual domains found in the mature protein. In all mammalian species analyzed to date, the selectins show tight chromosomal linkage, suggesting they arose by duplication events during evolution (see *Ensembl* genomic database). In humans and in mice, the selectin cluster is located on chromosome 1 [56]. Although many alternatively spliced mRNAs have been described for other adhesion molecules, such as VCAM-1, PECAM-1, and ICAM-1, only a few reports have documented selectin isoforms generated by this mechanism [57–60]. An alternatively spliced form of P-selectin has been described that lacks the transmembrane domain and results in a secreted form of the protein [61, 62], while a variant mRNA deleted for the fifth CR domain of the rat E-selectin gene has also been identified [63].

Soluble forms of P- and E-selectin, which most likely arise by shedding or cleavage from the cell membrane, have also been identified in serum. A number of studies have reported higher circulating levels of these soluble adhesion molecules in patients with different inflammatory diseases, including cardiovascular disorders (for reviews see [64–66]). Thrombin activation of platelets has been shown to result in the rapid cleavage of P-selectin and its appearance in the plasma [67, 68], and this may serve as one mechanism that leads to increased levels of the soluble form of this adhesion molecule in patients with active disease. The potential roles of these soluble forms in regulating inflammatory responses are not fully understood, and they may act to both block selectin ligand interactions and induce signaling events in leukocytes. Andre et al. [69] analyzed this question using a line of gene targeted mice that express a mutant form of P-selectin lacking the cytoplasmic domain, which results in high circulating levels of soluble P-selectin in the plasma. These mice showed an accelerated hemostasis phenotype, and were characterized by a high concentration of pro-coagulant microparticles containing tissue factor. These findings suggest that both soluble P-selectin and microparticles containing this adhesion molecule are important in coagulation events, and that they may contribute to the development of thrombotic and cardiovascular disorders in humans [70].

Regulation of P- and E-selectin expression

The mechanisms responsible for the regulation of P- and E-selectin expression and function have been actively studied. In general, inflammatory mediators that promote leukocyte/endothelial cell adhesion stimulate rapid expression of these proteins at the cell surface. These include mediators such as TNF-α, IL-1β, LPS, complement components, immune complexes, oxygen radicals, and others [71]. P-selectin expression is also stimulated by histamine, thrombin, N-formyl-methionyl-leucyl-phenylalanine (fMLP), and adenosine diphosphate (ADP) (platelets) [72]. However, several key differences exist between the molecular mechanisms and intracellular signaling pathways that control expression of these two genes. As briefly described above, P-selectin, unlike E-selectin, is constitutively produced and packaged in the α granules of the platelet precursors, megakaryocytes, and Weibel-Palade bodies of endothelial cells. This allows the rapid induction of cell surface expression on platelets, within seconds, and on endothelial cells, within several minutes following their activation. In contrast, E-selectin is not found on the surface of resting endothelial cells, and expression requires promoter activation and new transcription. Following translation in endothelial cells, both P- and E-selectin are modified with high mannose N-linked oligosaccharides as they move from the endoplasmic reticulum through the Golgi network [73]. P-selectin then moves into the secretory granules, which are thought to form at the *trans* Golgi network (TGN) [53], while E-selectin moves directly from the TGN to the cell surface [73]. Efficient trafficking of P-selectin requires sequences in both the cytoplasmic and luminal domains of this adhesion molecule, although the specific extracellular regions involved in this process have not been determined [49, 50, 53].

Both P- and E-selectin are removed from the endothelial cell surface by endocytosis in clathrin-coated pits, which move through the endosomal pathway to lysosomes for degradation [51, 52, 74–76]. P-selectin can also be redirected back into secretory granules, and can be re-expressed on the endothelial cell surface [52].

Multiple regions in the cytoplasmic domain of P-selectin are responsible for both internalization and recycling through secretory vesicles in endothelial cells [50, 52, 73]. In contrast to endothelial cells, deletion of the entire cytoplasmic domain does not affect the localization of P-selectin in the α granules in platelets, or its expression following thrombin stimulation, and does not result in reduced platelet-leukocyte interactions [49].

The transcription factors and promoter elements that control E-selectin transcription in endothelial cells following cytokine stimulation have been well studied (for review see [73]). The transcription factors nuclear factor-κB (NF-κB) p50/p65, activating transcription factor-2 (ATF-2), and c-JUN, as well as the high mobility group I(Y) [HMG-I(Y)] are involved in mediating cytokine induction of the E-selectin promoter, and these proteins have all been shown to bind to DNA elements in the first 160 bp upstream of the 5′ end of the transcription start site [77–83]. Mutational analyses have identified a number of different DNA elements in the E-selectin promoter, including 3 NF-κB sites, which are essential for maximal induction of E-selectin transcription following stimulation with mediators such as TNF-α [77, 84, 85]. The mechanisms responsible for repression of E-selectin transcription have also been examined. E-selectin mRNA production is rapidly diminished following removal of TNF-α stimulation, and transcriptional repression involves the regulator, IκBα, and loss of the NF-κB subunits p50 and p65 from the nucleus [86, 87]. In addition, a recent study found that the POU domain transcription factor Oct-1, but not Oct-2, is involved in the repression of E-selectin transcription in HUVEC [88]. Further investigations suggested that Oct-1 acts through binding of the NF-κB p65 subunit, and that IL-6 can induce expression of Oct-1 and lead to suppression of E-selectin transcription [88].

P-selectin expression has also been shown to be regulated at the transcriptional level, although the specific proteins and DNA elements that control mRNA production are not as extensively characterized as those for E-selectin. The human P-selectin promoter contains multiple transcriptional start sites and lacks a TATA box sequence [73]. Several potential regulatory DNA elements have been identified in the 5' end of the P-selectin gene in humans, including a novel NF-κB site that only binds to p50 and p52 homodimers, and not p50/p65 heterodimers, a GATA element, several Ets motifs, two Stat6 binding sites, a HOX element, and a sequence similar to the GT-IIC element of the SV40 enhancer [89–92]. Pan and McEver [90] showed that the protein Bcl-3 interacts with p52 homodimers and the κB element to induce transcription of P-selectin in endothelial cells in response to PMA, while interactions of the p50 homodimers with this DNA element lead to transcriptional repression. Mutations in the GATA element have also been shown to significantly reduce P-selectin transcription [89].

It is interesting to note that differences exist in the transcriptional regulation of the P-selectin gene in mammals. Although inflammatory mediators such as IL-3, IL-4, and oncostatin-M have been shown to induce P-selectin transcription in HUVEC

[93, 94], LPS or TNF-α stimulation does not lead to increased mRNA production. Similarly, injection of baboons with *E. coli*, which leads to rapid increases in the levels of circulating TNF-α and LPS, does not result in acute increases in P-selectin transcription [94, 95]. In contrast, these activators significantly stimulate mRNA production *in vivo* in mice and rats, and also in cultured bovine and murine endothelial cell lines [96–98]. In humans and in mice, the differential regulation of the P-selectin gene has been associated with the presence of several different regulatory elements in the promoter sequences [91, 99].

E- and P-selectin ligands

Despite an extensive number of studies, few physiological ligands for P- or E-selectin have been identified that interact with these adhesion molecules to promote leukocyte rolling and emigration *in vivo*. All three selectins recognize specific glycoproteins on the surface of leukocytes, endothelial cells, or platelets. These ligands undergo extensive post-translational modifications by a number of different enzymes, including glycosyltransferases and sulfotransferases, whose expression or activity levels vary depending on the specific leukocyte or endothelial cell population [32, 100]. The majority of ligand studies to date have focused on defining the roles of PSGL-1, which serves as the major ligand for P-selectin on both leukocytes and platelets, and also interacts with both E- and L-selectin [101–109]. Investigations in different inflammatory models have shown, in many cases, that loss or inhibition of PSGL-1 interactions with P-selectin significantly inhibits leukocyte rolling, and can lead to reduced recruitment *in vivo* [107, 110–112]. Although fewer studies have examined the specific contributions of PSGL-1 binding to E-selectin for mediating leukocyte rolling, several published reports suggest that interactions between these adhesion molecules are important for recruitment in organs such as the skin [113–115].

PSGL-1-independent ligands for P- and E-selectin have been reported; however, very little functional information is available regarding the importance of these interactions. CD24, also known as the heat-stable antigen (HSA), is a mucin-type glycosylphosphatidylinositol-linked cell surface molecule expressed on neutrophils, monocytes, and some B cell populations, as well as on many tumor cells [116]. CD24 has been shown to bind to P-selectin, but not E-selectin on both endothelial cells and platelets [117, 118]. An important role for P-selectin/CD24 interactions was demonstrated by Aigner et al. [119], who showed that rolling of non-PSGL-1-expressing breast carcinoma cells was mediated to a large extent by these adhesion molecules. Both E-selectin ligand-1 (ESL-1) and CD44 have been shown to bind to E-selectin on a variety of leukocyte subtypes and on tumor lines [120–125]. ESL-1 was cloned in 1995, and has a high degree of amino acid identity with the chicken cysteine-rich fibroblast growth factor receptor [120]. Functional studies, however, demonstrating an important role for ESL-1/E-selectin interactions in mediating rolling or recruitment are lacking. CD44 expressed on neutrophils has been found to bind to E-selectin, and mediate slow rolling and short-term recruitment through interactions with this selectin in mice. Hematopoietic cell E- and L-selectin ligand (HCELL) is a sialofucosylated glycoform of CD44, which binds to both E- and L-selectin [122–125]. HCELL interactions with E-selectin have been shown to mediate rolling of myeloid cells and colon carcinoma cell lines using *in vitro* assays, although *in vivo* studies have not been reported [122, 124, 125].

Roles of P- and E-selectin in acute inflammatory responses

Selectin function has now been examined in a wide variety of inflammatory model systems, and a comprehensive review of all of the different published studies is not possible here. Investigations performed in both acute and chronic models illustrate the different roles of the endothelial selectins in regulating immune and inflammatory responses. The contributions of P- and E-selectin have been addressed primarily through use of inhibitory antibodies, or using mice containing gene targeted mutations in the genes encoding for these adhesion molecules. Both methods for analyses of selectin function have been informative, although mutant mice have been particularly valuable for long-term or chronic studies, since continual dosing with an inhibitory antibody can elicit an anti-isotype immune response. Mice with mutations in the individual selectin genes, as well as double and triple mutations have now been reported. Initial studies of P- or E-selectin mutant mice did not show any evidence of spontaneous disease [126]. P-selectin mutant mice also presented with a slight, but significant increase in circulating neutrophil counts compared to wild-type mice, although this phenotype was not observed in mice lacking E-selectin expression. Minor defects in the coagulation response were noted in P-selectin-deficient mice, which showed a small, but significant increase in bleeding times using a tail tip assay [127]. P-selectin mutant platelets also showed decreased rosette formation with neutrophils following thrombin activation [128]. Using intravital microscopy, P-selectin mutant mice were found to have a significant reduction in early leukocyte rolling (0–2 h) compared to non-mutants [27, 129, 130], while in peritonitis experiments, early (0–4 h) peritoneal neutrophil emigration was also reduced in these mutants, but not at 24 h [27, 131]. Early analyses of E-selectin mutant mice failed to detect any defects in the inflammatory response, unless either P- or L-selectin was additionally blocked with a monoclonal antibody [130, 132].

Mice with null mutations in both E- and P-selectin presented with a severe inflammatory phenotype, characterized by increased susceptibility to mucocutaneous infections, periodontitis, highly elevated leukocyte counts, hypergammaglobulinemia, reduced L-selectin expression on peripheral blood leukocytes, and cervical lymphadenopathy with plasmacytosis [133, 134]. The spontaneous inflammatory manifestations in these E/P-selectin double mutant mice are strikingly similar to those seen in patients with leukocyte adhesion deficiency type II (LAD type II), a genetic disease that affects fucosylation of glycoconjugates, and consequently the ability of these ligands to bind to the selectins (for review see [135]). Patients with this syndrome suffer from recurrent bacterial infections, neutrophilia, and show decreases in neutrophil rolling. Interestingly, PSGL-1 deficiency in mice does not lead to a similar clinical phenotype, despite the fact that this adhesion molecule serves as a ligand for both P- and E-selectin [111]. This strongly suggests that loss of interaction with other non-PSGL-1 ligands, especially those for E-selectin, contribute to the inflammatory phenotype in E- and P-selectin double-mutant mice. Finally, further analyses of different lines of E-/P-selectin-deficient mice, in which the double mutation was backcrossed onto several different inbred strain backgrounds, or studies of mice housed in different animal facilities, have shown a high degree of variability in the clinical phenotype, especially with regard to the development and extent of skin lesions (D. C. Bullard et al., unpublished observations). These findings suggest that both genetic and environmental influences can modulate the phenotype of E- and P-selectin-deficient mice, although these factors have not yet been identified.

Further investigations of these double-mutant mice revealed a total absence of trauma or TNF-α-induced leukocyte rolling up to 2 h, complete inhibition of peritoneal emigration of neutrophils after 4 h, and a significant inhibition of oxazolone-induced delayed-type contact hypersensitivity (DTH) [133, 134, 136]. Interestingly, leukocyte emigration in E-/P-selectin double-mutant mice is not completely inhibited, and recruitment can occur at later time points following the induction of the inflammatory response. For example, equivalent numbers of neutrophils were found in the peritoneal cavity of mutant and non-mutant mice, 24 h after the induction of peritonitis [133]. In addition, extravascular leukocytes were also observed in skin sections from mice with chronic dermal inflammation [133, 137].

Roles in inflammatory diseases

Several lines of experimental evidence strongly suggest that both P- and E-selectin are actively involved in regulating the development of different inflammatory diseases. Although P- and E-selectin expression has been shown to be rapidly induced, and declines generally within the first 24 hours following stimulation in both *in vitro* and *in vivo* inflammatory models, chronic induction of these adhesion molecules has also been reported, especially in inflamed tissues of patients with inflammatory disorders. Immunohistochemical analyses using selectin monoclonal antibodies have documented vascular expression in diseases such as atherosclerosis, rheumatoid arthritis (RA), diabetes, systemic lupus erythematosus (SLE), psoriasis, vasculitic diseases, and others (reviewed in [138–140]). Many of the inflammatory mediators associated with the pathogenesis of these diseases can up-regulate P- and E-selectin expression in cultured endothelial cells or on platelets (see discussion above). Further indirect evidence for selectin involvement has also come from genetic studies of patients, with a number of different reports showing significant associations between specific selectin gene polymorphisms and the development of different inflammatory diseases [141–144].

Functional analyses of the selectins have also been performed using different animal models for chronic inflammatory diseases, especially in murine systems. For example, an important role for both P- and E-selectin in the development of atherosclerosis has been documented using the C57BL/6 high-fat diet, low-density lipoprotein receptor mutant (LDLR), or ApoE-deficient mouse models of atherosclerosis. Loss of P-selectin expression in all three of these models led to decreased atherosclerotic lesions compared to control mice, suggesting that this adhesion molecule significantly contributes to monocyte rolling and adhesion occurring at the sites of fatty steak formation [145–148]. In the ApoE-deficient model, lesion area was also shown to be reduced in E-selectin-deficient mice, although not to the same extent as that of P-selectin mutants [148]. E/P-selectin double-mutant mice exhibited a significant decrease in early (8–22 week) and late (37 week) atherosclerotic lesion formation in the LDLR model when compared to wild-type mice [149]. Investigations of the selectins in the experimental autoimmune encephalomyelitis (EAE) model of multiple sclerosis (MS), have not documented a primary role for P- or E-selectin in the initiation or progression of CNS inflammation leading to the development of demyelination. Engelhardt et al. [150] showed that the development of EAE was not inhibited by treatment of mice with inhibitory anti-P-selectin or anti-E-selectin monoclonal antibodies. Similar observations were made using E-/P-selectin double-mutant mice and PSGL-1-deficient mice in this same model [151–153]. These studies suggest selectin interactions with PSGL-1 are not required for T cell emigration or monocyte/macrophage recruitment events during the initiation or progression of EAE, and that other adhesion molecules play a more dominant role in mediating leukocyte rolling in the CNS vasculature.

Studies of P- and E-selectin in other inflammatory disease model systems have, in several cases, produced surprising and sometimes conflicting results. Previously, we found that mice lacking P-selectin, E-selectin, or E-/P-selectin double-mutant mice showed accelerated and more severe collagen-induced arthritis (CIA), a model for human RA [154, 155]. In E-/P-selectin double-mutant mice, this phenotype was associated with increased production of macrophage inflammatory protein-1α (CCL3) and IL-1β in joint tissue [155]. These findings suggested that the endothelial selectins were not required for mediating leukocyte recruitment in this model, and that their expression may instead serve to influence cytokine and chemokine expression during the development of arthritis. In contrast to these observations, Sumariwalla et al. [156] found that treatment of arthritic mice with a recombinant PSGL-1–Ig fusion protein suppressed the progression of inflammation and protected against joint damage in a similar CIA model. Studies of selectin function in other rodent arthritis models have reported mixed results [157–160], with several papers showing that loss or inhibition of P- or E-selectin decreased the incidence or severity of joint inflammation, while others found no effect on the development of arthritis.

Investigations of selectin function in animal models of SLE-associated tissue inflammation suggest that P- and E-selectin may also play regulating roles in the control of SLE development. MRL/MpJ-*Fas^lpr* mice develop a systemic inflammatory disease characterized by autoantibody formation, immune complex-mediated glomerulonephritis, vasculitis, and dermatitis [161]. P-selectin and PSGL-1 mutant MRL/MpJ-*Fas^lpr* mice were not protected against the development of autoimmune-mediated inflammation, but showed accelerated forms of glomerulonephritis and dermatitis [162]. These observations were similar to studies in the anti-glomerular basement membrane model (anti-GBM), where P-selectin-deficient mice presented with increased mortality and a more severe form of glomerulonephritis [163]. Interestingly, the rapid progression of glomerulonephritis in both P-selectin and PSGL-1 MRL/MpJ-*Fas^lpr* mutant mice was associated with increased expression of the chemokine CCL2 in kidney tissue and in purified endothelial cells from P-selectin mutant mice [162]. Thus, these studies suggest that P-selectin and PSGL-1 regulate the development of inflammatory disease in this model, possibly through inhibition of CCL2 expression. Furthermore, these findings collectively indicate that both P- and E-selectin play complex roles in regulating inflammatory responses in the joint and kidney, and suggest that further studies are necessary to define the contributions of these molecules in RA, SLE, as well as other diseases.

P-selectin function in platelets

Although P-selectin promotes leukocyte interactions with platelets, other studies now suggest it is also important for platelet/endothelial cell adhesion events, which may serve to promote thrombus formation, as well as mediate leukocyte recruitment events during immune or inflammatory responses in different tissues. Using intravital microscopy, Frenette et al. [164] showed that platelets roll on activated endothelial cells through a P-selectin-dependent mechanism. Transfer of wild-type or P-selectin-deficient platelets into P-selectin mutant mice resulted in a significant decrease in rolling compared to similar transfers into wild-type mice. Platelet-expressed P-selectin mediates lymphocyte rolling in lymph nodes through interactions with peripheral lymph node addressin (PNAd), which can serve as an L-selectin-independent mechanism for lymphocyte trafficking to high endothelial venules (HEV) during the initiation of T cell-dependent immune responses [165, 166]. Platelet-expressed P-selectin is involved in the recruitment of leukocytes in response to different inflammatory stimuli, as well as for triggering signaling events in leukocytes through engagement of PSGL-1 [167–172].

Selectin-based therapies

Anti-adhesion molecule therapies have now been tested in clinical trials for a number of different inflammatory disorders. However, despite the development of several different selectin-based inhibitors, few studies have demonstrated efficacy in clinical trials (reviewed in [173–177]). Recently, recombinant PSGL-1–Ig was tested in a clinical trial involving patients with acute myocardial infarction [178]. This inhibitor was previously shown to inhibit damage to the myocardium in several different animal models [176]. However, the clinical trial was prematurely stopped due to a lack of efficacy [178]. More promising results have recently been reported using the pan-selectin antagonist, bimosiamose, in clinical trials involving psoriasis patients [179]. In these studies, a small number of patients with moderate to severe psoriasis were treated bimosiamose for a period of 14 days, with the majority showing clinical and histological signs of disease improvement [179]. Larger studies, however, are necessary to further evaluate the efficacy of this selectin inhibitor for the treatment of psoriasis, as well as other inflammatory diseases.

Conclusions

P- and E-selectin share several unique features that allow each protein to mediate leukocyte interactions during the early stages of inflammatory responses. Their structure permits rapid bond association and dissociation rates with their ligands, as well as high tensile strength, which are required for stable leukocyte rolling interactions on endothelial cells. In addition, their induction mechanisms for expression at the cell surface allow for the rapid onset of leukocyte or platelet rolling and adhesion following stimulation. Numerous *in vivo* studies have shown that loss or inhibition of P- and E-selectin can inhibit tissue damage, although these investigations have also highlighted the significant redundancy that exists among the selectins and other adhesion molecules in mediating rolling interactions. More work is necessary, however, to specifically define their roles in the regulating immune and inflammatory responses, especially in inflammatory diseases. The contributions of P- and E-selectin to Th17 effector and T regulatory cell recruitment and functions in the development of normal immune responses, as well as during the initiation of inflammatory disorders need to be defined. These studies would provide valuable information regarding the functions of these adhesion molecules in the pathogenesis of RA, MS, psoriasis, and inflammatory bowel diseases, and may highlight other diseases where selectin-based inhibitors could potentially be used for therapeutic purposes. Also, the signaling pathways in endothelial cells, elicited by engagement of these adhesion molecules, remain to be investigated to determine the mechanisms by which P- and E-selectin control inflammatory responses. Additional studies of PSGL-1-independent ligands for E-selectin are necessary to specifically define their contributions in mediating rolling and emigration of different leukocyte populations.

References

1   Hsu-Lin S, Berman CL, Furie BC, August D, Furie B (1984) A platelet membrane protein expressed during platelet activation and secretion. Studies using a monoclonal antibody specific for thrombin-activated platelets. *J Biol Chem* 259: 9121–9126

2   McEver RP, Martin MN (1984) A monoclonal antibody to a membrane glycoprotein binds only to activated platelets. *J Biol Chem* 259: 9799–9804

3   Stenberg PE, McEver RP, Shuman MA, Jacques YV, Bainton DF (1985) A platelet alpha-granule membrane protein (GMP-140) is expressed on the plasma membrane after activation. *J Cell Biol* 101: 880–886

4   Berman CL, Yeo EL, Wencel-Drake JD, Furie BC, Ginsberg MH, Furie B (1986) A platelet alpha granule membrane protein that is associated with the plasma membrane after activation. Characterization and subcellular localization of platelet activation-dependent granule-external membrane protein. *J Clin Invest* 78: 130–137

5   McEver RP, Beckstead JH, Moore KL, Marshall-Carlson L, Bainton DF (1989) GMP-140, a platelet alpha-granule membrane protein, is also synthesized by vascular endothelial cells and is localized in Weibel-Palade bodies. *J Clin Invest* 84: 92–99

6   Hattori R, Hamilton KK, McEver RP, Sims PJ (1989) Complement proteins C5b-9 induce secretion of high molecular weight multimers of endothelial von Willebrand factor and translocation of granule membrane protein GMP-140 to the cell surface. *J Biol Chem* 264: 9053–9060

7   Hattori R, Hamilton KK, Fugate RD, McEver RP, Sims PJ (1989) Stimulated secretion of endothelial von Willebrand factor is accompanied by rapid redistribution to the cell surface of the intracellular granule membrane protein GMP-140. *J Biol Chem* 264: 7768–7771

8   Bonfanti R, Furie BC, Furie B, Wagner DD (1989) PADGEM (GMP140) is a component of Weibel-Palade bodies of human endothelial cells. *Blood* 73: 1109–1112

9   Johnston GI, Cook RG, McEver RP (1989) Cloning of GMP-140, a granule membrane protein of platelets and endothelium: sequence similarity to proteins involved in cell adhesion and inflammation. *Cell* 56: 1033–1044

10 Bevilacqua MP, Stengelin S, Gimbrone MA Jr, Seed B (1989) Endothelial leukocyte adhesion molecule 1: an inducible receptor for neutrophils related to complement regulatory proteins and lectins. *Science* 243: 1160–1165

11 Hamburger SA, McEver RP (1990) GMP-140 mediates adhesion of stimulated platelets to neutrophils. *Blood* 75: 550–554

12 Larsen E, Celi A, Gilbert GE, Furie BC, Erban JK, Bonfanti R, Wagner DD, Furie B (1989) PADGEM protein: a receptor that mediates the interaction of activated platelets with neutrophils and monocytes. *Cell* 59: 305–312

13 Geng JG, Bevilacqua MP, Moore KL, McIntyre TM, Prescott SM, Kim JM, Bliss GA, Zimmerman GA, McEver RP (1990) Rapid neutrophil adhesion to activated endothelium mediated by GMP-140. *Nature* 343: 757–760

14 Gamble JR, Skinner MP, Berndt MC, Vadas MA (1990) Prevention of activated neutrophil adhesion to endothelium by soluble adhesion protein GMP140. *Science* 249: 414–417

15 Palabrica T, Lobb R, Furie BC, Aronovitz M, Benjamin C, Hsu YM, Sajer SA, Furie B (1992) Leukocyte accumulation promoting fibrin deposition is mediated *in vivo* by P-selectin on adherent platelets. *Nature* 359: 848–851

16 Falati S, Liu Q, Gross P, Merrill-Skoloff G, Chou J, Vandendries E, Celi A, Croce K, Furie BC, Furie B (2003) Accumulation of tissue factor into developing thrombi *in vivo* is dependent upon microparticle P-selectin glycoprotein ligand 1 and platelet P-selectin. *J Exp Med* 197: 1585–1598

17 Bevilacqua MP, Pober JS, Mendrick DL, Cotran RS, Gimbrone MA Jr (1987) Identification of an inducible endothelial-leukocyte adhesion molecule. *Proc Natl Acad Sci USA* 84: 9238–9242

18 Rice GE, Bevilacqua MP (1989) An inducible endothelial cell surface glycoprotein mediates melanoma adhesion. *Science* 246: 1303–1306

19 Graber N, Gopal TV, Wilson D, Beall LD, Polte T, Newman W (1990) T cells bind to cytokine-activated endothelial cells *via* a novel, inducible sialoglycoprotein and endothelial leukocyte adhesion molecule-1. *J Immunol* 145: 819–830

20 Weller PF, Rand TH, Goelz SE, Chi-Rosso G, Lobb RR (1991) Human eosinophil adherence to vascular endothelium mediated by binding to vascular cell adhesion molecule 1 and endothelial leukocyte adhesion molecule 1. *Proc Natl Acad Sci USA* 88: 7430–7433

21 Lobb RR, Chi-Rosso G, Leone DR, Rosa MD, Bixler S, Newman BM, Luhowskyj S, Benjamin CD, Dougas IG, Goelz SE et al (1991) Expression and functional characterization of a soluble form of endothelial-leukocyte adhesion molecule 1. *J Immunol* 147: 124–129

22 Carlos T, Kovach N, Schwartz B, Rosa M, Newman B, Wayner E, Benjamin C, Osborn L, Lobb R, Harlan J (1991) Human monocytes bind to two cytokine-induced adhesive ligands on cultured human endothelial cells: endothelial-leukocyte adhesion molecule-1 and vascular cell adhesion molecule-1. *Blood* 77: 2266–2271

23 Springer TA (1994) Traffic signals for lymphocyte recirculation and leukocyte emigration: the multistep paradigm. *Cell* 76: 301–314

24 Ley K (1996) Molecular mechanisms of leukocyte recruitment in the inflammatory process. *Cardiovasc Res* 32: 733–742

25 Lawrence MB, Springer TA (1991) Leukocytes roll on a selectin at physiologic flow rates: distinction from and prerequisite for adhesion through integrins. *Cell* 65: 859–873

26 Jones DA, Abbassi O, McIntire LV, McEver RP, Smith CW (1993) P-selectin mediates neutrophil rolling on histamine-stimulated endothelial cells. *Biophys J* 65: 1560–1569

27 Mayadas TN, Johnson RC, Rayburn H, Hynes RO, Wagner DD (1993) Leukocyte rolling and extravasation are severely compromised in P selectin-deficient mice. *Cell* 74: 541–554

28 Abbassi O, Kishimoto TK, McIntire LV, Smith CW (1993) Neutrophil adhesion to endothelial cells. *Blood* Cells 19: 245–259

29 Abbassi O, Kishimoto TK, McIntire LV, Anderson DC, Smith CW (1993) E-selectin supports neutrophil rolling *in vitro* under conditions of flow. *J Clin Invest* 92: 2719–2730

30 Lawrence MB, Springer TA (1993) Neutrophils roll on E-selectin. *J Immunol* 151: 6338–6346

31 Huang KS, Graves BJ, Wolitzky BA (1997) Functional analysis of selectin structure. In: D Vestweber (ed): *The selectins: Initiators of leukocyte endothelial adhesion*. Harwood Academic, Amsterdam, 1–29

32 Kansas GS (1996) Selectins and their ligands: Current concepts and controversies. *Blood* 88: 3259–3287

33 Zelensky AN, Gready JE (2005) The C-type lectin-like domain superfamily. *FEBS J* 272: 6179–6217

34 Erbe DV, Wolitzky BA, Presta LG, Norton CR, Ramos RJ, Burns DK, Rumberger JM, Rao BN, Foxall C, Brandley BK et al (1992) Identification of an E-selectin region critical for carbohydrate recognition and cell adhesion. *J Cell Biol* 119: 215–227

35 Geng JG, Heavner GA, McEver RP (1992) Lectin domain peptides from selectins interact with both cell surface ligands and Ca^{2+} ions. *J Biol Chem* 267: 19846–19853

36 Erbe DV, Watson SR, Presta LG, Wolitzky BA, Foxall C, Brandley BK, Lasky LA (1993) P- and E-selectin use common sites for carbohydrate ligand recognition and cell adhesion. *J Cell Biol* 120: 1227–1235

37 Hollenbaugh D, Bajorath J, Stenkamp R, Aruffo A (1993) Interaction of P-selectin (CD62) and its cellular ligand: analysis of critical residues. *Biochemistry* 32: 2960–2966

38 Pigott R, Needham LA, Edwards RM, Walker C, Power C (1991) Structural and functional studies of the endothelial activation antigen endothelial leucocyte adhesion molecule-1 using a panel of monoclonal antibodies. *J Immunol* 147: 130–135

39 Kansas GS, Spertini O, Stoolman LM, Tedder TF (1991) Molecular mapping of functional domains of the leukocyte receptor for endothelium, LAM-1. *J Cell Biol* 114: 351–358

40 Gibson RM, Kansas GS, Tedder TF, Furie B, Furie BC (1995) Lectin and epidermal growth factor domains of P-selectin at physiologic density are the recognition unit for leukocyte binding. *Blood* 85: 151–158

41 Revelle BM, Scott D, Beck PJ (1996) Single amino acid residues in the E- and P-selectin epidermal growth factor domains can determine carbohydrate binding specificity. *J Biol Chem* 271: 16160–16170

42 Murphy JF, McGregor JL (1994) Two sites on P-selectin (the lectin and epidermal growth factor-like domains) are involved in the adhesion of monocytes to thrombin-activated endothelial cells. *Biochem J* 303: 619–624

43 Graves BJ, Crowther RL, Chandran C, Rumberger JM, Li S, Huang KS, Presky DH, Familletti PC, Wolitzky BA, Burns DK (1994) Insight into E-selectin/ligand interaction from the crystal structure and mutagenesis of the lec/EGF domains. *Nature* 367: 532–538

44 Somers WS, Tang J, Shaw GD, Camphausen RT (2000) Insights into the molecular basis of leukocyte tethering and rolling revealed by structures of P- and E-selectin bound to SLe(X) and PSGL-1. *Cell* 103: 467–479

45 Li SH, Burns DK, Rumberger JM, Presky DH, Wilkinson VL, Anostario M Jr., Wolitzky BA, Norton CR, Familletti PC, Kim KJ et al (1994) Consensus repeat domains of E-selectin enhance ligand binding. *J Biol Chem* 269: 4431–4437

46 Patel KD, Nollert MU, McEver RP (1995) P-selectin must extend a sufficient length from the plasma membrane to mediate rolling of neutrophils. *J Cell Biol* 131: 1893–1902

47 Jutila MA, Watts G, Walcheck B, Kansas GS (1992) Characterization of a functionally important and evolutionarily well-conserved epitope mapped to the short consensus repeats of E-selectin and L-selectin. *J Exp Med* 175: 1565–1573

48 Kansas GS, Pavalko FM (1996) The cytoplasmic domains of E- and P-selectin do not constitutively interact with alpha-actinin and are not essential for leukocyte adhesion. *J Immunol* 157: 321–325

49 Hartwell DW, Mayadas TN, Berger G, Frenette PS, Rayburn H, Hynes RO, Wagner DD (1998) Role of P-selectin cytoplasmic domain in granular targeting *in vivo* and in early inflammatory responses. *J Cell Biol* 143: 1129–1141

50 Disdier M, Morrissey JH, Fugate RD, Bainton DF, McEver RP (1992) Cytoplasmic domain of P-selectin (CD62) contains the signal for sorting into the regulated secretory pathway. *Mol Biol Cell* 3: 309–321

51 Green SA, Setiadi H, McEver RP, Kelly RB (1994) The cytoplasmic domain of P-selectin contains a sorting determinant that mediates rapid degradation in lysosomes. *J Cell Biol* 124: 435–448

52 Subramaniam M, Koedam JA, Wagner DD (1993) Divergent fates of P- and E-selectins after their expression on the plasma membrane. *Mol Biol Cell* 4: 791–801

53 Harrison-Lavoie KJ, Michaux G, Hewlett L, Kaur J, Hannah MJ, Lui-Roberts WW, Norman KE, Cutler DF (2006) P-selectin and CD63 use different mechanisms for delivery to Weibel-Palade bodies. *Traffic* 7: 647–662

54 Fujimoto T, McEver RP (1993) The cytoplasmic domain of P-selectin is phosphorylated on serine and threonine residues. *Blood* 82: 1758–1766

55 Yoshida M, Szente BE, Kiely JM, Rosenzweig A, Gimbrone MA Jr. (1998) Phosphorylation of the cytoplasmic domain of E-selectin is regulated during leukocyte-endothelial adhesion. *J Immunol* 161: 933–941

56 Watson ML, Kingsmore SF, Johnston GI, Siegelman MH, Le Beau MM, Lemons RS, Bora NS, Howard TA, Weissman IL, McEver RP et al (1990) Genomic organization of the selectin family of leukocyte adhesion molecules on human and mouse chromosome 1. *J Exp Med* 172: 263–272

57 King PD, Sandberg ET, Selvakumar A, Fang P, Beaudet AL, Dupont B (1995) Novel

isoforms of murine intercellular adhesion molecule-1 generated by alternative RNA splicing. *J Immunol* 154: 6080–6093

58   Osborn L, Vassallo C, Benjamin CD (1992) Activated endothelium binds lymphocytes through a novel binding site in the alternately spliced domain of vascular cell adhesion molecule-1. *J Exp Med* 176: 99–107

59   Yan HC, Baldwin HS, Sun J, Buck CA, Albelda SM, DeLisser HM (1995) Alternative splicing of a specific cytoplasmic exon alters the binding characteristics of murine platelet/endothelial cell adhesion molecule-1 (PECAM-1). *J Biol Chem* 270: 23672–23680

60   Leung E, Berg RW, Langley R, Greene J, Raymond LA, Augustus M, Ni J, Carter KC, Spurr N, Choo KH et al (1997) Genomic organization, chromosomal mapping, and analysis of the 5′ promoter region of the human MAdCAM-1 gene. *Immunogenetics* 46: 111–119

61   Johnston GI, Bliss GA, Newman PJ, McEver RP (1990) Structure of the human gene encoding granule membrane protein-140, a member of the selectin family of adhesion receptors for leukocytes. *J Biol Chem* 265: 21381–21385

62   Ishiwata N, Takio K, Katayama M, Watanabe K, Titani K, Ikeda Y, Handa M (1994) Alternatively spliced isoform of P-selectin is present *in vivo* as a soluble molecule. *J Biol Chem* 269: 23708–23715

63   Billups KL, Sherley JL, Palladino MA, Tindall JW, Roberts KP (1995) Evidence for E-selectin complement regulatory domain mRNA splice variants in the rat. *J Lab Clin Med* 126: 580–587

64   Woollard KJ (2005) Soluble bio-markers in vascular disease: much more than gauges of disease? *Clin Exp Pharmacol Physiol* 32: 233–240

65   Hope SA, Meredith IT (2003) Cellular adhesion molecules and cardiovascular disease. Part I. Their expression and role in atherogenesis. *Intern Med J* 33: 380–386

66   Hope SA, Meredith IT (2003) Cellular adhesion molecules and cardiovascular disease. Part II. Their association with conventional and emerging risk factors, acute coronary events and cardiovascular risk prediction. *Intern Med J* 33: 450–462

67   Berger G, Hartwell DW, Wagner DD (1998) P-Selectin and platelet clearance. *Blood* 92: 4446–4452

68   Michelson AD, Barnard MR, Hechtman HB, MacGregor H, Connolly RJ, Loscalzo J, Valeri CR (1996) *In vivo* tracking of platelets: circulating degranulated platelets rapidly lose surface P-selectin but continue to circulate and function. *Proc Natl Acad Sci USA* 93: 11877–11882

69   Andre P, Hartwell D, Hrachovinova I, Saffaripour S, Wagner DD (2000) Pro-coagulant state resulting from high levels of soluble P-selectin in blood. *Proc Natl Acad Sci USA* 97: 13835–13840

70   Cambien B, Wagner DD (2004) A new role in hemostasis for the adhesion receptor P-selectin. *Trends Mol Med* 10: 179–186

71   Patel KD, Zimmerman GA, Prescott SM, McEver RP, McIntyre TM (1991) Oxygen radicals induce human endothelial cells to express GMP-140 and bind neutrophils. *J Cell Biol* 112: 749–759

72 Kubes P, Kanwar S (1994) Histamine induces leukocyte rolling in post-capillary venules. A P-selectin-mediated event. *J Immunol* 152: 3570–3577

73 McEver RP (1997) Regulation of expression of E-selectin and P-selectin. In: D Vestweber (ed): *The selectins – initiators of leukocyte endothelial adhesion*. Harwood Academic, Amsterdam, 31–48

74 Smeets EF, de Vries T, Leeuwenberg JF, van den Eijnden DH, Buurman WA, Neefjes JJ (1993) Phosphorylation of surface E-selectin and the effect of soluble ligand (sialyl Lewisx) on the half-life of E-selectin. *Eur J Immunol* 23: 147–151

75 von Asmuth EJ, Smeets EF, Ginsel LA, Onderwater JJ, Leeuwenberg JF, Buurman WA (1992) Evidence for endocytosis of E-selectin in human endothelial cells. *Eur J Immunol* 22: 2519–2526

76 Kuijpers TW, Raleigh M, Kavanagh T, Janssen H, Calafat J, Roos D, Harlan JM (1994) Cytokine-activated endothelial cells internalize E-selectin into a lysosomal compartment of vesiculotubular shape. A tubulin-driven process. *J Immunol* 152: 5060–5069

77 Whelan J, Ghersa P, Hooft van Huijsduijnen R, Gray J, Chandra G, Talabot F, DeLamarter JF (1991) An NF kappa B-like factor is essential but not sufficient for cytokine induction of endothelial leukocyte adhesion molecule 1 (ELAM-1) gene transcription. *Nucleic Acids Res* 19: 2645–2653

78 Kaszubska W, Hooft van Huijsduijnen R, Ghersa P, DeRaemy-Schenk AM, Chen BP, Hai T, DeLamarter JF, Whelan J (1993) Cyclic AMP-independent ATF family members interact with NF-kappa B and function in the activation of the E-selectin promoter in response to cytokines. *Mol Cell Biol* 13: 7180–7190

79 Lewis H, Kaszubska W, DeLamarter JF, Whelan J (1994) Cooperativity between two NF-kappa B complexes, mediated by high-mobility-group protein I(Y), is essential for cytokine-induced expression of the E-selectin promoter. *Mol Cell Biol* 14: 5701–5709

80 Schindler U, Baichwal VR (1994) Three NF-kappa B binding sites in the human E-selectin gene required for maximal tumor necrosis factor alpha-induced expression. *Mol Cell Biol* 14: 5820–5831

81 Whitley MZ, Thanos D, Read MA, Maniatis T, Collins T (1994) A striking similarity in the organization of the E-selectin and beta interferon gene promoters. *Mol Cell Biol* 14: 6464–6475

82 Read MA, Whitley MZ, Gupta S, Pierce JW, Best J, Davis RJ, Collins T (1997) Tumor necrosis factor alpha-induced E-selectin expression is activated by the nuclear factor-kappaB and c-JUN N-terminal kinase/p38 mitogen-activated protein kinase pathways. *J Biol Chem* 272: 2753–2761

83 Edelstein LC, Pan A, Collins T (2005) Chromatin modification and the endothelial-specific activation of the E-selectin gene. *J Biol Chem* 280: 11192–11202

84 Montgomery KF, Osborn L, Hession C, Tizard R, Goff D, Vassallo C, Tarr PI, Bomsztyk K, Lobb R, Harlan JM et al (1991) Activation of endothelial-leukocyte adhesion molecule 1 (ELAM-1) gene transcription. *Proc Natl Acad Sci USA* 88: 6523–6527

85 Read MA, Whitley MZ, Williams AJ, Collins T (1994) NF-kappa B and I kappa B alpha: an inducible regulatory system in endothelial activation. *J Exp Med* 179: 503–512

86 Read MA, Neish AS, Gerritsen ME, Collins T (1996) Postinduction transcriptional repression of E-selectin and vascular cell adhesion molecule-1. *J Immunol* 157: 3472–3479

87 Boyle EM, Jr., Sato TT, Noel RF, Jr., Verrier ED, Pohlman TH (1999) Transcriptional arrest of the human E-selectin gene. *J Surg Res* 82: 194–200

88 dela Paz NG, Simeonidis S, Leo C, Rose DW, Collins T (2007) Regulation of NF-kappaB-dependent gene expression by the POU domain transcription factor Oct-1. *J Biol Chem* 282: 8424–8434

89 Pan J, McEver RP (1993) Characterization of the promoter for the human P-selectin gene. *J Biol Chem* 268: 22600–22608

90 Pan J, McEver RP (1995) Regulation of the human P-selectin promoter by Bcl-3 and specific homodimeric members of the NF-kappa B/Rel family. *J Biol Chem* 270: 23077–23083

91 Pan J, Xia L, McEver RP (1998) Comparison of promoters for the murine and human P-selectin genes suggests species-specific and conserved mechanisms for transcriptional regulation in endothelial cells. *J Biol Chem* 273: 10058–10067

92 Khew-Goodall Y, Wadham C, Stein BN, Gamble JR, Vadas MA (1999) Stat6 activation is essential for interleukin-4 induction of P-selectin transcription in human umbilical vein endothelial cells. *Arterioscler Thromb Vasc Biol* 19: 1421–1429

93 Khew-Goodall Y, Butcher CM, Litwin MS, Newlands S, Korpelainen EI, Noack LM, Berndt MC, Lopez AF, Gamble JR, Vadas MA (1996) Chronic expression of P-selectin on endothelial cells stimulated by the T-cell cytokine, interleukin-3. *Blood* 87: 1432–1438

94 Yao L, Pan J, Setiadi H, Patel KD, McEver RP (1996) Interleukin 4 or oncostatin M induces a prolonged increase in P-selectin mRNA and protein in human endothelial cells. *J Exp Med* 184: 81–92

95 Yao L, Setiadi H, Xia L, Laszik Z, Taylor FB, McEver RP (1999) Divergent inducible expression of P-selectin and E-selectin in mice and primates. *Blood* 94: 3820–3828

96 Weller A, Isenmann S, Vestweber D (1992) Cloning of the mouse endothelial selectins. Expression of both E- and P-selectin is inducible by tumor necrosis factor alpha. *J Biol Chem* 267: 15176–15183

97 Sanders WE, Wilson RW, Ballantyne CM, Beaudet AL (1992) Molecular cloning and analysis of *in vivo* expression of murine P-selectin. *Blood* 80: 795–800

98 Auchampach JA, Oliver MG, Anderson DC, Manning AM (1994) Cloning, sequence comparison and *in vivo* expression of the gene encoding rat P-selectin. *Gene* 145: 251–255

99 Pan J, Xia L, Yao L, McEver RP (1998) Tumor necrosis factor-alpha- or lipopolysaccharide-induced expression of the murine P-selectin gene in endothelial cells involves novel kappaB sites and a variant activating transcription factor/cAMP response element. *J Biol Chem* 273: 10068–10077

100 Sperandio M (2006) Selectins and glycosyltransferases in leukocyte rolling *in vivo*. *FEBS J* 273: 4377–4389

101 Moore KL, Stults NL, Diaz S, Smith DF, Cummings RD, Varki A, McEver RP (1992) Identification of a specific glycoprotein ligand for P-selectin (CD62) on myeloid cells. *J Cell Biol* 118: 445–456

102 Norgard KE, Moore KL, Diaz S, Stults NL, Ushiyama S, McEver RP, Cummings RD, Varki A (1993) Characterization of a specific ligand for P-selectin on myeloid cells. A minor glycoprotein with sialylated O-linked oligosaccharides. *J Biol Chem* 268: 12764–12774

103 Sako D, Chang XJ, Barone KM, Vachino G, White HM, Shaw G, Veldman GM, Bean KM, Ahern TJ, Furie B et al (1993) Expression cloning of a functional glycoprotein ligand for P-selectin. *Cell* 75: 1179–1186

104 Moore KL, Eaton SF, Lyons DE, Lichenstein HS, Cummings RD, McEver RP (1994) The P-selectin glycoprotein ligand from human neutrophils displays sialylated, fucosylated, O-linked poly-N-acetyllactosamine. *J Biol Chem* 269: 23318–23327

105 Alon R, Rossiter H, Wang X, Springer TA, Kupper TS (1994) Distinct cell surface ligands mediate T lymphocyte attachment and rolling on P and E selectin under physiological flow. *J Cell Biol* 127: 1485–1495

106 Asa D, Raycroft L, Ma L, Aeed PA, Kaytes PS, Elhammer AP, Geng JG (1995) The P-selectin glycoprotein ligand functions as a common human leukocyte ligand for P- and E-selectins. *J Biol Chem* 270: 11662–11670

107 Moore KL, Patel KD, Bruehl RE, Li F, Johnson DA, Lichenstein HS, Cummings RD, Bainton DF, McEver RP (1995) P-selectin glycoprotein ligand-1 mediates rolling of human neutrophils on P-selectin. *J Cell Biol* 128: 661–671

108 Walcheck B, Moore KL, McEver RP, Kishimoto TK (1996) Neutrophil-neutrophil interactions under hydrodynamic shear stress involve L-selectin and PSGL-1. A mechanism that amplifies initial leukocyte accumulation of P-selectin *in vitro. J Clin Invest* 98: 1081–1087

109 Guyer DA, Moore KL, Lynam EB, Schammel CM, Rogelj S, McEver RP, Sklar LA (1996) P-selectin glycoprotein ligand-1 (PSGL-1) is a ligand for L-selectin in neutrophil aggregation. *Blood* 88: 2415–2421

110 Snapp KR, Wagers AJ, Craig R, Stoolman LM, Kansas GS (1997) P-selectin glycoprotein ligand-1 is essential for adhesion to P-selectin but not E-selectin in stably transfected hematopoietic cell lines. *Blood* 89: 896–901

111 Yang J, Hirata T, Croce K, Merrill-Skoloff G, Tchernychev B, Williams E, Flaumenhaft R, Furie BC, Furie B (1999) Targeted gene disruption demonstrates that P-selectin glycoprotein ligand 1 (PSGL-1) is required for P-selectin-mediated but not E-selectin-mediated neutrophil rolling and migration. *J Exp Med* 190: 1769–1782

112 Haddad W, Cooper CJ, Zhang Z, Brown JB, Zhu Y, Issekutz A, Fuss I, Lee HO, Kansas GS, Barrett TA (2003) P-selectin and P-selectin glycoprotein ligand 1 are major determinants for Th1 cell recruitment to nonlymphoid effector sites in the intestinal lamina propria. *J Exp Med* 198: 369–377

113 Fuhlbrigge RC, Kieffer JD, Armerding D, Kupper TS (1997) Cutaneous lymphocyte

antigen is a specialized form of PSGL-1 expressed on skin-homing T cells. *Nature* 389: 978–981

114 Hirata T, Merrill-Skoloff G, Aab M, Yang J, Furie BC, Furie B (2000) P-Selectin glycoprotein ligand 1 (PSGL-1) is a physiological ligand for E-selectin in mediating T helper 1 lymphocyte migration. *J Exp Med* 192: 1669–1676

115 Zanardo RC, Bonder CS, Hwang JM, Andonegui G, Liu L, Vestweber D, Zbytnuik L, Kubes P (2004) A down-regulatable E-selectin ligand is functionally important for PSGL-1-independent leukocyte-endothelial cell interactions. *Blood* 104: 3766–3773

116 Aigner S, Sthoeger ZM, Fogel M, Weber E, Zarn J, Ruppert M, Zeller Y, Vestweber D, Stahel R, Sammar M et al (1997) CD24, a mucin-type glycoprotein, is a ligand for P-selectin on human tumor cells. *Blood* 89: 3385–3395

117 Sammar M, Aigner S, Hubbe M, Schirrmacher V, Schachner M, Vestweber D, Altevogt P (1994) Heat-stable antigen (CD24) as ligand for mouse P-selectin. *Int Immunol* 6: 1027–1036

118 Aigner S, Ruppert M, Hubbe M, Sammar M, Sthoeger Z, Butcher EC, Vestweber D, Altevogt P (1995) Heat stable antigen (mouse CD24) supports myeloid cell binding to endothelial and platelet P-selectin. *Int Immunol* 7: 1557–1565

119 Aigner S, Ramos CL, Hafezi-Moghadam A, Lawrence MB, Friederichs J, Altevogt P, Ley K (1998) CD24 mediates rolling of breast carcinoma cells on P-selectin. *FASEB J* 12: 1241–1251

120 Steegmaier M, Levinovitz A, Isenmann S, Borges E, Lenter M, Kocher HP, Kleuser B, Vestweber D (1995) The E-selectin-ligand ESL-1 is a variant of a receptor for fibroblast growth factor. *Nature* 373: 615–620

121 Dimitroff CJ, Descheny L, Trujillo N, Kim R, Nguyen V, Huang W, Pienta KJ, Kutok JL, Rubin MA (2005) Identification of leukocyte E-selectin ligands, P-selectin glycoprotein ligand-1 and E-selectin ligand-1, on human metastatic prostate tumor cells. *Cancer Res* 65: 5750–5760

122 Hanley WD, Burdick MM, Konstantopoulos K, Sackstein R (2005) CD44 on LS174T colon carcinoma cells possesses E-selectin ligand activity. *Cancer Res* 65: 5812–5817

123 Katayama Y, Hidalgo A, Chang J, Peired A, Frenette PS (2005) CD44 is a physiological E-selectin ligand on neutrophils. *J Exp Med* 201: 1183–1189

124 Burdick MM, Chu JT, Godar S, Sackstein R (2006) HCELL is the major E- and L-selectin ligand expressed on LS174T colon carcinoma cells. *J Biol Chem* 281: 13899–13905

125 Dagia NM, Gadhoum SZ, Knoblauch CA, Spencer JA, Zamiri P, Lin CP, Sackstein R (2006) G-CSF induces E-selectin ligand expression on human myeloid cells. *Nat Med* 12: 1185–1190

126 Bullard DC, Beaudet AL (1997) Analysis of selectin deficient mice. In: D Vestweber (ed): *The selectins – initiators of leukocyte endothelial adhesion*. Harwood Academic, Amsterdam, 133–142

127 Subramaniam M, Frenette PS, Saffaripour S, Johnson RC, Hynes RO, Wagner DD (1996) Defects in hemostasis in P-selectin-deficient mice. *Blood* 87: 1238–1242

128 Johnson RC, Mayadas TN, Frenette PS, Mebius RE, Subramaniam M, Lacasce A,

Hynes RO, Wagner DD (1995) Blood cell dynamics in P-selectin-deficient mice. *Blood* 86: 1106–1114

129 Ley K, Bullard DC, Arbones ML, Bosse R, Vestweber D, Tedder TF, Beaudet AL (1995) Sequential contribution of L- and P-selectin to leukocyte rolling *in vivo*. *J Exp Med* 181: 669–675

130 Kunkel EJ, Jung U, Bullard DC, Norman KE, Wolitzky BA, Vestweber D, Beaudet AL, Ley K (1996) Absence of trauma-induced leukocyte rolling in mice deficient in both P-selectin and intercellular adhesion molecule 1. *J Exp Med* 183: 57–65

131 Bullard DC, Qin L, Lorenzo I, Quinlin WM, Doyle NA, Bosse R, Vestweber D, Doerschuk CM, Beaudet AL (1995) P-selectin/ICAM-1 double mutant mice: acute emigration of neutrophils into the peritoneum is completely absent but is normal into pulmonary alveoli. *J Clin Invest* 95: 1782–1788

132 Labow MA, Norton CR, Rumberger JM, Lombard-Gillooly KM, Shuster DJ, Hubbard J, Bertko R, Knaack PA, Terry RW, Harbison ML et al (1994) Characterization of E-selectin-deficient mice: demonstration of overlapping function of the endothelial selectins. *Immunity* 1: 709–720

133 Bullard DC, Kunkel EJ, Kubo H, Hicks MJ, Lorenzo I, Doyle NA, Doerschuk CM, Ley K, Beaudet AL (1996) Infectious susceptibility and severe deficiency of leukocyte rolling and recruitment in E-selectin and P-selectin double mutant mice. *J Exp Med* 183: 2329–2336

134 Frenette PS, Mayadas TN, Rayburn H, Hynes RO, Wagner DD (1996) Susceptibility to infection and altered hematopoiesis in mice deficient in both P- and E- selectins. *Cell* 84: 563–574

135 Wild MK, Luhn K, Marquardt T, Vestweber D (2002) Leukocyte adhesion deficiency II: therapy and genetic defect. *Cells Tissues Organs* 172: 161–173

136 Staite ND, Justen JM, Sly LM, Beaudet AL, Bullard DC (1996) Inhibition of delayed-type contact hypersensitivity in mice deficient in both E-selectin and P-selectin. *Blood* 88: 2973–2979

137 Mizgerd JP, Bullard DC, Hicks MJ, Beaudet AL, Doerschuk CM (1999) Chronic inflammatory disease alters adhesion molecule requirements for acute neutrophil emigration in mouse skin. *J Immunol* 162: 5444–5448

138 Bevilacqua MP, Nelson RM, Mannori G, Cecconi O (1994) Endothelial-leukocyte adhesion molecules in human diseases. *Annu Rev Med* 45: 361–378

139 McMurray RW (1996) Adhesion molecules in autoimmune disease. *Semin Arthritis Rheum* 25: 215–233

140 Ley K (2003) The role of selectins in inflammation and disease. *Trends Mol Med* 9: 263–268

141 El-Magadmi M, Alansari A, Teh LS, Ordi J, Gul A, Inanc M, Bruce I, Hajeer A (2001) Association of the A561C E-selectin polymorphism with systemic lupus erythematosus in 2 independent populations. *J Rheumatol* 28: 2650–2652

142 Tregouet DA, Barbaux S, Escolano S, Tahri N, Golmard JL, Tiret L, Cambien F (2002)

Specific haplotypes of the P-selectin gene are associated with myocardial infarction. *Hum Mol Genet* 11: 2015–2023

143 Watanabe Y, Inoue T, Okada H, Kotaki S, Kanno Y, Kikuta T, Suzuki H (2006) Impact of selectin gene polymorphisms on rapid progression to end-stage renal disease in patients with IgA nephropathy. *Intern Med* 45: 947–951

144 Bourgain C, Hoffjan S, Nicolae R, Newman D, Steiner L, Walker K, Reynolds R, Ober C, McPeek MS (2003) Novel case-control test in a founder population identifies P-selectin as an atopy-susceptibility locus. *Am J Hum Genet* 73: 612–626

145 Nageh MF, Sandberg ET, Marotti KR, Lin AH, Melchior EP, Bullard DC, Beaudet AL (1997) Deficiency of inflammatory cell adhesion molecules protects against atherosclerosis in mice. *Arterioscler Thromb Vasc Biol* 17: 1517–1520

146 Johnson RC, Chapman SM, Dong ZM, Ordovas JM, Mayadas TN, Herz J, Hynes RO, Schaefer EJ, Wagner DD (1997) Absence of P-selectin delays fatty streak formation in mice. *J Clin Invest* 99: 1037–1043

147 Dong ZM, Brown AA, Wagner DD (2000) Prominent role of P-selectin in the development of advanced atherosclerosis in ApoE-deficient mice. *Circulation* 101: 2290–2295

148 Collins RG, Velji R, Guevara NV, Hicks MJ, Chan L, Beaudet AL (2000) P-Selectin or intercellular adhesion molecule (ICAM)-1 deficiency substantially protects against atherosclerosis in apolipoprotein E-deficient mice. *J Exp Med* 191: 189–194

149 Dong ZM, Chapman SM, Brown AA, Frenette PS, Hynes RO, Wagner DD (1998) The combined role of P- and E-selectins in atherosclerosis. *J Clin Invest* 102: 145–152

150 Engelhardt B, Vestweber D, Hallmann R, Schulz M (1997) E- and P-selectin are not involved in the recruitment of inflammatory cells across the blood-brain barrier in experimental autoimmune encephalomyelitis. *Blood* 90: 4459–4472

151 Osmers I, Bullard DC, Barnum SR (2005) PSGL-1 is not required for development of experimental autoimmune encephalomyelitis. *J Neuroimmunol* 166: 193–196

152 Engelhardt B, Kempe B, Merfeld-Clauss S, Laschinger M, Furie B, Wild MK, Vestweber D (2005) P-selectin glycoprotein ligand 1 is not required for the development of experimental autoimmune encephalomyelitis in SJL and C57BL/6 mice. *J Immunol* 175: 1267–1275

153 Kerfoot SM, Norman MU, Lapointe BM, Bonder CS, Zbytnuik L, Kubes P (2006) Reevaluation of P-selectin and alpha 4 integrin as targets for the treatment of experimental autoimmune encephalomyelitis. *J Immunol* 176: 6225–6234

154 Bullard DC, Mobley JM, Justen JM, Sly LM, Chosay JG, Dunn CJ, Lindsey JR, Beaudet AL, Staite ND (1999) Acceleration and increased severity of collagen-induced arthritis in P-selectin mutant mice. *J Immunol* 163: 2844–2849

155 Ruth JH, Amin MA, Woods JM, He X, Samuel S, Yi N, Haas CS, Koch AE, Bullard DC (2005) Accelerated development of arthritis in mice lacking endothelial selectins. *Arthritis Res Ther* 7: R959–970

156 Sumariwalla PF, Malfait AM, Feldmann M (2004) P-selectin glycoprotein ligand 1 therapy ameliorates established collagen-induced arthritis in DBA/1 mice partly through the suppression of tumour necrosis factor. *Clin Exp Immunol* 136: 67–75

157 Seiler KP, Ma Y, Weis JH, Frenette PS, Hynes RO, Wagner DD, Weis JJ (1998) E and P selectins are not required for resistance to severe murine lyme arthritis. *Infect Immun* 66: 4557–4559

158 Schimmer RC, Schrier DJ, Flory CM, Dykens J, Tung DKL, Jacobson PB, Friedl HP, Conroy MC, Schimmer BB, Ward PA (1997) Streptococcal cell wall-induced arthritis: Requirements for neutrophils, P-selectin, intercellular adhesion molecule-1, and macrophage-inflammatory protein-2. *J Immunol* 159: 4103–4108

159 Walter UM, Issekutz AC (1997) The role of E- and P-selectin in neutrophil and monocyte migration in adjuvant-induced arthritis in the rat. *Eur J Immunol* 27: 1498–1505

160 Issekutz AC, Mu JY, Liu G, Melrose J, Berg EL (2001) E-selectin, but not P-selectin, is required for development of adjuvant-induced arthritis in the rat. *Arthritis Rheum* 44: 1428–1437

161 Theofilopoulos AN, Dixon FJ (1985) Murine models of systemic lupus erythematosus. *Adv Immunol* 37: 269–391

162 He X, Schoeb TR, Panoskaltsis-Mortari A, Zinn KR, Kesterson RA, Zhang J, Samuel S, Hicks MJ, Hickey MJ, Bullard DC (2006) Deficiency of P-selectin or P-selectin glycoprotein ligand-1 leads to accelerated development of glomerulonephritis and increased expression of CC chemokine ligand 2 in lupus-prone mice. *J Immunol* 177: 8748–8756

163 Rosenkranz AR, Mendrick DL, Cotran RS, Mayadas TN (1999) P-selectin deficiency exacerbates experimental glomerulonephritis: a protective role for endothelial P-selectin in inflammation. *J Clin Invest* 103: 649–659

164 Frenette PS, Johnson RC, Hynes RO, Wagner DD (1995) Platelets roll on stimulated endothelium *in vivo*: an interaction mediated by endothelial P-selectin. *Proc Natl Acad Sci USA* 92: 7450–7454

165 Diacovo TG, Puri KD, Warnock RA, Springer TA, von Andrian UH (1996) Platelet-mediated lymphocyte delivery to high endothelial venules. *Science* 273: 252–255

166 Diacovo TG, Catalina MD, Siegelman MH, von Andrian UH (1998) Circulating activated platelets reconstitute lymphocyte homing and immunity in L-selectin-deficient mice. *J Exp Med* 187: 197–204

167 Singbartl K, Forlow SB, Ley K (2001) Platelet, but not endothelial, P-selectin is critical for neutrophil-mediated acute postischemic renal failure. *FASEB J* 15: 2337–2344

168 Kuligowski MP, Kitching AR, Hickey MJ (2006) Leukocyte recruitment to the inflamed glomerulus: a critical role for platelet-derived P-selectin in the absence of rolling. *J Immunol* 176: 6991–6999

169 Zarbock A, Singbartl K, Ley K (2006) Complete reversal of acid-induced acute lung injury by blocking of platelet-neutrophil aggregation. *J Clin Invest* 116: 3211–3219

170 Zarbock A, Polanowska-Grabowska RK, Ley K (2007) Platelet-neutrophil-interactions: Linking hemostasis and inflammation. *Blood* Rev 21: 99–111

171 Evangelista V, Manarini S, Sideri R, Rotondo S, Martelli N, Piccoli A, Totani L, Piccardoni P, Vestweber D, de Gaetano G et al (1999) Platelet/polymorphonuclear leukocyte

interaction: P-selectin triggers protein-tyrosine phosphorylation-dependent CD11b/CD18 adhesion: role of PSGL-1 as a signaling molecule. *Blood* 93: 876–885

172 Piccardoni P, Sideri R, Manarini S, Piccoli A, Martelli N, de Gaetano G, Cerletti C, Evangelista V (2001) Platelet/polymorphonuclear leukocyte adhesion: a new role for SRC kinases in Mac-1 adhesive function triggered by P-selectin. *Blood* 98: 108–116

173 Ehrhardt C, Kneuer C, Bakowsky U (2004) Selectins-an emerging target for drug delivery. *Adv Drug Deliv Rev* 56: 527–549

174 Kaila N, Thomas BE (2002) Design and synthesis of sialyl Lewis(x) mimics as E- and P-selectin inhibitors. *Med Res Rev* 22: 566–601

175 Kaneider NC, Leger AJ, Kuliopulos A (2006) Therapeutic targeting of molecules involved in leukocyte-endothelial cell interactions. *FEBS J* 273: 4416–4424

176 Romano SJ (2005) Selectin antagonists: therapeutic potential in asthma and COPD. *Treat Respir Med* 4: 85–94

177 Zollner TM, Asadullah K, Schon MP (2007) Targeting leukocyte trafficking to inflamed skin: still an attractive therapeutic approach? *Exp Dermatol* 16: 1–12

178 Mertens P, Maes A, Nuyts J, Belmans A, Desmet W, Esplugas E, Charlier F, Figueras J, Sambuceti G, Schwaiger M et al (2006) Recombinant P-selectin glycoprotein ligand-immunoglobulin, a P-selectin antagonist, as an adjunct to thrombolysis in acute myocardial infarction. The P-Selectin Antagonist Limiting Myonecrosis (PSALM) trial. *Am Heart J* 152: 125 e121–128

179 Friedrich M, Bock D, Philipp S, Ludwig N, Sabat R, Wolk K, Schroeter-Maas S, Aydt E, Kang S, Dam TN et al (2006) Pan-selectin antagonism improves psoriasis manifestation in mice and man. *Arch Dermatol Res* 297: 345–351

Firm adhesion

Endothelial ICAM-1 functions in adhesion and signaling during leukocyte recruitment

Scott D. Auerbach, Lin Yang and Francis W. Luscinskas

Center for Excellence in Vascular Biology, Departments of Pathology, Brigham and Women's Hospital and Harvard Medical School, 77 Avenue Louis Pasteur, Boston, MA 02115, USA

Introduction

Intercellular adhesion molecule-1 (ICAM-1, CD54) was identified more than 20 years ago as a cytokine-inducible adhesion molecule [1]. Endothelial cell ICAM-1 surface expression is elevated at sites of endothelial cell activation *in vivo* and *in vitro*, and contributes to stable adhesion and transmigration of circulating blood leukocytes through its interaction with leukocyte β_2 integrins. This chapter briefly reviews ICAM-1 structure and function, and then address the mechanisms through which ICAM-1-dependent adhesion and signaling control leukocyte transmigration.

ICAM-1 tissue distribution, structure and regulation of expression

Tissue distribution

The extracellular domain of ICAM-1 functions in cell-cell adhesion and cell-matrix interactions. ICAM-1 has a broad tissue distribution and is constitutively expressed on the surface of vascular endothelial cells, epithelium, smooth muscle cells, fibroblasts, astrocytes, keratinocytes, and by immunocompetent cells such as activated T and B cells, monocytes and macrophages. The expression of ICAM-1 is regulated primarily at the level of gene transcription [2]. ICAM-1 surface expression is markedly augmented by inflammatory mediators, including TNF-α, IL-1α and β, IFN-γ and certain Gram bacterial endotoxins, and a combination of cytokines leads to synergistic enhancement of ICAM-1 expression in endothelium. Interestingly, this inducible expression can be modulated by anti-inflammatory cytokines such as transforming growth factor β (TGF-β), IL-10 and steroidal anti-inflammatory glucocorticoids (reviewed in [2]).

Adhesion Molecules: Function and Inhibition, edited by Klaus Ley
© 2007 Birkhäuser Verlag Basel/Switzerland

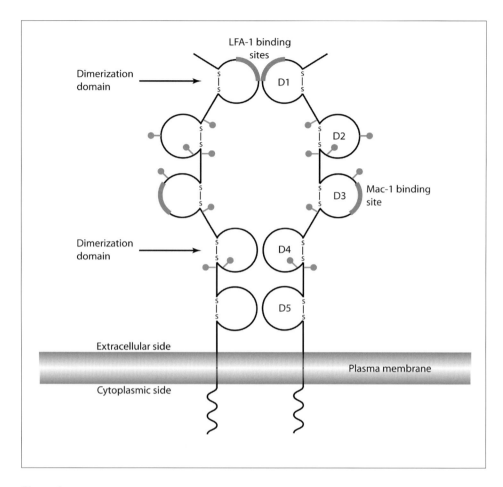

Figure 1
Model of ICAM-1 structure.
Human ICAM-1 is 505 amino acids and predicted to contain five Ig-like domains (D1–D5). Binding sites for the β₂ integrins LFA-1 (on D1) Mac-I (on D3) are indicated. ICAM-1 is depicted in the "O-dimer" conformation, with intimate contact made primarily between with D1 and D4 based on the report of Yang and coworkers [7]. The potential O- and N-linked glycosylation sites are identified by lollipops.

Structure

ICAM-1 is a transmembrane glycoprotein and has a mass that ranges from 85 to 110 kDa depending on the extent of post-translational glycosylation (see Fig. 1 for cartoon). It is a member of the immunoglobulin (Ig)-like gene superfamily and consists of an extracellular domain composed of five Ig-like domains that contain glycosylation sites, a predicted transmembrane region and a cytoplasmic tail of 28 amino acids [3]. In addition, four other ICAM family members (ICAM-2, ICAM-3, ICAM-4, ICAM-5) have been identified and are known ligands for leukocyte integrins.

In ICAM-1, domains 1 and 3 are critical for functional interactions with leukocyte β_2 integrins; domain 1 binds LFA-1, while domain 3 binds Mac-1 [4]. In addition, earlier studies found that surface-expressed ICAM-1 dimerizes and that the dimer form is the predominant species expressed by cytokine-activated vascular endothelium [5, 6]. Subsequent studies have explored the structural features that underlie ICAM-1 dimerization. X-ray crystallography suggests that ICAM-1 forms "O-dimers" with domains 1, 4 and the D4–5 stem being essential for dimer formation [7]. Self assembly occurs *via* intimate dimerization between D1 and D4, with D4–D5 (proximal to the membrane) forming a rigid stem structure. The authors of this study proposed that a dramatic conformational change may occur in ICAM-1 when LFA-1 binds to D1, based on their structural data, as well as an observation by Grakoui and coworkers [8] that ICAM-1 can form one-dimensional clusters upon LFA-1 binding. The proposed conformational change, illustrated below (see Fig. 2) transforms a linear array of "O-dimers" into concatenated "W-dimers" that assume a one-dimensional chain. It is not clear at present precisely how the proposed array of W-dimers enhances/facilitates leukocyte adhesion or transmigration and/or outside-in signaling but this idea certainly merits further study.

Regulation of ICAM-1 expression by vascular endothelium

The regulation of ICAM-1 expression has been well characterized by several groups (see the comprehensive review by Roebuck [2] and references therein) and is beyond the scope of this article. The key signal transduction pathways that regulate its expression include NF-κB, protein kinase C (PKC) and the mitogen-activated protein kinases (MAPK) ERK, JNK and p38. There are several transcription factors known to be involved in activation of gene expression including NF-κB family members, AP-1, Sp1, Ets.

ICAM-1 cytoplasmic domain associates with cytoplasmic linker proteins

General signals initiated by ICAM-1 occupancy

The signaling through ICAM-1 has been studied in more detail than other molecules and multiple reports document that the cytoplasmic tail is essential for transmitting signals that promote leukocyte transmigration and cytoskeletal remodeling. Engage-

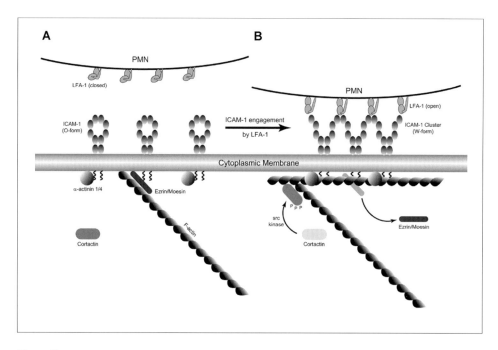

Figure 2
Cytoskeletal remodeling resulting from ICAM-1 engagement by LFA-1.
(A) Unoccupied ICAM-1 dimers adopt the O-form [7]. ERM proteins (ezrin and/or moesin) and α-actinin (1 and/or 4) associate with the ICAM-1 tail, while cortactin is free. In this scheme, only ezrin/moesin links ICAM-1 to F-actin prior to LFA-1 engagement. (B) Neutrophil LFA-1 has engaged ICAM-1 for transmigration, resulting in a conformational change in both LFA-1 and ICAM-1, the latter adopting the W-form clusters. Ezrin/moesin dissociates from the ICAM-1 tail, while α-actinin remains bound. New bundles of F-actin may be formed along a scaffold provided by α-actinin under these conditions, perhaps due to Ca^{2+} influx, PIP2 enrichment, and/or PIP3 synthesis, as well as other ICAM-1 dependent and independent pathways converging (i.e., PECAM-1, CD47, CD99). Cortactin is recruited to the ICAM-1 cluster and becomes phosphorylated by src family kinases. In this manner, ICAM-1 clustering may trigger cytoskeletal remodeling.

ment of ICAM-1 in cytokine-activated endothelium by adherent leukocytes triggers elevations in intracellular free Ca^{2+}, myosin contractility, activation of PKC and of small GTPases, in particular, members of the Rho family [9–11], p38 MAPK [12], and the tyrosine kinase p60[Src] (src kinases) [13]. A prominent substrate for p60[Src] in human and murine endothelium is the actin-binding protein cortactin [14]. Most of these functions have been shown to be dependent upon signaling through the

ICAM-1 cytoplasmic tail [15–17]. How these above signals converge to control the cytoskeletal remodeling that accompanies leukocyte transmigration is poorly understood despite a significant effort on the part of many investigators. It is likely that the small GTPases (Rho, Rac, Cdc42, Rap1) are critical for accomplishing the cytoskeletal remodeling that accompanies leukocyte transmigration. The reader is directed to several recent more detailed reviews on small GTPases in endothelium and leukocytes during adhesion [18, 19]. An approach that several labs have taken is to study the role of known molecules that bind to the cytoplasmic domain of ICAM-1 and examine their role in endothelial cell cytoskeletal remodeling and leukocyte transendothelial migration. ICAM-1 is predicted to have 28 C-terminal residues on the cytoplasmic side of the membrane. It is a remarkable feature of ICAM-1 that its single transmembrane domain and small cytoplasmic tail are capable of transmitting any information at all through the cytoplasmic membrane. In the following sections of this review, the interactions that are known to occur between ICAM-1 and endothelial cell cytoplasmic proteins are described.

α-Actinins

The α-actinins are a family of four highly homologous proteins [20] (α-actinin 1–4) that function as dimers and assemble F-actin into bundles of parallel and anti-parallel filaments [21]. The α-actinin isoforms 1 and 4, specific to non-muscle cells, are expressed in endothelium have a demonstrated affinity for the ICAM-1 cytoplasmic tail; isoforms 2 and 3, found in the Z-disks of muscle cells, are unlikely to play a role in this process and were not detected by PCR screens in endothelium (S. Auerbach and F. W. Luscinskas; unpublished results). The non-muscle α-actinins can be distinguished by their localization within cells. α-Actinin isoform 1 is localized in focal adhesions [22], while isoform 4 is present in membrane ruffles [23]. In endothelial cells, α-actinin 4 (but not α-actinin 1) is enriched at endothelial cell-cell junctions (S Auerbach and F. W. Luscinsksas; unpublished observations). The importance of the α-actinins in regulating the endothelial cell's participation in leukocyte transmigration is highlighted by their *in vivo* co-localization with ICAM-1 [24]. ICAM-1 is not unique in its affinity for the α-actinins. Non-muscle cell α-actinins have numerous potential binding partners besides ICAM-1, suggesting multiple roles for these proteins in cytoskeletal organization beyond that of leukocyte transmigration (reviewed in [25]). Known binding partners for α-actinins include molecules associated with adhesion and leukocyte targeting (ICAM-1, ICAM-2, β_2 integrins), focal adhesions (vinculin, palladin, β_1 and β_3 integrins) and signaling (iNOS, PI3 kinase). In addition to its affinity for various membrane-associated proteins, α-actinin may associate directly to the membrane phospholipids PIP2 and PIP3 (see [26] for review). The regulatory influences that these two phospholipids may exert on α-actinins, as well as cytoskeleton-associated proteins of the ezrin/radixin/moesin (ERM) family (see below) are likely to be major components of the leukocyte transmigratory mechanism.

The participation of α-actinins in leukocyte transmigration occurs, at least in part, *via* its binding to the ICAM-1 cytoplasmic tail. This interaction has been demonstrated *in vitro*, in the absence of a phospholipid membrane, using purified α-actinin and a synthetic ICAM-1 tail peptide [27]. The binding of α-actinin to ICAM-1 is therefore likely to be constitutive, and probably does not require accessory or adapter molecules, or small regulatory molecules. Additionally, a yeast two-hybrid screen for ICAM-1 tail binding partners identified a small set of proteins with affinities for the bait peptide sequence, among which were α-actinin 1 and 4 [24]. A tentative model describing the role of ICAM-1 in leukocyte transmigration should therefore include the constitutive affinity of α-actinin for the ICAM-1 tail. The α-actinins may not be regulated by their recruitment to ICAM-1 clusters that form during transmigration, but instead by changes in the local environment near the cytoplasmic membrane that result from ICAM-1 clustering (see Fig. 2). Although both α-actinin 1 and 4 bind to the ICAM-1 tail, Celli and co-workers [24] suggested that α-actinin 4 plays the more dominant role in transmigration, since siRNA knockdown of α-actinin 4 (but not α-actinin 1) significantly reduced neutrophil transmigration efficiency.

If α-actinins are constitutively bound to ICAM-1, how might ICAM-1 clustering regulate their activities, and what activities are regulated? As mentioned previously, α-actinins mediate the bundling of parallel (and anti-parallel) F-actin filaments to form stress fibers, or higher-order structures resembling stress fibers. This activity is regulated by cytosolic Ca^{2+} as well as by the membrane phospholipids PIP2 and PIP3 [26]. Recent structural evidence based on cryo-electron microscopy of α-actinin suggests more complex effects of Ca^{2+} [21]. Ca^{2+} ions may have an asymmetric effect on the affinity of the α-actinin dimmer for F-actin, whereby one actin-binding site is inhibited, but the other is not. Reports of PIP2 regulation of α-actinin are similarly inconclusive: *In vitro*, both PIP2 and PIP3 exerted inhibitory effects on the ability of α-actinin to bundle F-actin using purified protein–protein approaches [28], while a stimulatory role for PIP2 was identified in a different study [29, 30]. These results, however, conflict somewhat with the general observation presented by Insall [31] that PIP3, produced by PI3-kinase, acts as a second messenger that favors F-actin polymerization. More recently, fluorescent reporter molecules such as pleckstrin homology (PH) domains fused to green fluorescent protein (GFP) were used to demonstrate that both PIP2 and PIP3 are enriched in the endothelial cell docking structure that forms transiently around stably adherent lymphoblasts [32]. The possible involvement in transmigration of membrane phospholipids as regulatory molecules for α-actinins as well as ERMs is an attractive possibility because it would represent a mechanism of spatial regulation of proteins that come into contact with the cytoplasmic membrane. Nevertheless, while it is likely that Ca^{2+}, PIP2, and PIP3 are important regulators of the α-actinins and play roles in leukocyte transmigration, it is not yet clear what those regulatory roles might be.

Cortactin

Cortactin was initially identified as a target of v-Src tyrosine kinase phosphorylation [33] and subsequently as molecules that participate in F-actin filament organization [14]. Like α-actinin, cortactin is a multifunctional scaffold protein that has multiple binding partners, many of which (*e.g.*, paxillin, ZO-1) link F-actin filaments to the cytoplasmic membrane *via* cortactin. Cortactin functions as part of a large protein complex that can nucleate the formation of new F-actin filaments that branch from existing ones to form characteristic y-shaped (dendritic) branch points. The role of cortactin in this complex appears to be threefold [34]: (1) cortactin activates Arp2/3, itself a 7-member protein complex that is directly responsible for the synthesis of new dendritic actin filaments upon pre-existing ones (reviewed in [35]; (2) it stabilize new branch points assembled by Arp2/3; (3) it can anchor the complex to other cellular structures. Indeed, cortactin accumulates in various motility-associated structures including lamellipodia [34] and invadopodia [36], and is also present in cell adhesion structures (tight junctions and adherens junctions) [37].

Cortactin function in endothelial cells is required for leukocyte transmigration. Ligation of ICAM-1 on endothelial cells by leukocyte binding, or by ICAM-1-specific monoclonal antibody cross-linking, recruits cortactin to sites of clustered ICAM-1, and stimulates the src-mediated tyrosine phosphorylation of cortactin [11, 13, 38]. This mechanism of cortactin activation by recruitment contrasts with that of α-actinin, which is likely to be constitutively bound to the ICAM-1 tail independent of the extracellular environment of ICAM-1, or whether ICAM-1 is clustered or not. Supporting the proposed requirement of cortactin for leukocyte transmigration, inhibition of cortactin expression in endothelial cells using siRNA interference was found to reduce the efficiency of leukocyte transmigration, as does pharmacological inhibition of src kinase by PP2 [17, 38, 39]. A working model for cortactin's activation by recruitment to sites of clustered ICAM-1 is represented in Figure 2.

It is not clear what the specific effects of phosphorylation have on cortactin activity, although it is clear that endothelial cells must phosphorylate cortactin to support leukocyte transmigration. It has been suggested that phosphorylation may affect the binding of cortactin to other proteins, either by changing the electrostatic charge of the protein, or modifying its conformational state [40]. Tyrosine- and serine/threonine protein kinases can affect actin dynamics through their effects on the state of phosphorylation of cortactin [40]. Kinases may act upon cortactin with complementary or counteractive effects; for example, one study suggested that the Erk serine/threonine kinase might activate cortactin and stimulate F-actin branch formation, but that this effect is nullified by the action of src kinase on cortactin

[41]. Cortactin is also a substrate for non-receptor-type subgroup kinases Fyn, Syk, and Fer [42, 43].

Ezrin/radixin/moesin

The ERMs are a class of proteins whose dominant feature, and one shared by cortactin and the α-actinins, is their ability to link the F-actin cytoskeleton to the plasma membrane by serving as an adapter molecule between F-actin and membrane-associated proteins. Ezrin, radixin, and moesin can bind to the cytoplasmic tails of CD43, CD44, and ICAM-2, L-selectin (reviewed in [44]), as well as to that of ICAM-1 [45]. Unlike cortactin and the α-actinins, the ERM proteins appear to lack any F-actin cross-linking or branch forming abilities. Several studies have identified the membrane phospholipids PIP2 as a regulatory molecule for ERM proteins, affecting both actin binding and cellular distribution [46, 47]. It is, therefore likely that ERMs and α-actinins might respond to similar signals at the cytoplasmic membrane during leukocyte transmigration. The participation of ezrin and moesin in this process is supported by observations that both proteins co-localize with ICAM-1 in endothelial cells. However, ezrin and moesin differ from cortactin and the α-actinins in one important respect. While clustering of ICAM-1 on the endothelial cell surface appears to recruit cortactin to the ICAM-1 tail and retain α-actinin, it may result in a loss of affinity for ezrin and moesin. This conclusion may be drawn from the observation that, in brain endothelial cells, co-localization of both ezrin and moesin with ICAM-1 is disrupted by antibody-mediated ICAM-1 clustering [48]. The model presented in Figure 2 depicts ezrin/moesin as being associated with ICAM-1 prior to clustering, but dissociating afterwards. Interestingly, the affinity of the ICAM-1 cytoplasmic tail for ezrin and moesin in this system did not permit the co-immunoprecipitation of ICAM-1 with either binding partner, suggesting a weak affinity, or an interaction that depends on one or more structures of the intact cell.

In summary, several endothelial cytosolic proteins bind to the ICAM-1 tail including three classes of proteins (cortactin, the α-actinins, ERMs) that also bind F-actin. These proteins serve to link the F-actin cytoskeleton to ICAM-1 in the plasma membrane, to create branch points in the F-actin network, and to bundle parallel F-actin strands. Engagement of ICAM-1, either by antibody cross-linking or leukocyte binding, causes localized cytoskeletal remodeling due, in part, to "outside-in" signaling mediated by ICAM-1. Cortactin, α-actinins, ezrin, and moesin, in turn, are responsible for the F-actin reorganization that results from ICAM-1 cross-linking, as well as anchoring ICAM-1 to the cytoskeleton. That these adapter proteins play distinct roles in the process of cytoskeletal remodeling is substantiated by their distinct modes of binding to the ICAM-1 tail: ICAM-1 clustering causes cortactin recruitment, α-actinin retention, and ERM dissociation.

ICAM-1 mobility in endothelial cells and association with cytoskeleton

It is clear from work done by Yang, Shaw and colleagues [17, 49] using live cell fluorescence-imaging approaches that neutrophil adhesion and transmigration is accompanied by dramatic changes in ICAM-1 localization and leukocyte integrin redistribution, whereby ICAM-1 and LFA-1 become redistributed to a ring-shaped structure at the interface between neutrophil and endothelium. This raises several interesting questions about this important process: How does engagement of ICAM-1 by LFA-1 trigger ICAM-1 redistribution? What force-generating process within the endothelial cell redistributes ICAM-1? How is the ring-shaped ICAM-1 structure maintained during leukocyte transmigration? Using fluorescence recovery after photobleaching (FRAP), Yang and colleagues [50] compared the mobility of ICAM-1 tagged with GFP within the endothelial cell membrane and an ICAM-1GFP mutant lacking its cytoplasmic tail. ICAM-1 mobility in the membrane was reduced fourfold upon engagement by cross-linking antibodies, while the tailless mutant did not display this effect (Fig. 3). This defect is in addition to the significantly higher diffusion coefficient of ICAM-1GFP tailless over wild type in the unengaged (ligand free) state. The authors suggest that ICAM-1 becomes less mobile upon antibody cross-linking or LFA-1 engagement due to a tighter association with the actin cytoskeleton. The mobility of ICAM-1 is directly mediated by proteins that bind simultaneously to the ICAM-1 cytoplasmic tail and the cytoskeleton (*i.e.*, α-actinin, cortactin, ERMs). For example, the *in vitro* affinity between α-actinin and the ICAM-1 tail measured by Carpen and co-workers [27], as well as the ability of α-actinin to co-immunoprecipitate with ICAM-1 [24] suggests a constitutive interaction exists between these two proteins. Cortactin, by contrast, is only recruited to the ICAM-1 tail during leukocyte diapedesis or after cross-linking with anti-ICAM-1 mAb [39]. Thus, cortactin may serve to anchor ICAM-1 to the cytoskeleton only as a consequence of its recruitment to the site of leukocyte adhesion, whereas the α-actinins may weakly anchor ICAM-1 to the underlying cytoskeleton prior to the cytoskeletal remodeling that accompanies leukocyte diapedesis. Given the fact that the α-actinins are regulated by Ca^{2+}, PIP2 and PIP3, three signaling molecules that influence the cytoskeleton, they may also function in the rapid modulation of cytoskeletal dynamics that occur during leukocyte transmigration as observed by Yang et al. [17].

Outside-in signaling by ICAM-1

Outside-in signaling refers to the ability of ICAM-1 to transmit information about extracellular conditions into the cell cytoplasm. Engagement of an endothelial cell monolayer by T cells, neutrophils, or monocytes triggers localized and transient remodeling of the actin cytoskeleton, and antibody cross-linking is thought to mimic

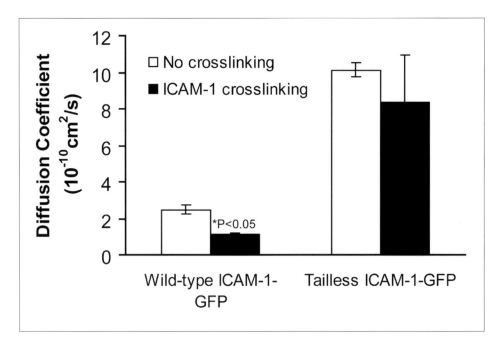

Figure 3
ICAM-1 mobility in vascular endothelium is regulated by interaction of its cytoplasmic tail with the cytoskeleton.
Data are diffusion coefficient determinations of wild-type and tailless ICAM-1GFP molecules in endothelial cells [17] under resting conditions or after 30 min of anti-ICAM-1 cross-linking using the methods described by Yang and coworkers [17].

this effect globally, across the entire apical surface of an endothelial cell, causing the appearance of F-actin stress fibers [11]. Cytoskeletal remodeling is a generalized consequence of ICAM-1 clustering. ICAM-1 is not unique in this capacity because cross-linking of VCAM-1 and E-selectin also leads to outside-in signaling and some actin cytoskeletal remodeling [51].

When a leukocyte transmigrates through an endothelial cell monolayer, its influence on ICAM-1 begins with the direct binding of LFA-1. ICAM-1 clustering as a result of engagement by a counter-receptor was first noticed by Sanders and Vitetta in interacting B and T cells [52]. More recent reports provide highly detailed descriptions of structures formed between leukocytes and endothelial cells using confocal imaging of fixed cells [32, 53]. These studies provided striking snapshots of transient "docking structures" that appear as transmigration is initiated, but do not address how those structures arise. The first account of how both ICAM-1 and LFA-1 redistribute over time during leukocyte transmigration was provided by Shaw and colleagues [49]. They monitored the time-dependent distribution of both proteins on the surfaces of the endothelial cell and neutrophil, respectively, and transmigration was observed in a parallel plate flow chamber, which attempts to duplicate the shear forces present in the microvasculature. Staining of both LFA-1 and ICAM-1 with fluorescently tagged non-function blocking antibodies permitted their visualization in live cells during the process of transmigration across vascular endothelium. A change from a uniform distribution to orderly, ring-shaped structures in which both co-localizing proteins were enriched was seen in the early stage of transmigration. Later studies in which a cortactin-GFP was expressed in endothelial cells showed that this protein also co-localizes with ICAM-1 during transmigration [39]. These measurements of protein redistribution during transmigration can serve as a model for how the assembly of the components of the docking structure described by Barreiro et al. [32] might be monitored in real time during transmigration.

Tilghman and Hoover [38] found that phosphorylated cortactin localized to ICAM-1 clusters in endothelial cells after engagement by leukocytes (THP-1 monocytic cells). Two additional findings further illuminate the process: First, that cortactin appears to be phosphorylated by src kinase after (most likely as a result of) recruitment to the cytoplasmic membrane. Second, the phosphorylation of cortactin is, in turn, required for the further maintenance of ICAM-1 clusters. Thus, a simple picture of a possible feedback loop between ICAM-1 and cortactin emerges, in which ICAM-1 clustering activates cortactin, which in turn, stabilizes the ICAM-1 clusters, perhaps as a result of F-actin cross-linking and recruitment of additional stabilizing molecules. Independently, Barreiro and colleagues [32] reported the participation of the ERM proteins ezrin and moesin in endothelial cells in ICAM-1 clustering mentioned briefly above. These authors described a "docking structure", consisting of ICAM-1, VCAM-1, ezrin and moesin, which was present at the endothelial cell-leukocyte interface, enveloping the bound leukocyte in a cup-like structure. Several other endothelial cell cytoskeleton-associated proteins (VASP, vinculin, talin, and paxillin) were likewise recruited, although it remains to be resolved what their functions are in such a complex. Ezrin and moesin have been shown to co-localize with ICAM-1 in brain endothelial cells, although their association with ICAM-1 is either indirect, or too weak to permit the co-sedimentation of an ICAM-1-ERM complex in a sucrose gradient [48]. Although the identities of all proteins that co-localize with ICAM-1 during leukocyte transmigration have not been elucidated, some of these molecules have been shown to bind to the ICAM-1 tail. With the exception of cortactin and the α-actinins, however, it has not been conclusively demonstrated that any of these proteins have indispensable functions in the process, and future studies are needed to address this gap in knowledge.

Working model of endothelial cell ICAM-1 behavior during engagement by LFA-1

How do endothelial cell cytoplasmic proteins "sense" conformational changes in the extracellular domain of ICAM-1? Ideally, a conformational change in one protein will affect the way in which other proteins interact with it. Unfortunately, there are no structural studies of the ICAM-1 transmembrane domain or cytoplasmic tail that might supply details about how conformation of these domains is affected by engagement of the extracellular domain by LFA-1. The affinities of cytoplasmic proteins for the ICAM-1 tail could be coupled to conformational changes in the ICAM-1 extracellular domain by either of two potential mechanisms. First, there may be changes in the conformation of the ICAM-1 tail concomitant with conformational changes in the extracellular domain. This 28-residue-long structure may undergo a partial or complete transition from α-helical to random coil, or *vice versa*, upon ICAM-1 engagement at the other side of the phospholipid membrane; it is impossible to know, given the absence of structural data for the ICAM-1 tail. Second, there may be no change in the secondary structure of the ICAM-1 tail upon ICAM-1 engagement; instead, signal transmission through the plasma membrane may achieved by clustering alone, which raises the density and alters the distribution of ICAM-1 tail domains at the cytoplasmic leaflet of the membrane. This is certainly the simpler explanation of the two because it requires no intramolecular rearrangements, only the well-documented redistribution of ICAM-1. The model presented in Figure 2 assumes the second mechanism, only by default, because clustering is known to occur, but nothing is known about the conformation of the ICAM-1 tail.

What effect does ICAM-1 engagement by LFA-1 ultimately have on the endothelial cell? The endpoint of leukocyte engagement and ICAM-1 clustering is transmigration. The model presented in Figure 2 makes no attempt to describe the entire cascade of events that conclude with the leukocyte having completely penetrated the endothelial monolayer; instead, it attempts to portray some of the better-understood effects of ICAM-1 clustering on the cytoskeleton, with the understanding that cytoskeletal remodeling is an early step in the process of transmigration.

What might the cytoskeleton-ICAM-1 interactions look like prior binding leukocyte LFA-1?

ICAM-1 dimers assume the O-form, and move independently of each other (*i.e.*, their extracellular domains are not linked). Yet, Yang and colleagues [17, 50] showed that ICAM-1 is linked to something *via* the cytoplasmic tail. Panel A of Figure 2 shows ICAM-1 linked *via* ezrin/moesin to an F-actin filament. Ezrin and moesin are described as co-localizing with ICAM-1 in resting endothelial cells.

They could serve to link F-actin with the cytoplasmic membrane *via* ICAM-1, although they are not known to mediate the formation of branched or cross-linked actin networks. α-Actinin is depicted as bound to ICAM-1, while cortactin is not bound. This portrayal is consistent with the affinities of these proteins for ICAM-1 described in an earlier section of this review. We speculate that cortactin does not link ICAM-1 to the F-actin cytoskeleton under these conditions, even though cortical actin is present in resting endothelial cells. α-Actinin, although associated with ICAM-1, may not yet have become tightly associated with F-actin filaments. Both cortactin and α-actinin will mediate the eventual link between ICAM-1 and the F-actin cytoskeleton upon ligation of ICAM-1 by LFA-1 (Fig. 2B).

What might the cytoskeleton of the endothelial cell look like after ligation of ICAM-1 by LFA-1?

Figure 2B depicts ICAM-1 engaged by leukocyte LFA-1. As a result, ICAM-1 dimers are in the W-form, which favors clustering. ICAM-1 clustering has triggered the release of ezrin/moesin from the tail, as suggested by the results of Romero and co-workers [48]. Of the two remaining proteins, cortactin's behavior is the most understood. It is recruited to the site of ICAM-1 clustering, and is predicted to initiate, with Arp2/3, the formation of new dendritic F-actin filaments. The role of α-actinin is less clear. Because antibody cross-linking of ICAM-1 triggers the formation of stress fibers (structures with which α-actinin is associated) in endothelial cells, our model depicts the F-actin bundling activity of α-actinin as being activated by ICAM-1 clustering, although the mechanism by which this might occur is uncertain. The α-actinins are capable of bundling F-actin fibers, as well as anchoring them to ICAM-1, but this activity is almost certain to be under spatial and temporal regulation by localized elevations of cytosolic Ca^{2+} and PIP2 and PIP3. The latter two phospholipids only regulate membrane-associated target proteins, and may constitute an important form of spatial regulation for ICAM-1 when not bound to ligands [26]. In summary, ICAM-1 engagement by LFA-1 triggers the remodeling of F-actin near the cell membrane, at the site of eventual leukocyte transmigration.

What role does ICAM-1 clustering and cytoskeletal remodeling play in the eventual transmigration of the bound leukocyte?

Grakoui and colleagues [8] suggested that the immunological synapse between a T cell and an antibody-presenting cell (APC) creates a stable fulcrum against which the cells may apply force. Like the leukocyte-endothelial cell junction, the APC-T cell synapse is held together by LFA-1 and ICAM-1, which form a ring-like structure. It is tempting to think of the ICAM-1-LFA-1 ring formed at the initiation of leukocyte transmigration as playing a similar role. To provide mechanical advantage, ligated ICAM-1 dimers (W-form) also serve as anchor points for F-actin. In the model, both α-actinin and cortactin are envisioned as creating a three-dimensional network of F-actin, anchored to the endothelial cell membrane. This network may serve as a corridor through which the leukocyte will eventually pass. The corridor would be dynamic and transient, forming and then dissipating as the leukocyte passes through.

Acknowledgements
The authors wish to thank Dr. Michael Gimbrone, Brigham and Women's Hospital for helpful discussion and continued support, and members of the Center for Excellent in Vascular Biology, Brigham and Women's Hospital for numerous helpful discussions and advice for this article. Our apologies in advance to the authors of papers we have failed to mention due to space limitations. We are indebted to funding provided by the National Institutes of Health to F.W.L. (HL36028, HL53993 and HL 56985) and a N.I.H. NRSA award HL 86217 to S.D.A.

References

1   Rothlein R, Dustin ML, Marlin SD, Springer TA (1986) A human intercellular adhesion molecule (ICAM-1) distinct from LFA-1. *J Immunol* 137: 1270–1274
2   Roebuck KA, Finnegan A (1999) Regulation of intercellular adhesion molecule-1 (CD54) gene expression. *J Leukoc Biol* 66: 876–888
3   Makgoba MW, Sanders ME, Luce GEG, Dustin ML, Springer TA, Clark EA, Mannoni P, Shaw S (1988) ICAM-1 a ligand for LFA-1 dependent adhesion to B, T and myeloid cells. *Nature* 331: 86–88
4   Springer TA (1994) Traffic signals for lymphocyte recirculation and leukocyte emigration: the multistep paradigm. *Cell* 76: 301–314
5   Miller J, Knorr R, Ferrone M, Houdei R, Carron CP, Dustin ML (1995) Intercellular adhesion molecule-1 dimerization and its consequences for adhesion mediated by lymphocyte function associated-1. *J Exp Med* 172: 1231–1241
6   Reilly PL, Waska JR, Jeanfavre DD, McNally E, Rothlein R, Bormann BJ (1995) The native structure of intercellular adhesion molecule-1 (ICAM-1) is a dimer. *J Immunol* 155: 529–532
7   Yang Y, Jun CD, Liu JH, Zhang R, Joachimiak A, Springer TA, Wang JH (2004) Structural basis for dimerization of ICAM-1 on the cell surface. *Mol Cell* 14: 269–276
8   Grakoui A, Bromley SK, Sumen C, Davis MM, Shaw AS, Allen PM, Dustin ML (1999) The immunological synapse: a molecular machine controlling T cell activation. *Science* 285: 221–227

9   Etienne S, Adamson P, Greenwood J, Strosberg AD, Cazaubon S, Couraud PO (1998) ICAM-1 signaling pathways associated with Rho activation in microvascular brain endothelial cells. *J Immunol* 161: 5755–5761

10   Adamson P, Etienne S, Couraud PO, Calder V, Greenwood J (1999) Lymphocyte migration through brain endothelial cell monolayers involves signaling through endothelial ICAM-1 *via* a rho-dependent pathway. *J Immunol* 162: 2964–2973

11   Etienne-Manneville S, Manneville JB, Adamson P, Wilbourn B, Greenwood J, Couraud PO (2000) ICAM-1-coupled cytoskeletal rearrangements and transendothelial lymphocyte migration involve intracellular calcium signaling in brain endothelial cell lines. *J Immunol* 165: 3375–3383

12   Wang Q, Doerschuk CM (2001) The p38 mitogen-activated protein kinase mediates cytoskeletal remodeling in pulmonary microvascular endothelial cells upon intracellular adhesion molecule-1 ligation. *J Immunol* 166: 6877–6884

13   Durieu-Trautmann O, Chaverot N, Cazaubon S, Strosberg AD, Couraud PO (1994) Intercellular adhesion molecule 1 activation induces tyrosine phosphorylation of the cytoskeleton-associated protein cortactin in brain microvessel endothelial cells. *J Biol Chem* 269: 12536–12540

14   Weed SA, Parsons JT (2001) Cortactin: coupling membrane dynamics to cortical actin assembly. *Oncogene* 20: 6418–6434

15   Sans E, Delachanal E, Duperray A (2001) Analysis of the roles of ICAM-1 in neutrophil transmigration using a reconstituted mammalian cell expression model: implication of ICAM-1 cytoplasmic domain and Rho-dependent signaling pathway. *J Immunol* 166: 544–551

16   Greenwood J, Amos CL, Walters CE, Couraud PO, Lyck R, Engelhardt B, Adamson P (2003) Intracellular domain of brain endothelial intercellular adhesion molecule-1 is essential for T lymphocyte-mediated signaling and migration. *J Immunol* 171: 2099–2108

17   Yang L, Kowalski JR, Yacono P, Bajmoczi M, Shaw SK, Froio RM, Golan DE, Thomas SM, Luscinskas FW (2006) Endothelial cell cortactin coordinates intercellular adhesion molecule-1 clustering and actin cytoskeleton remodeling during polymorphonuclear leukocyte adhesion and transmigration. *J Immunol* 177: 6440–6449

18   Cernuda-Morollon E, Ridley AJ (2006) Rho GTPases and leukocyte adhesion receptor expression and function in endothelial cells. *Circ Res* 98: 757–767

19   Imhof BA, Aurrand-Lions M (2004) Adhesion mechanisms regulating the migration of monocytes. *Nat Rev Immunol* 4: 432–444

20   Dixson JD, Forstner MJ, Garcia DM (2003) The alpha-actinin gene family: a revised classification. *J Mol Evol* 56: 1–10

21   Liu J, Taylor DW, Taylor KA (2004) A 3-D reconstruction of smooth muscle alpha-actinin by CryoEm reveals two different conformations at the actin-binding region. *J Mol Biol* 338: 115–125

22   Tsuruta D, Gonzales M, Hopkinson SB, Otey C, Khuon S, Goldman RD, Jones JC

(2002) Microfilament-dependent movement of the beta3 integrin subunit within focal contacts of endothelial cells. *FASEB J* 16: 866–868

23 Araki N, Hatae T, Yamada T, Hirohashi S (2000) Actinin-4 is preferentially involved in circular ruffling and macropinocytosis in mouse macrophages: analysis by fluorescence ratio imaging. *J Cell Sci* 113: 3329–3340

24 Celli L, Ryckewaert JJ, Delachanal E, Duperray A (2006) Evidence of a functional role for interaction between ICAM-1 and nonmuscle α-actinins in leukocyte diapedesis. *J Immunol* 177: 4113–4121

25 Otey CA, Carpen O (2004) Alpha-actinin revisited: a fresh look at an old player. *Cell Motil Cytoskeleton* 58: 104–111

26 Janmey PA, Lindberg U (2004) Cytoskeletal regulation: rich in lipids. *Nat Rev Mol Cell Biol* 5: 658–666

27 Carpen O, Pallai P, Staunton DE, Springer TA (1992) Association of intercellular adhesion molecule-1 (ICAM-1) with actin-containing cytoskeleton and alpha-actinin. *J Cell Biol* 118: 1223–1234

28 Corgan AM, Singleton C, Santoso CB, Greenwood JA (2004) Phosphoinositides differentially regulate alpha-actinin flexibility and function. *Biochem J* 378: 1067–1072

29 Fukami K, Sawada N, Endo T, Takenawa T (1996) Identification of a phosphatidylinositol 4,5-bisphosphate-binding site in chicken skeletal muscle alpha-actinin. *J Biol Chem* 271: 2646–2650

30 Fukami K, Furuhashi K, Inagaki M, Endo T, Hatano S, Takenawa T (1992) Requirement of phosphatidylinositol 4,5-bisphosphate for alpha-actinin function. *Nature* 359: 150–152

31 Insall RH, Weiner OD (2001) PIP3, PIP2, and cell movement – similar messages, different meanings? *Dev Cell* 1: 743–747

32 Barreiro O, Yanez-Mo M, Serrador JM, Montoya MC, Vicente-Manzanares M, Tejedor R, Furthmayr H, Sanchez-Madrid F (2002) Dynamic interaction of VCAM-1 and ICAM-1 with moesin and ezrin in a novel endothelial docking structure for adherent leukocytes. *J Cell Biol* 157: 1233–1245

33 Kanner SB, Reynolds AB, Parsons JT (1991) Tyrosine phosphorylation of a 120-kilodalton pp60[src] substrate upon epidermal growth factor and platelet-derived growth factor stimulation and in polyomavirus middle-T-antigen-transformed cells. *Mol Cell Biol* 11: 713–720

34 Weed SA, Karginov AV, Schafer DA, Weaver AM, Kinley AW, Cooper JA, Parsons JT (2000) Cortactin localization to sites of actin assembly in lamellipodia requires interactions with F-actin and the Arp2/3 complex. *J Cell Biol* 151: 29–40

35 Goley ED, Welch MD (2006) The ARP2/3 complex: an actin nucleator comes of age. *Nat Rev Mol Cell Biol* 7: 713–726

36 Bowden ET, Barth M, Thomas D, Glazer RI, Mueller SC (1999) An invasion-related complex of cortactin, paxillin and PKCmu associates with invadopodia at sites of extracellular matrix degradation. *Oncogene* 18: 4440–4449

37 Ivanov AI, McCall IC, Parkos CA, Nusrat A (2004) Role for actin filament turnover and a myosin II motor in cytoskeleton-driven disassembly of the epithelial apical junctional complex. *Mol Biol Cell* 15: 2639–2651

38 Tilghman RW, Hoover RL (2002) The Src-cortactin pathway is required for clustering of E-selectin and ICAM-1 in endothelial cells. *FASEB J* 16: 1257–1259

39 Yang L, Kowalski JR, Zhan X, Thomas SM, Luscinskas FW (2006) Endothelial cell cortactin phosphorylation by Src contributes to polymorphonuclear leukocyte transmigration *in vitro*. *Circ Res* 98: 394–402

40 Daly RJ (2004) Cortactin signalling and dynamic actin networks. *Biochem J* 382: 13–25

41 Martinez-Quiles N, Ho HY, Kirschner MW, Ramesh N, Geha RS (2004) Erk/Src phosphorylation of cortactin acts as a switch on-switch off mechanism that controls its ability to activate N-WASP. *Mol Cell Biol* 24: 5269–5280

42 Fan L, Ciano-Oliveira C, Weed SA, Craig AW, Greer PA, Rotstein OD, Kapus A (2004) Actin depolymerization-induced tyrosine phosphorylation of cortactin: the role of Fer kinase. *Biochem J* 380: 581–591

43 Huang J, Asawa T, Takato T, Sakai R (2003) Cooperative roles of Fyn and cortactin in cell migration of metastatic murine melanoma. *J Biol Chem* 278: 48367–48376

44 Tsukita S, Yonemura S (1999) Cortical actin organization: lessons from ERM (ezrin/radixin/moesin) proteins. *J Biol Chem* 274: 34507–34510

45 Heiska L, Alfthan K, Gronholm M, Vilja P, Vaheri A, Carpen O (1998) Association of ezrin with intercellular adhesion molecule-1 and -2 (ICAM-1 and ICAM-2). Regulation by phosphatidylinositol 4, 5-bisphosphate. *J Biol Chem* 273: 21893–21900

46 Barret C, Roy C, Montcourrier P, Mangeat P, Niggli V (2000) Mutagenesis of the phosphatidylinositol 4,5-bisphosphate (PIP$_2$) binding site in the NH$_2$-terminal domain of ezrin correlates with its altered cellular distribution. *J Cell Biol* 151: 1067–1080

47 Bompard G, Martin M, Roy C, Vignon F, Freiss G (2003) Membrane targeting of protein tyrosine phosphatase PTPL1 through its FERM domain *via* binding to phosphatidylinositol 4,5-biphosphate. *J Cell Sci* 116: 2519–2530

48 Romero IA, Amos CL, Greenwood J, Adamson P (2002) Ezrin and moesin co-localise with ICAM-1 in brain endothelial cells but are not directly associated. *Brain Res Mol Brain Res* 105: 47–59

49 Shaw SK, Ma S, Kim M, Rao RM, Hartman CU, Froio R, Liu Y, Yang L, Jones T., Nusrat A et al (2004) Coordinated redistribution of leukocyte LFA-1 and endothelial cell ICAM-1 accompanies neutrophil transmigration. *J Exp Med* 200: 1571–1580

50 Yang L, Froio RM, Sciuto TE, Dvorak AM, Alon R, Luscinskas FW (2005) ICAM-1 regulates neutrophil adhesion and transcellular migration of TNF-alpha-activated vascular endothelium under flow. *Blood* 106: 584–592

51 Lorenzon P, Vecile E, Nardon E, Ferrero E, Harlan JM, Tedesco F, Dobrina A (1998) Endothelial cell E- and P-selectin and vascular cell adhesion molecule-1 function as signaling receptors. *J Cell Biol* 142: 1381–1391

52 Sanders VM, Vitetta ES (1991) B cell-associated LFA-1 and T cell-associated ICAM-1 transiently cluster in the area of contact between interacting cells. *Cell Immunol* 132: 45–55

53 Carman CV, Springer TA (2004) A transmigratory cup in leukocyte diapedesis both through individual vascular endothelial cells and between them. *J Cell Biol* 167: 377–388

α_4-integrins: structure, function and secrets

Britta Engelhardt

Theodor Kocher Institute, University of Bern, Freiestrasse 1, 3012 Bern, Switzerland

Discovery of $\alpha_4\beta_1$-integrin

The initial description of $\alpha_4\beta_1$-integrin dates back 30 years, when in 1987, based on biochemical studies, Martin Hemler and co-workers [1] described it as a distinct Mr 150 000/130 000 $\alpha_4\beta$ heterodimer and thus a new member of the VLA-protein family on human T lymphoblastoid cells and peripheral blood T cells. Interestingly, some of the most widely used and characterized antibodies against α_4-integrin were originally described by Sanchez-Madrid in 1986 against VLA-3 in error [2]. The name VLA (very late activation) antigen was originally coined to describe a novel set of antigens that appeared in a very late stage of activated T cell differentiation [3]. VLA antigens were found to be composed of non-covalently associated heterodimers with a common β-subunit but unique α-subunits, and were readily recognized as members of the larger integrin family of adhesion receptors. These receptors are involved in a multitude of distinct cell-matrix and cell-cell adhesion functions. Based on their discovery, β_1-integrins are often referred to as VLA antigens, therefore $\alpha_4\beta_1$-integrin is still mostly referred to as VLA-4 until today. The corresponding term of the CD nomenclature for $\alpha_4\beta_1$-integrin is CD49dCD29.

The first hints regarding the possible function of VLA-4 on T cells came from independent discoveries in several laboratories. An antibody called L25, later recognized to bind to VLA-4 [4] was originally found to interfere with the cytotoxic activity of human T cells [5]. VLA-4 was identified as a novel fibronectin receptor on lymphoid cells mediating binding to the connecting segment (CS)-1 region, which is generated by alternative splicing of pre-mRNA of fibronectin [6, 7]. At the same time VLA-4 was discovered to mediate lymphocyte binding to vascular cell adhesion molecule (VCAM)-1 [8], which is up-regulated on endothelial cells upon inflammation. Thus, VLA-4 was quickly recognized to play an important role in T cell effector functions and in the recruitment of circulating immune cells into tissue during inflammation.

Adhesion Molecules: Function and Inhibition, edited by Klaus Ley
© 2007 Birkhäuser Verlag Basel/Switzerland

α_4 association with an alternate, the β_P or β_7-integrin subunit

In 1989, the mouse homologue of human α_4-integrin was characterized by Bernhard Holzman in Irving Weissman's laboratory [9], and found to associate with either mouse β_1-integrin or a novel integrin β-subunit then called β_P [10]. β_P was found to be biochemically and immunohistochemically distinct from β_1, and $\alpha_4\beta_P$ was shown to mediate lymphocyte adhesion to high endothelial venules (HEVs) in gut-associated lymphoid tissue [as later shown, through interaction with mucosal addressin cell adhesion molecule (MAdCAM)-1]. Upon publication of a novel human β-integrin subunit named β_7 [11], β_P was found to be the mouse homologue to the human β_7-integrin subunit and to form heterodimers with the α_4 integrin subunit [12]. $\alpha_4\beta_7$ was originally described as lamina propria-associated molecule (LPAM)-1 [9, 12]. In contrast to the ubiquitously expressed β_1-integrin subunit, distribution of the β_7-integrin subunit was found to be restricted to subsets of lymphocytes only.

Expression of α_4-integrins

$\alpha_4\beta_1$- and $\alpha_4\beta_7$-integrins are constitutively expressed on naive T and B cells independent of their localization within the blood stream or within lymphoid organs. Very high expression of $\alpha_4\beta_7$-integrin is found on circulating B cells and on a gut homing subset of $CD4^+$ memory T cells, which is distinct from the skin homing subset of T cells [13, 14]. $\alpha_4\beta_7$-integrin expression on T cells and B cells was shown to be enhanced by the vitamin A metabolite, retinoic acid, which is produced from retinol by dendritic cells in gut-associated lymphoid tissues [15, 16]. In addition to expression of the chemokine receptor CCR9, elevated levels of $\alpha_4\beta_7$-integrin ensure gut tropism of T and B cells. Besides on cells of the adaptive immune system, α_4-integrins are also expressed on cells of the innate immune system such as eosinophils, basophils and natural killer cells, while monocytes express moderate to high levels of $\alpha_4\beta_1$ and no $\alpha_4\beta_7$. Upon stimulation, however, monocytes can up-regulate surface expression of $\alpha_4\beta_7$. Even though neutrophils have originally been reported to have no α_4-integrins, in the meanwhile several studies have documented involvement of α_4-integrins in neutrophil recruitment into inflamed tissues (reviewed in [17]), suggesting that low levels of α_4-integrins on neutrophils suffice for functional activity. α_4-integrins are also expressed on hematopoietic stem cells [18]. Furthermore, expression of α_4-integrins – mostly with unspecified β-chain – has been reported in a developmentally regulated fashion on non-hematopoietic cells in a number of embryonic tissues including somites, heart, vascular smooth muscle and skeletal muscle, on differentiated neuroepithelium of the embryonic retina and neural crest-derived cells including melanoblasts [19–21]. Finally, α_4-integrin was described on various tumor cells including metastatic melanoma [22].

Structure of $\alpha_4\beta_1$- and $\alpha_4\beta_7$-integrins

α_4-integrins are comprised of two non-covalently associated type I transmembrane glycoproteins. Each subunit contains a large extracellular domain, a single transmembrane domain and a short cytoplasmic tail. The α_4-integrin subunit was cloned and sequenced in Martin Hemler's laboratory in 1989 [4]. The sequence of the β_1-integrin subunit was already known at that time [23], whereas the integrin β_7-subunit was characterized, cloned and sequenced after the characterization of α_4-integrin [10, 11].

The α_4-integrin subunit is expressed at the cell surface as a M_r 150 000 protein. It is unique in that α_4-integrin unlike any other integrin α-subunit has a proteolytic cleavage site between Lys557 and Arg 558 or Ser 559 yielding fragments of M_r 80 000 and 70 000, respectively. Conversion of the intact 150 000 form to the cleaved form can be observed upon activation of T lymphocytes. Therefore, $\alpha_4\beta_1$-integrin can be found in multiple forms on the cell surface, *i.e.*, solely as α_4^{150}; or $\alpha_4^{80,70}$ or a mixture of both, α_4^{150} and $\alpha_4^{80/70}$ [24]. Regardless of cleavage, both α_4-integrin subunits remain associated with each other and the β_1-integrin subunit, despite the lack of covalent bonds! This might explain why α_4-subunit cleavage has not been found to affect $\alpha_4\beta_1$-integrin mediated adhesion functions [24]. A third form of the α_4-integrin subunit with an M_r of 180 000 has been shown to represent a conformational variant of unknown nature of the M_r 150 000 form [25].

Amino acid sequence analysis of the N-terminal portion of α_4-integrin revealed the presence of seven conserved repeated domains and the lack of an inserted (I)-domain between the second and third repeat, which is found in the α-subunits of the leukocyte β_2-integrins and is known to mediate their binding to ligands. The seven N-terminal 60-amino acid repeats of α_4-integrin have been predicted to be organized into a large domain with the appearance of a seven-bladed propeller [26]. This prediction was fully confirmed by the elucidation of the crystal structure of the extracellular domain of $\alpha_V\beta_3$, which also lacks the I domain in its α-subunit [27]. The N-terminal propeller domain of the α_V-integrin subunit is attached to an elongated leg formed of three β-sandwich domains, which have been termed thigh, calf 1 and calf 2, respectively (Fig. 1, and reviewed in [28]). The β-integrin domain organization is completely different. The I-like domain of the β_3-subunit, although it is not localized at the extreme N terminus of the β-chain, forms the head domain. The I-like domain of β-integrin subunits harbors a metal-ion-dependent-adhesion-site (MIDAS), which binds the divalent cations necessary for integrin binding to its ligand. The leg of the β-subunit is formed by four tandem arrays of cysteine-rich repeats, which fold into EGF-like folds and are highly characteristic of all β-integrin subunits (reviewed in [28]). The most unexpected feature of the crystal structure of $\alpha_V\beta_3$-integrin was that it showed the extracellular parts of $\alpha_V\beta_3$-integrin in a bent conformation, with the "knee" between the thigh and the calf 1 domain of the α-subunit and the EGF-2 and EFG-3 in the β-subunit, respectively (Fig. 1). In apparent

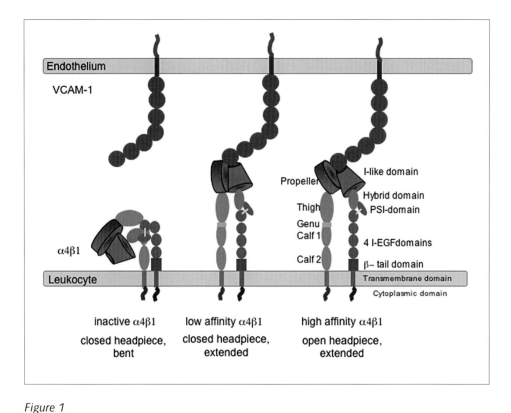

Figure 1

Models for the proposed activity states of $\alpha_4\beta_1$-integrin on circulating leukocytes.
In the bent conformation α_4-integrin is supposedly inactive. Straightening of $\alpha_4\beta_1$-integrin might allow for low-affinity interaction of the propeller and the I-like domain with the endothelial ligand VCAM-1 mediating leukocyte rolling. Separation of the cytoplasmic tails allows for intracellular signaling. Altered orientation of the propeller and the I-like domain in the extended form – probably induced by inside-out-signalling – will mediate high-affinity binding to VCAM-1 required for adhesion strengthening of the leukocyte in the blood stream. In the absence of ligand, the different forms of $\alpha_4\beta_1$-integrin are thought to exist in an allosteric equilibrium on the leukocyte surface (modified after [28]).

contrast to the crystal structure, electron microscopical pictures of integrins bound to ligand had shown integrins in an extended form [29]. As the crystal structure of $\alpha_V\beta_3$-integrin was determined in the absence of ligand, and as integrins can exist in an inactive and active state, the bent conformation of integrins is presumed to be the inactive and the extended form the active state, which has been supported by functional studies (see below).

The cytoplasmic tails of integrins are very short, yet strictly required for integrin function as they are linked to the cytoskeletal and signaling partners of integrins (summarized in [28, 30–32]). In the context of this chapter, it is particularly relevant that α_4-integrin cytoplasmic domains can regulate the activation of α_4-integrins, $i.e.$, affect the structure and function of their extracellular domains.

Based on our present knowledge including the structural analyses, it can be concluded that the propeller domain of α_4-integrin together with the I-like domain of the β-chain mediate ligand binding (Fig. 1).

$\alpha_4\beta_1$-integrin and its ligands

Very early after its discovery, $\alpha_4\beta_1$-integrin was found to bind to vascular cell adhesion molecule (VCAM)-1 [8], which is up-regulated on vascular endothelial cells during inflammation. $\alpha_4\beta_1$-integrin binds to sites within the first and fourth immunoglobulin (Ig)-like domain of the full-length seven-domain form of VCAM-1 [33]. Within domain one of VCAM-1 a dominant acidic peptide motif QIDSPL was shown to be critical for α_4-integrin recognition. Interestingly, binding of $\alpha_4\beta_1$ to domain 4 of VCAM-1 requires integrin activation, whereas adhesion to VCAM-1 domain 1 does not [34]. Similarly, $\alpha_4\beta_7$ can also bind to VCAM-1 by interaction with the first or fourth Ig-like domain of the full-length seven domain form of VCAM-1 [35], although with much lower affinity than $\alpha_4\beta_1$. A related cell adhesion molecule, MAdCAM-1, which is constitutively expressed on HEVs in mucosa-associated lymphatic tissue, such as in the mesenteric lymph nodes and the Peyer's Patches, is recognized by $\alpha_4\beta_7$, but is a poor ligand for $\alpha_4\beta_1$ [36]. It was shown that the first domain of MAdCAM-1 mediates binding of $\alpha_4\beta_7$-integrin, while the second Ig domain of MAdCAM-1 strengthens this interaction [37].

Equally soon after its discovery, $\alpha_4\beta_1$-integrin was recognized to serve as an alternative fibronectin receptor on lymphoid cells, in addition to the already known fibronectin receptor $\alpha_5\beta_1$-integrin [6, 7]. When the ability of functionally defined monoclonal antibodies directed against the two integrins to inhibit lymphocyte adhesion to purified tryptic fragments of plasma fibronectin was examined, it was found that antibodies blocking either $\alpha_5\beta_1$ or $\alpha_4\beta_1$ only partially inhibited T lymphocyte adhesion to intact fibronectin [6]. In contrast, cocktails of both antibodies led to complete inhibition of T lymphocyte adhesion to fibronectin, suggesting that these receptors recognize independent sites on intact fibronectin. Using recombinant fibronectins, this observation was elegantly confirmed by demonstrating that $\alpha_4\beta_1$-integrin-dependent spreading of lymphoid cells can only be observed on fibronectin containing the alternatively spliced V region [7]. It has been well documented that $\alpha_5\beta_1$-integrin interacts with the cell binding domain of fibronectin containing the well-known critical acidic peptide motif Arg-Gly-Asp (RGD) [38]. In contrast, these studies demonstrated that $\alpha_4\beta_1$-integrin interacts with the acidic

LDVP sequence on the Type III connecting-segment-1 (CS-1) of fibronectin within an alternatively spliced V region of fibronectin. Several reports have shown that $\alpha_4\beta_7$-integrin recognizes the same site within the alternatively spliced connecting segment of fibronectin as $\alpha_4\beta_1$ [35]. Thus, both $\alpha_4\beta_1$-integrin and $\alpha_4\beta_7$-integrin mediated binding to fibronectin is molecularly distinct from that mediated by $\alpha_5\beta_1$-integrin.

Furthermore, in one report $\alpha_4\beta_1$ and $\alpha_4\beta_7$-integrins were shown to exhibit homophilic interactions with α_4-integrin subunits *in vitro*, suggesting that α_4-integrins might bind to α_4 subunits on adjacent cells thus providing an additional mechanism for α_4-integrin-mediated functions [39]. These findings have, however, never been confirmed with other methods or other cells.

As an additional ligand for $\alpha_4\beta_1$-integrin, the secreted highly acidic glycoprotein osteopontin, originally isolated from bone, has been identified [40]. Amongst other functions osteopontin is involved in inflammation and tumor progression. The precise binding sites of $\alpha_4\beta_1$-integrin have been mapped to the N-terminal fragment of osteopontin [41]. As further ligands for $\alpha_4\beta_1$-integrin the bacterial protein invasin [42], the secreted platelet protein thrombospondin [43] and the tight junction protein junctional adhesion molecule (JAM)-B [44] have been described. Thrombospondin binding requires activation of $\alpha_4\beta_1$-integrin, and $\alpha_4\beta_1$-integrin engagement of JAM-B is only enabled following prior adhesion of JAM-B with JAM-C and is not detectable in cells where JAM-C expression is absent. $\alpha_4\beta_1$- and $\alpha_4\beta_7$-integrins also bind to the disintegrin domains of ADAM7 and ADAM27 [45]. Finally, very recently both α_4-integrins were shown to bind to CD14 [46]. While these and still other ligand interactions are intriguing, the functional relevance of α_4-interactions with ligands other than VCAM-1, MAdCAM-1 and fibronectin *in vivo* still needs to be investigated in more detail.

Activation of $\alpha_4\beta_1$-integrin

The regulation of α_4-integrin activity at least on circulating immune cells appears to be different from that of other leukocyte integrins, which are expressed in an inactive state on the cell surface of leukocytes and need cellular activation prior to binding of their ligands.

However, several pathways of $\alpha_4\beta_1$-integrin activation have been defined. In T cells $\alpha_4\beta_1$-integrin can be activated from within the cell after engagement of the T cell receptor or after binding of a chemokine to its G-protein coupled chemokine receptor. In these cases $\alpha_4\beta_1$-integrin is activated through its cytoplasmic tail by the downstream signaling cascades induced by the T cell receptor or the G protein-coupled chemokine receptors. This process has been called inside-out-signaling and includes integrin cross-talk, where different integrins on the same cell have been shown to intercommunicate and thereby modulate their activities [47].

Involvement of both the α_4 and the β_1 subunit in ligand binding was suggested by conformational changes detectable in both chains, which can be monitored using select anti-α_4-integrin and anti-β_1-integrin antibodies exclusively binding to so-called ligand induced binding sites (LIBS) and ligand-attenuated epitopes on $\alpha_4\beta_1$ [48, 49]. Employing a large panel of anti-α_4-integrin antibodies, it was shown that binding of $\alpha_4\beta_1$ to VCAM-1, MAdCAM-1 or fibronectin induces different conformational changes [49], suggesting that distinct binding mechanisms may exist for $\alpha_4\beta_1$-binding to its ligands and that $\alpha_4\beta_1$-integrin is able to transduce ligand-specific signals into a cell. This process is commonly referred to as outside-in-signaling.

Functional epitope mapping of the α_4-integrin subunit with a large panel of anti-α_4-integrin antibodies defined three topographically distinct epitopes [50–54]. Antibodies binding to functional epitope A trigger α_4-integrin-mediated homophilic aggregation of lymphocytes, and partially block adhesion to fibronectin but not to VCAM-1. Antibodies recognizing the functional epitope B are potent inhibitors of α_4-mediated adhesion to fibronectin, VCAM-1 and MAdCAM-1. Amongst the epitope B binding antibodies, those inducing homophilic aggregation are assigned to functional epitope B2, those that do not to functional epitope B1. Antibodies directed against functional epitope C inhibit neither fibronectin nor VCAM-1 binding, but interfere with homophilic aggregation induced by epitope A or B antibodies [50]. Epitope mapping of some of these functional antibodies localized domain A at the N terminus of α_4-integrin with Arg89 and Asp90 involved in fibronectin binding [54], domain B to the third blade of the propeller domain, and domain C outside of the propeller structure to the calf region. In accordance with these studies amino acid exchange analyses have localized bindings sites for fibronectin and VCAM-1 within the third blade of the propeller domain of α_4-integrin [55]. As $\alpha_4\beta_1$- and $\alpha_4\beta_7$-integrins have overlapping but distinct ligands, the β-integrin chains must contribute to ligand binding. Amino acid exchange and domain swap analysis of the β-chains demonstrated that the MIDAS domain within the I-like domain of β_1- and β_7-integrin is required for binding to the respective ligands [56, 57], suggesting that both, the α_4 and the β-subunits physically interact with the ligand. This view is supported by structural analysis of $\alpha_V\beta_3$ integrin in the presence of ligand (cyclic RGDF-peptide), which showed the peptide-ligand bound at the $\alpha\beta$-interface in between the propeller domain of the α-subunit and the divalent cation associated with the MIDAS site of the I-like-domain of the β-subunit [58]. This finding is consistent with previous observations demonstrating that all integrin-ligand interactions are divalent cation dependent.

When these findings are combined it can be postulated that activation of α_4-integrins presumably leads to unfolding of the inactive bent $\alpha\beta$-heterodimer at the cell surface, placing the propeller domain of the α_4-subunit and the I-like domain of the β-subunit in the appropriate position for high-affinity binding to the respective ligands (reviewed by [28]). This model does not explain, however, how circulating immune cells mediate α_4-integrin-dependent adhesion to their endothelial ligands VCAM-1 and MAdCAM-1 without prior cellular activation in the presence of shear forces (see below). Studies with an anti-β_1-integrin antibody, which only recognizes a high-affinity or ligand-occupied epitope on the β_1-integrin chain suggested that circulating leukocytes express a characteristic third subset of $\alpha_4\beta_1$-integrins on their surface that is responsive to ligand with either low affinity or transient activity [59]. Circulating immune cells might therefore keep a characteristic allosteric balance between these three α_4-integrin activities – namely, inactive (probably bent conformation), a low affinity (likely multiple transient conformations) and the high affinity (extended conformation) on their cell surface allowing for rapid low- and regulated high-affinity ligand binding (Fig. 1).

None of the above explanations takes into account that, in addition to conformational changes, integrin-mediated cell adhesion has been suggested to be regulated by clustering of integrin molecules on the cell surface, thus increasing overall avidity for ligand binding. In case of α_4-integrins, this notion is supported by the finding that the cytoplasmic tail of α_4-integrin makes strong positive contributions to integrin-mediated cell adhesion by mediating integrin clustering and thereby by increasing overall avidity of binding [60].

Functions of $\alpha_4\beta_1$-integrin

$\alpha_4\beta_1$-integrin is crucial in embryonic development

Ablation of the α_4-integrin subunit leads to embryonic lethality due to a failure in two cell-cell adhesion events in placental and cardiac development. The early defect in α_4-null embryos is characterized by a failure of the allantois to fuse with the chorion during placental development and these embryos die at around embryonic day 11 [21]. In few α_4-null embryos the fusion of the allantois and the chorion is successful. However, those embryos still die at around embryonic day 11.5 due to severe hemorrhage in the heart region caused by a failure in epicardium-myocardium attachment during cardiac development [21]. The same placental and cardiac defects have been observed in VCAM-1-deficient embryos, suggesting that these defects are probably due to $\alpha_4\beta_1$/VCAM-1-mediated adhesive interactions during embryonic development [61, 62]. A role for α_4-integrin and its counter-receptor VCAM-1 has also been shown in skeletal muscle development at the stage of secondary myogenesis [63], whereas another study using α_4-null somatic chimeras failed to detect an involvement of α_4-integrin in skeletal muscle development [64].

When screening for antibodies inhibiting lymphopoiesis by disturbing progenitor cell interaction with stromal cells *in vitro*, an anti-α_4-integrin antibody was found to be most potent, suggesting an important role for α_4-integrins in the development and maintenance of hematopoietic stem cells [65].

Gene inactivation of β_7-integrin led to viable offspring, which, although the gut-associated lymphoid tissue was severely impaired, showed unaltered development of mature T and B cells [66]. These findings demonstrate that β_7-integrin does not play a crucial role for lymphocyte development or organogenesis in the gut associated tissue. In this context it is interesting to note that $\alpha_4\beta_1$-integrin rather than $\alpha_4\beta_7$-integrin is required to trigger the development of Peyer's patches by mediating the adhesion of CD4$^+$CD3$^-$ hematopoietic cells to VCAM-1 on stromal cells in the fetal intestine [67].

Thus, $\alpha_4\beta_1$-integrin seems to be the pivotal integrin during hematopoiesis. Ablation of the β_1-integrin results in arrest of peri-implantation development [68]. This phenomenon combined with the early embryonic lethality of α_4-null embryos prevents a straightforward analysis of $\alpha_4\beta_1$-integrin in hematopoiesis *in vivo*. Several laboratories have therefore addressed this question using different models of genetically modified mice. Analyzing hematopoiesis in α_4-null somatic chimeric mice demonstrated more or less an involvement of α_4-integrins in the development of all hematopoietic lineages in fetal liver, bone marrow and spleen *in vivo* [69, 70]. In these studies, α_4-integrins were found, although dispensable for hematopoietic precursor cell homing to hematopoietic organs, to be required for hematopoietic precursor cell interaction with bone marrow stromal cells [70]. Lack of α_4-integrin obviously leads to premature detachment of hematopoietic progenitors, which by skipping their expansion phase would shift directly towards differentiation, explaining the low yields of mature cells in peripheral tissues observed in these chimeric mice. It was therefore concluded that α_4-integrins act as a retention signal for hematopoietic progenitors to remain in the bone marrow compartment.

Analysis of hematopoiesis in β_1-integrin-deficient somatic chimeric mice demonstrated a requirement of β_1-integrins in hematopoietic stem cell homing to the fetal liver but surprisingly not for their differentiation [71]. The failure of β_1-integrin-deficient hematopoietic stem cells to engraft lethally irradiated mice further supported a key role for β_1-integrins in hematopoiesis [72]. It was therefore unexpected, when β_1-integrin mutant bone marrow chimeras with induced deletion of β_1-integrins in the adult hematopoietic system *in vivo* showed no defects in blood cell development [73].

One explanation at the time was that $\alpha_4\beta_1$ and $\alpha_4\beta_7$ might have redundant functions in blood cell development and that only the absence of both α_4-integrin-receptors leads to the described hematopoietic defects previously observed in α_4-null somatic chimeric mice. However, inducible deletion of α_4-integrin in adult mice produced only subtle effects on hematopoiesis characterized by a mild reduction of B cell and T cell populations in the bone marrow of these animals [74]. To unravel the apparently discrepant findings about the involvement of $\alpha_4\beta_1$ *versus* $\alpha_4\beta_7$ in hematopoiesis, recently mice with a constitutive knockout for β_7-integrin and a blood cell-restricted inducible knockout for β_1-integrin were generated [75]. After induced deletion of β_1-integrins in adult mice, these animals lack $\alpha_4\beta_1$, $\alpha_4\beta_7$

and all additional β_1 integrins on their blood cells, but still express α_4- and β_1-integrins in non-hematopoietic cells. Unexpectedly, this approach failed to demonstrate an essential function for $\alpha_4\beta_1$- and $\alpha_4\beta_7$-integrins in blood cell development or in progenitor cell retention in the bone marrow in adult mice.

Direct comparison of the data obtained in the animal models with induced deletion of either the β_1-integrin or the α_4-integrin subunit is difficult, because residual integrin expression is observed in these systems and needs to be considered when interpreting the observations made in each model system. Inducible deletion of β_1-integrin was reported to lead to the rapid loss of β_1-integrin expression on 95% of bone marrow cells and thymocytes [73]. However, even 12 weeks after induced deletion of β_1-integrin expression in hematopoietic cells, up to 15% of cells in the spleen and the Peyer's Patches express β_1-integrin on their cell surface [73]. Similarly, induced deletion of the α_4-integrin subunit was reported to result in the rapid loss of α_4-integrin expression on more than 95% of bone marrow cells; however, expression of α_4-integrin in other tissues has not been analyzed [74].

Taken together, those studies, which detected an involvement of $\alpha_4\beta_1$-integrin in hematopoiesis have in common that the α_4-subunit or the β_1-subunit were absent already before the onset of hematopoiesis during embryonic development and/or that the integrin subunits were additionally deleted on non-hematopoietic cells. In contrast, lack of both α_4-integrins solely on hematopoietic cells in the adult does not seem to hamper hematopoiesis. Thus, based on the available data a requisite role for $\alpha_4\beta_1$-integrins in hematopoiesis cannot be postulated.

α_4-integrins as costimulatory molecules on T cells

One of the very early observations has been that certain antibodies against the α_4-integrin subunit inhibit effector activity of human cytotoxic T cells *in vitro* [4, 5]. The relevant mechanism of action remained unknown. Target cells in this assay did not express VCAM-1, suggesting that cytotoxic T cells recognize either fibronectin or an unidentified ligand. In this regard it is interesting to note that T cells can produce fibronectin [76]. As one of the anti-α_4-integrin antibodies used in these initial assays does not inhibit $\alpha_4\beta_1$-integrin binding either to fibronectin or to VCAM-1, it was thought that this effect might either be mediated by homophilic binding to α_4-integrin [39] or by yet another ligand for $\alpha_4\beta_1$-integrin.

Ligation of the α_4-integrin subunit on human T lymphoblastoid cell lines or resting peripheral blood T cells either *via* the CS-1 domain of fibronectin or by specific anti-α_4-integrin antibodies can stimulate tyrosine phosphorylation of a unique set of cellular proteins including phospholipase Cγ (ppl40), ppl25 focal adhesion kinase (ppl20), paxillin (pp70 and pp50), p59^{fyn}/p56^{lck} (pp60-55) and mitogen-activated protein kinase (pp45) [77–79]. These studies suggested that α_4-integrin on T cells may play a role in the transduction of activation signals into T cells. This observation is further supported by the observation that soluble recombinant VCAM-1 can promote CD3-dependent T cell proliferation [80, 81]. Other studies demonstrated that certain anti-α_4-integrin antibodies can interfere with the antigen-specific proliferation of CD4+ T cells *in vitro* [82]. Thus, one might speculate that, depending on the epitope they recognize, different anti-α_4-integrin antibodies might have opposing effects on T cell activation. This notion is supported by the observation that different anti-α_4-integrin antibodies induce distinct tyrosine phosphorylation patterns in T cells [79]. As engagement of α_4-integrin on T cells has been shown to trigger β_2-integrin-mediated adhesion to ICAM-1 in a PI3-kinase-dependent manner [83], the recent observation that $\alpha_4\beta_1$-integrin colocalizes with LFA-1 at the peripheral supramolecular activation complex (pSMAC) in the immunological synapses on T cells further supports a function of α_4-integrin as a costimulatory molecule on T cells [84].

Last but not least, lymphocyte adhesion *via* α_4-integrins was shown to promote T and B cell survival [85, 86].

α_4-integrins play a unique dual role in the multi-step leukocyte recruitment from blood into tissue

Leukocyte recruitment from the blood into tissue is regulated by the sequential interactions of adhesion and signaling molecules on leukocytes and endothelial cells [87, 88]. The multi-step interaction starts with an initial transient contact of the circulating leukocyte with the vascular endothelium, leading to leukocyte rolling along the vascular wall with greatly reduced velocity. The leukocyte is then exposed to chemotactic factors of the family of chemokines presented on the endothelial surface, which bind to G protein-coupled receptors on the leukocyte surface. G protein-mediated "inside-out-signals" lead to the activation of integrins on the leukocyte surface, which can mediate the firm adhesion of the leukocytes to the vascular endothelium and trigger leukocyte diapedesis across the endothelium. As already pointed out above, α_4-integrins have been shown to be unique among the leukocyte integrins as they can initiate this multistep cascade by mediating leukocyte rolling by low-affinity interaction with the endothelial ligands VCAM-1 and MAdCAM-1 under the influence of shear forces [89, 90] and continue by mediating G protein-dependent high-affinity adhesion strengthening of the leukocyte to the endothelium.

Under physiological conditions $\alpha_4\beta_7$-integrin mediates the homing of naive lymphocytes to the gut-associated lymphoid tissues by binding to MAdCAM-1 on the HEVs in Peyer's patches and mesenteric lymph nodes. Intravital microscopy studies have shown that $\alpha_4\beta_7$ initiates lymphocyte rolling and mediates G protein-dependent firm adhesion [91], confirming the unique dual role of α_4-integrins in the multi-step recruitment cascade.

Right after its discovery, $\alpha_4\beta_1$-integrin was suggested to play an important role in the recruitment of circulating leukocytes from blood into inflamed tissue as its ligand VCAM-1 was found to be up-regulated on inflamed endothelial cells. We only began to understand the multi-step adhesion cascade of leukocyte recruitment into tissues at that time when the first studies performed by Thomas and Andrew Issekutz [92, 93] demonstrated that the migration of ^{111}In-labeled rat lymphocytes into the inflamed skin or into inflamed joints was dependent on α_4-integrin. Inhibition of α_4-integrin-mediated leukocyte recruitment using blocking antibodies proved beneficial in many different animal models of inflammation including organ transplantation (summarized in [94]). Two different anti-α_4-integrin antibodies were efficient in preventing a murine experimental model of contact hypersensitivity [95, 96]. Furthermore, blocking α_4-integrins resulted in inhibition of spontaneously developing insulitis and prevention of diabetes in nonobese diabetic mice [97]. Similarly, antibodies blocking α_4-integrin were found to attenuate colitis in the cotton top tamarin, a New World primate that experiences a spontaneous acute and chronic colitis resembling ulcerative colitis [98].

α_4-integrins have been most extensively studied regarding their involvement in the recruitment of circulating immune cells into the central nervous system (CNS) in experimental autoimmune encephalomyelitis (EAE), an animal model of multiple sclerosis (MS). After the seminal study by Ted Yednock and colleagues [99], numerous following studies confirmed and extended the predominant involvement of α_4-integrin/VCAM-1 in inflammatory cell recruitment into the CNS in different EAE models in a number of species [82, 100, 101]. Additionally, successful transfer of EAE was shown to depend on the surface expression levels of α_4-integrins on autoaggressive T cell clones [102]. Recently, intravital microscopy studies have provided direct evidence for α_4 integrin-mediated inflammatory cell interaction with the CNS microvasculature *in vivo*. In inflamed pial venules of mice suffering from EAE, α_4-integrins mediate rolling and G protein-dependent arrest of endogenous leukocytes [103]. Recruitment of encephalitogenic T cell blasts across the spinal cord white matter microvasculature was shown to depend on α_4-integrin-mediated initial capture and G protein-dependent arrest [104]. Lack of rolling of encephalitogenic T cell blasts might be due to the high expression level of high-affinity α_4-integrins on their cell surface, which might prevent the low-affinity interactions required for rolling [82]. In any case, firm adhesion of encephalitogenic T cells to the spinal cord microvasculature still required G protein-dependent activation of α_4-integrins, suggesting that more high-affinity α_4-integrin molecules are required for adhesion strengthening. Taken together, these date demonstrate that α_4-integrins play a predominant role in leukocyte recruitment into the CNS, as they mediate both the initial low-affinity interaction of circulating leukocytes with the CNS microvasculature and the subsequent G protein-dependent arrest requiring high-affinity binding of α_4-integrins.

In most of the studies described above, a role for $\alpha_4\beta_7$-integrin in leukocyte recruitment was not formally ruled out. In EAE, the $\alpha_4\beta_1$-integrin ligand

VCAM-1, but not the $\alpha_4\beta_7$-integrin ligand MAdCAM-1, is expressed on inflamed CNS microvessels [100]. Therefore, it was concluded that leukocyte recruitment into the CNS is mediated by $\alpha_4\beta_1$-VCAM-1 interaction, disregarding that $\alpha_4\beta_7$-integrin can also bind to VCAM-1 [13]. Despite the fact that encephalitogenic T cells express both α_4-integrins at equal levels, neutralizing $\alpha_4\beta_7$-antibodies do not inhibit EAE [82]. These data demonstrate that $\alpha_4\beta_7$-integrin is at least not required for EAE pathogenesis. In apparent contrast, an EAE study performed in β_7-integrin-deficient mice showed very mild clinical EAE, when compared to wild-type mice [105]. However, whether α_4-integrin levels on leukocytes are generally lower in β_7-integrin-deficient mice, which would explain these varying findings, was not investigated.

Another issue that has rarely been addressed is whether the therapeutic efficacy of blocking α_4-integrins in inflammatory disease models is truly due to the inhibition of α_4-integrin-mediated immune cell recruitment to the inflamed organs. Many studies performed in mouse models have exchangeably used two different monoclonal rat anti mouse α_4-integrin antibodies named PS/2 and R1.2. When used within the same studies, their therapeutic efficacy was indistinguishable [82, 96]. Surprisingly, when characterized in more detail, the inhibitory characteristics of both antibodies are not alike. PS/2 recognizes the functional epitope B2 on the α_4-integrin subunit and blocks $\alpha_4\beta_1$-mediated binding to VCAM-1 and fibronectin as well as $\alpha_4\beta_7$-dependent lymphocyte adhesion to VCAM-1 and MAdCAM-1 *in vitro* [36, 51]. *In vivo*, PS/2 efficiently blocks $\alpha_4\beta_7$-dependent lymphocyte homing to gut-associated lymphoid tissue and $\alpha_4\beta_1$-dependent T cell homing to the CNS *in vivo* [104, 106]. R1-2, which binds to the functional epitope B1 on the α_4-integrin subunit, was originally shown to block $\alpha_4\beta_7$-integrin-dependent lymphocyte binding to HEVs in Peyer's patches *in vitro* [9]. Surprisingly, R1.2 neither blocks $\alpha_4\beta_7$-mediated binding to purified MAdCAM-1 *in vitro* [36] nor inhibits $\alpha_4\beta_7$-dependent lymphocyte homing to gut-associated lymphoid tissue *in vivo* [106], suggesting that R1.2 is a poor inhibitor of α_4-integrin-mediated lymphocyte migration *in vivo*. This notion is further supported by the finding that, although both R1.2 and PS/2 were shown to inhibit T cell-dependent murine contact hypersensitivity, R1.2 blocked ear swelling by 80% without inhibiting the overall emigration of either nonimmune or immune T cells, suggesting that in this model R1.2 does not function by inhibiting immune cell recruitment [96]. As R1.2 was shown to block antigen-dependent activation of $CD4^+$ T cells *in vitro* [82], it can be assumed that equal therapeutic efficacies of R1.2 and PS/2 in mouse inflammatory disease models is due to the inhibition of maybe overlapping but distinct α_4-integrin-dependent pathomechanisms including T cell activation and immune cell migration.

In the context of cell migration it should finally be mentioned that $\alpha_4\beta_1$-integrins seem to play a role in tumor progression and metastasis. $\alpha_4\beta_1$-integrin has been detected on both benign and malignant cells of the melanocytic lineage and was shown to mediate adhesion to VCAM-1. Interestingly, in this setting $\alpha_4\beta_1$-integrin is found to suppress formation of metastasis in experimental tumor models probably by promoting adhesion of tumor cells within the tissue [107].

Therapeutic targeting of α_4-integrins in human disease

It stands to reason that therapies aimed at blockade of α_4-integrin-mediated inflammatory cell immigration into the affected organs might represent an effective therapeutic approach especially in those organs where α_4-integrins mediate the initial contact of immune cells with the vascular endothelium. Based on the dominant role of α_4-integrins in leukocyte recruitment into the CNS in an animal model of MS this therapeutic approach was readily translated into the clinic. The mouse monoclonal antibody AN100226m was selected for its ability to inhibit lymphocyte adhesion to TNF-α-stimulated brain endothelium and was shown to reverse disease in a chronic model of EAE [101]. AN100226 (TY21-6) was found to be a potent inhibitor of α_4-mediated binding to fibronectin and VCAM-1, but did not induce homophilic aggregation, which characterizes the antibody as an epitope B1 antibody [49], most probably binding to the third blade of the propeller domain in the α_4-integrin subunit (as described above). The antibody was humanized and placed on the IgG4 framework [108], and assigned the name Natalizumab according to the Federal Drug Administration (USA) nomenclature guidelines. In clinical trials for MS, Natalizumab proved to be extremely successful and demonstrated a highly significant benefit of treatment on both clinical parameters as well as magnetic resonance imaging measurements of disease activity [109, 110].

As Natalizumab binds the α_4-integrin subunit irrespective of its associated β chain, it blocks both $\alpha_4\beta_1$- and $\alpha_4\beta_7$-integrin-mediated actions. Early studies of Holzman and Weissman [10] implicated $\alpha_4\beta_7$-integrin in homing of lymphocytes to the intestine, and subsequent studies in an animal model of colitis demonstrated that antibodies to $\alpha_4\beta_7$-integrin or MAdCAM-1 attenuated T cell-mediated intestinal inflammation [111, 112]. Therefore, therapeutic efficacy of Natalizumab was additionally tested for the treatment of Crohn's disease. The results of previous and ongoing studies suggest that Natalizumab may be effective for the treatment of patients with moderately to severely active Crohn's disease (summarized in [113]).

Due to the overall encouraging results, Natalizumab was approved for the treatment of relapsing-remitting MS in the United States in November 2004. Shortly after release of Natalizumab, however, three of several thousand patients who had received the antibody during the clinical trials developed progressive multifocal leukoencephalopathy (PML), a reactivation of the human polyoma JC virus within CNS glial cells with mostly fatal outcome [114, 115]. These unexpected events led to voluntary suspension of the drug from the market and rapidly raised general concerns about the safety of a therapy targeting α_4-integrin-mediated immune cell migration into the CNS in MS [116, 117]. Although reactivation of JC virus in these patients is clearly treatment-associated, we can only speculate about the mechanisms involved, especially as JC virus resides in the kidney, the gut and probably also in the bone marrow of healthy individuals. A subsequent detailed review of possible cases of PML in more than 3000 patients treated with Natalizumab over an average period of 17.9 months found no new cases of PML [118]. In light of its great benefits in the treatment of MS, Natalizumab has been re-approved for the treatment of relapsing-remitting MS in the United States and in Europe in summer 2006. The clinical efficacy of Natalizumab in light of a recent observation made in a small number of patients, *i.e.*, decreases the ratio of CD4$^+$/CD8$^+$ T cells in the cerebrospinal fluid of MS patients [119], supports a critical role of CD4$^+$ T cells in the inflammatory cascade of MS. Whether Natalizumab targets α$_4$-integrin-mediated pathomechanism distinct from α$_4$-integrin-mediated immune cell entry into the CNS remains to be investigated. Further thorough evaluation of MS patients under long-term treatment with Natalizumab will allow for a better assessment of its therapeutic mechanisms and its benefit/risk ratio.

Outlook

The availability of a drug inhibiting α$_4$-integrin-mediated pathomechanisms in human diseases would suggest that we have come to understand the biological function of α$_4$-integrins *in vivo*. Having read this chapter, one might come to realize that, although we know the primary structure of α$_4$-integrin, its cellular distribution and its major ligands, we have only begun to understand the elegant allosteric regulation of α$_4$-integrin activity. Similarly, although there is no doubt that α$_4$-integrins mediate immune cell migration into gut-associated lymphoid tissue and into inflamed organs, the contribution of α$_4$-integrins in other biological processes such as embryogenesis, hematopoiesis or T cell costimulation have, despite tremendous efforts, not yet been solved. To understand the structure-function relationship of α$_4$-integrins and to obtain a complete picture of α$_4$-integrin-mediated biological functions *in vivo* are challenges for the future.

Acknowledgments
My special thanks go to Ted Yednock for his valuable expert comments on this chapter. Additional thanks go to Urban Deutsch for his productive criticism of the chapter.

References

1   Hemler ME, Huang C, Schwarz L (1987) The VLA protein family. Characterization of five distinct cell surface heterodimers each with a common 130,000 molecular weight beta subunit. *J Biol Chem* 262: 3300–3309
2   Sanchez-Madrid F, De Landazuri MO, Morago G, Cebrian M, Acevedo A, Bernabeu C

(1986) VLA-3: a novel polypeptide association within the VLA molecular complex: cell distribution and biochemical characterization. *Eur J Immunol* 16: 1343–1349

3   Hemler ME, Jacobson JG, Brenner MB, Mann D, Strominger JL (1985) VLA-1: a T cell surface antigen which defines a novel late stage of human T cell activation. *Eur J Immunol* 15: 502–508

4   Takada Y, Elices MJ, Crouse C, Hemler ME (1989) The primary structure of the alpha 4 subunit of VLA-4: homology to other integrins and a possible cell-cell adhesion function. *EMBO J* 8: 1361–1368

5   Clayberger C, Krensky AM, McIntyre BW, Koller TD, Parham P, Brodsky F, Linn DJ, Evans EL (1987) Identification and characterization of two novel lymphocyte function-associated antigens, L24 and L25. *J Immunol* 138: 1510–1514

6   Wayner EA, Garcia-Pardo A, Humphries MJ, McDonald JA, Carter WG (1989) Identification and characterization of the T lymphocyte adhesion receptor for an alternative cell attachment domain (CS-1) in plasma fibronectin. *J Cell Biol* 109: 1321–1330

7   Guan JL, Hynes RO (1990) Lymphoid cells recognize an alternatively spliced segment of fibronectin *via* the integrin receptor alpha 4 beta 1. *Cell* 60: 53–61

8   Elices MJ, Osborn L, Takada Y, Crouse C, Luhowskyj S, Hemler ME, Lobb RR (1990) VCAM-1 on activated endothelium interacts with the leukocyte integrin VLA-4 at a site distinct from the VLA-4/fibronectin binding site. *Cell* 60: 577–584

9   Holzmann B, McIntyre BW, Weissman IL (1989) Identification of a murine Peyer's patch-specific lymphocyte homing receptor as an integrin molecule with an alpha chain homologous to human VLA-4 alpha. *Cell* 56: 37–46

10   Holzmann B, Weissman IL (1989) Peyer's patch-specific lymphocyte homing receptors consist of a VLA-4-like alpha chain associated with either of two integrin beta chains, one of which is novel. *EMBO J* 8: 1735–1741

11   Erle DJ, Ruegg C, Sheppard D, Pytela R (1991) Complete amino acid sequence of an integrin beta subunit (beta 7) identified in leukocytes. *J Biol Chem* 266: 11009–11016

12   Kilshaw PJ, Murant SJ (1991) Expression and regulation of beta 7(beta p) integrins on mouse lymphocytes: relevance to the mucosal immune system. *Eur J Immunol* 21: 2591–2597

13   Erle DJ, Briskin MJ, Butcher EC, Garcia-Pardo A, Lazarovits AI, Tidswell M (1994) Expression and function of the MAdCAM-1 receptor, integrin alpha 4 beta 7, on human leukocytes. *J Immunol* 153: 517–528

14   Rott LS, Briskin MJ, Andrew DP, Berg EL, Butcher EC (1996) A fundamental subdivision of circulating lymphocytes defined by adhesion to mucosal addressin cell adhesion molecule-1. Comparison with vascular cell adhesion molecule-1 and correlation with beta 7 integrins and memory differentiation. *J Immunol* 156: 3727–3736

15   Iwata M, Hirakiyama A, Eshima Y, Kagechika H, Kato C, Song SY (2004) Retinoic acid imprints gut-homing specificity on T cells. *Immunity* 21: 527–538

16   Mora JR, Iwata M, Eksteen B, Song SY, Junt T, Senman B, Otipoby KL, Yokota A, Takeuchi H, Ricciardi-Castagnoli P et al (2006) Generation of gut-homing IgA-secreting B cells by intestinal dendritic cells. *Science* 314: 1157–1160

17   Johnston B, Kubes P (1999) The alpha4-integrin: an alternative pathway for neutrophil recruitment? *Immunol Today* 20: 545–550

18   Williams DA, Rios M, Stephens C, Patel VP (1991) Fibronectin and VLA-4 in haemato-poietic stem cell-microenvironment interactions. *Nature* 352: 438–441.

19   Sheppard AM, Onken MD, Rosen GD, Noakes PG, Dean DC (1994) Expanding roles for alpha 4 integrin and its ligands in development. *Cell Adhes Commun* 2: 27–43

20   Stepp MA, Urry LA, Hynes RO (1994) Expression of alpha 4 integrin mRNA and protein and fibronectin in the early chicken embryo. *Cell Adhes Commun* 2: 359–375

21   Yang JT, Rayburn H, Hynes RO (1995) Cell adhesion events mediated by alpha 4 integrins are essential in placental and cardiac development. *Development* 121: 549–560

22   Qian F, Vaux DL, Weissman IL (1994) Expression of the integrin alpha 4 beta 1 on melanoma cells can inhibit the invasive stage of metastasis formation. *Cell* 77: 335–347

23   Argraves WS, Suzuki S, Arai H, Thompson K, Pierschbacher MD, Ruoslahti E (1987) Amino acid sequence of the human fibronectin receptor. *J Cell Biol* 105: 1183–1190

24   Teixido J, Parker CM, Kassner PD, Hemler ME (1992) Functional and structural analysis of VLA-4 integrin alpha 4 subunit cleavage. *J Biol Chem* 267: 1786–1791

25   Parker CM, Pujades C, Brenner MB, Hemler ME (1993) Alpha 4/180, a novel form of the integrin alpha 4 subunit. *J Biol Chem* 268: 7028–7035

26   Springer TA (1997) Folding of the N-terminal, ligand-binding region of integrin alpha-subunits into a beta-propeller domain. *Proc Natl Acad Sci USA* 94: 65–72

27   Xiong JP, Stehle T, Diefenbach B, Zhang R, Dunker R, Scott DL, Joachimiak A, Goodman SL, Arnaout MA (2001) Crystal structure of the extracellular segment of integrin alpha Vbeta3. *Science* 294: 339–345

28   Hynes RO (2002) Integrins: bidirectional, allosteric signaling machines. *Cell* 110: 673–687

29   Du X, Gu M, Weisel JW, Nagaswami C, Bennett JS, Bowditch R, Ginsberg MH (1993) Long range propagation of conformational changes in integrin alpha IIb beta 3 [Erratum in: *J Biol Chem* (1994) 269: 11673]. *J Biol Chem* 268: 23087–23092

30   Wiesner S, Lange A, Fassler R (2006) Local call: from integrins to actin assembly. *Trends Cell Biol* 16: 327–329

31   Legate KR, Montanez E, Kudlacek O, Fassler R (2006) ILK, PINCH and parvin: the tIPP of integrin signalling. *Nat Rev Mol Cell Biol* 7: 20–31

32   Luo BH, Carman CV, Springer TA (2007) Structural basis of integrin regulation and signaling. *Annu Rev Immunol* 25: 619–647

33   Osborn L, Vassallo C, Benjamin CD (1992) Activated endothelium binds lymphocytes through a novel binding site in the alternately spliced domain of vascular cell adhesion molecule-1. *J Exp Med* 176: 99–107

34   Kilger G, Needham LA, Nielsen PJ, Clements J, Vestweber D, Holzmann B (1995) Differential regulation of alpha-4 integrin-dependent binding to domains 1 and 4 of vascular cell adhesion molecule-1. *J Biol Chem* 270: 5979–5984

35   Ruegg C, Postigo AA, Sikorski EE, Butcher EC, Pytela R, Erle DJ (1992) Role of integrin

alpha 4 beta 7/alpha 4 beta P in lymphocyte adherence to fibronectin and VCAM-1 and in homotypic cell clustering. *J Cell Biol* 117: 179–189

36   Berlin C, Berg EL, Briskin MJ, Andrew DP, Kilshaw PJ, Holzmann B, Weissman IL, Hamann A, Butcher EC (1993) Alpha 4 beta 7 integrin mediates lymphocyte binding to the mucosal vascular addressin MAdCAM-1. *Cell* 74: 185–185

37   Briskin MJ, Rott L, Butcher EC (1996) Structural requirements for mucosal vascular addressin binding to its lymphocyte receptor alpha 4 beta 7. Common themes among integrin-Ig family interactions. *J Immunol* 156: 719–726

38   Pytela R, Pierschbacher MD, Ruoslahti E (1985) Identification and isolation of a 140 kd cell surface glycoprotein with properties expected of a fibronectin receptor. *Cell* 40: 191–198

39   Altevogt P, Hubbe M, Ruppert M, Lohr J, Hoegen P, Sammar M, Andrew DP, McEvoy LM, Humphries MJ, Butcher EC (1995) The α4-integrin chain is a ligand for α4β7 and α4β1. *J Exp Med* 182: 345–355

40   Bayless KJ, Meininger GA, Scholtz JM, Davis GE (1998) Osteopontin is a ligand for the alpha4beta1 integrin. *J Cell Sci* 111: 1165–1174

41   Bayless KJ, Davis GE (2001) Identification of dual alpha 4beta1 integrin binding sites within a 38 amino acid domain in the N-terminal thrombin fragment of human osteopontin. *J Biol Chem* 276: 13483–13489

42   Ennis E, Isberg RR, Shimizu Y (1993) Very late antigen 4-dependent adhesion and costimulation of resting human T cells by the bacterial beta 1 integrin ligand invasin. *J Exp Med* 177: 207–212

43   Yabkowitz R, Dixit VM, Guo N, Roberts DD, Shimizu Y (1993) Activated T-cell adhesion to thrombospondin is mediated by the alpha 4 beta 1 (VLA-4) and alpha 5 beta 1 (VLA-5) integrins. *J Immunol* 151: 149–158

44   Cunningham SA, Rodriguez JM, Arrate MP, Tran TM, Brock TA (2002) JAM2 interacts with alpha4beta1. Facilitation by JAM3. *J Biol Chem* 277: 27589–27592

45   Bridges LC, Sheppard D, Bowditch RD (2005) ADAM disintegrin-like domain recognition by the lymphocyte integrins alpha4beta1 and alpha4beta7. *Biochem J* 387: 101–108

46   Humphries JD, Humphries MJ (2007) CD14 is a ligand for the integrin alpha4beta1. *FEBS Lett* 581: 757–763

47   Porter JC, Hogg N (1997) Integrin cross talk: activation of lymphocyte function-associated antigen-1 on human T cells alters alpha4beta1- and alpha5beta1-mediated function. *J Cell Biol* 138: 1437–1447

48   Bazzoni G, shih D-T, Buck CA, Hemler ME (1995) Monoclonal antibody 9EG7 defines a novel β1 integrin epitope induced by soluble ligand and manganese, but inhibited by calcium. *J Biol Chem* 270: 25570–25577

49   Newham P, Craig SE, Clark K, Mould AP, Humphries MJ (1998) Analysis of ligand-induced and ligand-attenuated epitopes on the leukocyte integrin alpha4beta1: VCAM-1, mucosal addressin cell adhesion molecule-1, and fibronectin induce distinct conformational changes. *J Immunol* 160: 4508–4517

50 Pulido R, Elices MJ, Campanero MR, Osborn L, Schiffer S, Garcia-Pardo A, Lobb R, Hemler ME, Sanchez-Madrid F (1991) Functional evidence for three distinct and independently inhibitable adhesion activities mediated by the human integrin VLA-4. Correlation with distinct alpha 4 epitopes. *J Biol Chem* 266: 10241–10245

51 Kamata T, Puzon W, Takada Y (1995) Identification of putative ligand-binding sites of the integrin $\alpha 4\beta 1$ (VLA-4, CD49d/CD29). *Biochem J* 305: 945–951

52 Schiffer SG, Hemler ME, Lobb RR, Tizard R, Osborn L (1995) Molecular mapping of functional antibody binding sites of alpha 4 integrin. *J Biol Chem* 270: 14270–14273

53 Andrew DP, Berlin C, Honda S, Yoshino T, Hamann A, Holzmann B, Kilshaw PJ, Butcher EC (1994) Distinct but overlapping epitopes are involved in alpha 4 beta 7-mediated adhesion to vascular cell adhesion molecule-1, mucosal addressin-1, fibronectin, and lymphocyte aggregation. *J Immunol* 153: 3847–3861

54 Munoz M, Serrador J, Sanchez-Madrid F, Teixido J (1996) A region of the integrin VLA alpha 4 subunit involved in homotypic cell aggregation and in fibronectin but not vascular cell adhesion molecule-1 binding. *J Biol Chem* 271: 2696–2702

55 Irie A, Kamata T, Puzon-McLaughlin W, Takada Y (1995) Critical amino acid residues for ligand binding are clustered in a predicted beta-turn of the third N-terminal repeat in the integrin alpha 4 and alpha 5 subunits. *EMBO J* 14: 5550–5556

56 Takada Y, Ylanne J, Mandelman D, Puzon W, Ginsberg MH (1992) A point mutation of integrin beta 1 subunit blocks binding of alpha 5 beta 1 to fibronectin and invasin but not recruitment to adhesion plaques. *J Cell Biol* 119: 913–921

57 Higgins JM, Cernadas M, Tan K, Irie A, Wang J, Takada Y, Brenner MB (2000) The role of alpha and beta chains in ligand recognition by beta 7 integrins. *J Biol Chem* 275: 25652–25664

58 Xiong JP, Stehle T, Zhang R, Joachimiak A, Frech M, Goodman SL, Arnaout MA (2002) Crystal structure of the extracellular segment of integrin alpha Vbeta3 in complex with an Arg-Gly-Asp ligand. *Science* 296: 151–155

59 Yednock TA, Cannon C, Vandevert C, Goldbach EG, Shaw G, Ellis DK, Liaw C, Fritz LC, Tanner LI (1995) Alpha 4 beta 1 integrin-dependent cell adhesion is regulated by a low affinity receptor pool that is conformationally responsive to ligand. *J Biol Chem* 270: 28740–28750

60 Yauch RL, Felsenfeld DP, Kraeft SK, Chen LB, Sheetz MP, Hemler ME (1997) Mutational evidence for control of cell adhesion through integrin diffusion/clustering, independent of ligand binding. *J Exp Med* 186: 1347–1355

61 Kwee L, Baldwin HS, Shen HM, Stewart CL, Buc C, Buch CA, Labow MA (1995) Defective development of the embryonic and extraembryonic circulatory systems in vascular cell adhesion molecule (VCAM-1) deficient mice. *Development* 121: 489–503

62 Gurtner GC, Davis V, Li H, McCoy MJ, Sharpe A, Cybulsky MI (1995) Targeted disruption of the murine VCAM1 gene: essential role of VCAM-1 in chorioallantoic fusion and placentation. *Genes Dev* 9: 1–14

63 Rosen GD, Sanes JR, LaChance R, Cunningham JM, Roman J, Dean DC (1992) Roles

for the integrin VLA-4 and its counter receptor VCAM-1 in myogenesis. *Cell* 69: 1107–1119

64   Yang JT, Rando TA, Mohler WA, Rayburn H, Blau HM, Hynes RO (1996) Genetic analysis of alpha 4 integrin functions in the development of mouse skeletal muscle. *J Cell Biol* 135: 829–835

65   Miyake K, Weissman IL, Greenberger JS, Kincade PW (1991) Evidence for a role of the integrin VLA-4 in lympho-hemopoiesis. *J Exp Med* 173: 599–607

66   Wagner N, Lohler J, Kunkel EJ, Ley K, Leung E, Krissansen G, Rajewsky K, Muller W (1996) Critical role for beta7 integrins in formation of the gut-associated lymphoid tissue. *Nature* 382: 366–370

67   Finke D, Acha-Orbea H, Mattis A, Lipp M, Kraehenbuhl J (2002) CD4+CD3− cells induce Peyer's patch development: role of alpha4beta1 integrin activation by CXCR5. *Immunity* 17: 363–373

68   Fassler R, Meyer M (1995) Consequences of lack of beta 1 integrin gene expression in mice. *Genes Dev* 9: 1896–1908

69   Arroyo AG, Yang JT, Rayburn H, Hynes RO (1999) Alpha4 integrins regulate the proliferation/differentiation balance of multilineage hematopoietic progenitors *in vivo*. *Immunity* 11: 555–566

70   Arroyo AG, Yang JT, Rayburn H, Hynes RO (1996) Differential requirements for alpha4 integrins during fetal and adult hematopoiesis. *Cell* 85: 997–1008

71   Hirsch E, Iglesias A, Potocnik AJ, Hartmann U, Fassler R (1996) Impaired migration but not differentiation of haematopoietic stem cells in the absence of beta1 integrins. *Nature* 380: 171–175

72   Potocnik AJ, Brakebusch C, Fassler R (2000) Fetal and adult hematopoietic stem cells require beta1 integrin function for colonizing fetal liver, spleen, and bone marrow. *Immunity* 12: 653–663

73   Brakebusch C, Fillatreau S, Potocnik AJ, Bungartz G, Wilhelm P, Svensson M, Kearney P, Korner H, Gray D, Fassler R (2002) Beta1 integrin is not essential for hematopoiesis but is necessary for the T cell-dependent IgM antibody response. *Immunity* 16: 465–477

74   Scott LM, Priestley GV, Papayannopoulou T (2003) Deletion of alpha4 integrins from adult hematopoietic cells reveals roles in homeostasis, regeneration, and homing. *Mol Cell Biol* 23: 9349–9360

75   Bungartz G, Stiller S, Bauer M, Muller W, Schippers A, Wagner N, Fassler R, Brakebusch C (2006) Adult murine hematopoiesis can proceed without beta1 and beta7 integrins. *Blood* 108: 1857–1864

76   Godfrey HP, Canfield LS, Kindler HL, Angadi CV, Tomasek JJ, Goodman JW (1988) Production of a fibronectin-associated lymphokine by cloned mouse T cells. *J Immunol* 141: 1508–1515

77   Nojima Y, Humphries MJ, Mould AP, Komoriya A, Yamada KM, Schlossman SF, Morimoto C (1990) VLA-4 mediated CD3-dependent CD4+ T cell activation *via* the CS1 alternatively spliced domain of fibronectin. *J Exp Med* 172: 1185–1192

78 Nojima Y, Rothstein DM, Sugita K, Schlossman SF, Morimoto C (1992) Ligation of VLA-4 on T cells stimulates tyrosine phosphorylation of a 105-kD protein. *J Exp Med* 175: 1045–1053

79 Sato T, Tachibana K, Nojima Y, D'Avirro N, Morimoto C (1995) Role of VLA-4 molecule in T cell costimulation. *J Immunol* 155: 2938–2947

80 Burkly LC, Jakubowski A, Newman BM, Rosa MD, Chi-Rosso G, Lobb RR (1991) Signaling by vascular cell adhesion molecule-1 (VCAM-1) through VLA-4 promotes CD3-dependent T cell proliferation. *Eur J Immunol* 21: 2871–2875

81 Damle NK, Aruffo A (1991) Vascular cell adhesion molecule 1 induces T cell antigen receptor dependent activation of CD4+ T lymphocytes. *Proc Natl Acad Sci USA* 88: 6403–6407

82 Engelhardt B, Laschinger M, Schulz M, Samulowitz U, Vestweber D, Hoch G (1998) The development of experimental autoimmune encephalomyelitis in the mouse requires alpha4-integrin but not alpha4beta7-integrin. *J Clin Invest* 102: 2096–2105

83 Hyduk SJ, Cybulsky MI (2002) Alpha 4 integrin signaling activates phosphatidylinositol 3-kinase and stimulates T cell adhesion to intercellular adhesion molecule-1 to a similar extent as CD3, but induces a distinct rearrangement of the actin cytoskeleton. *J Immunol* 168: 696–704

84 Mittelbrunn M, Molina A, Escribese MM, Yanez-Mo M, Escudero E, Ursa A, Tejedor R, Mampaso F, Sanchez-Madrid F (2004) VLA-4 integrin concentrates at the peripheral supramolecular activation complex of the immune synapse and drives T helper 1 responses. *Proc Natl Acad Sci USA* 101: 11058–11063

85 Leussink VI, Zettl UK, Jander S, Pepinsky RB, Lobb RR, Stoll G, Toyka KV, Gold R (2002) Blockade of signaling *via* the very late antigen (VLA-4) and its counterligand vascular cell adhesion molecule-1 (VCAM-1) causes increased T cell apoptosis in experimental autoimmune neuritis. *Acta Neuropathol (Berl)* 103: 131–136

86 Koopman G, Keehnen RM, Lindhout E, Newman W, Shimizu Y, van Seventer GA, de Groot C, Pals ST (1994) Adhesion through the LFA-1 (CD11a/CD18)-ICAM-1 (CD54) and the VLA-4 (CD49d)-VCAM-1 (CD106) pathways prevents apoptosis of germinal center B cells. *J Immunol* 152: 3760–3767

87 Butcher EC (1991) Leukocyte-endothelial cell recognition: three (or more) steps to specificity and diversity. *Cell* 67: 1033–1036

88 Springer TA (1993) Signals on endothelium for lymphocyte recirculation and leukocyte emigration: the area code paradigm. *Harvey Lect* 89: 53–103

89 Berlin C, Bargatze RF, Campbell JJ, von Andrian UH, Szabo MC, Hasslen SR, Nelson RD, Berg EL, Erlandsen SL, Butcher EC (1995) Alpha 4 integrins mediate lymphocyte attachment and rolling under physiologic flow. *Cell* 80: 413–422

90 Alon R, Kassner PD, Carr MW, Finger EB, Hemler ME, Springer TA (1995) The integrin VLA-4 supports tethering and rolling in flow on VCAM-1. *J Cell Biol* 128: 1243–1253

91 Bargatze RF, Jutila MA, Butcher EC (1995) Distinct roles of L-selectin and integrins

α4β7 and LFA-1 in lymphocyte homing to Peyer's patch-HEV in situ: the multistep model confirmed and refined. *Immunity* 3: 99–108

92 Issekutz TB (1991) Inhibition of *in vivo* lymphocyte migration to inflammation and homing to lymphoid tissues by the TA-2 monoclonal antibody. A likely role for VLA-4 *in vivo*. *J Immunol* 147: 4178–4184

93 Issekutz TB, Issekutz AC (1991) T lymphocyte migration to arthritic joints and dermal inflammation in the rat: differing migration patterns and the involvement of VLA-4. *Clin Immunol Immunopathol* 61: 436–447

94 Lobb RR, Hemler ME (1994) The pathophysiological role of α4 integrins *in vivo*. *J Clin Invest* 94: 1722–1728

95 Chisholm PL, Williams CA, Lobb RR (1993) Monoclonal antibodies to the integrin alpha-4 subunit inhibit the murine contact hypersensitivity response. *Eur J Immunol* 23: 682–688

96 Ferguson TA, Kupper TS (1993) Antigen-independent processes in antigen-specific immunity. A role for alpha 4 integrin. *J Immunol* 150: 1172–1182

97 Yang XD, Karin N, Tisch R, Steinman L, McDevitt HO (1993) Inhibition of insulitis and prevention of diabetes in nonobese diabetic mice by blocking L-selectin and very late antigen 4 adhesion receptors. *Proc Natl Acad Sci USA* 90: 10494–10498

98 Podolsky DK, Lobb R, King N, Benjamin CD, Pepinsky B, Sehgal P, deBeaumont M (1993) Attenuation of colitis in the cotton-top tamarin by anti-alpha 4 integrin monoclonal antibody. *J Clin Invest* 92: 372–380

99 Yednock TA, Cannon C, Fritz LC, Sanchez Madrid F, Steinman L, Karin N (1992) Prevention of experimental autoimmune encephalomyelitis by antibodies against alpha 4 beta 1 integrin. *Nature* 356: 63–66

100 Steffen BJ, Butcher EC, Engelhardt B (1994) Evidence for involvement of ICAM-1 and VCAM-1 in lymphocyte interaction with endothelium in experimental autoimmune encephalomyelitis in the central nervous system in the SJL/J mouse. *Am J Pathol* 145: 189–201

101 Kent SJ, Karlik SJ, Cannon C, Hines DK, Yednock TA, Fritz LC, Horner HC (1995) A monoclonal antibody to alpha 4 integrin suppresses and reverses active experimental allergic encephalomyelitis. *J Neuroimmunol* 58: 1–10

102 Baron JL, Madri JA, Ruddle NH, Hashim G, Janeway CA Jr. (1993) Surface expression of alpha 4 integrin by CD4 T cells is required for their entry into brain parenchyma. *J Exp Med* 177: 57–68

103 Kerfoot S, Kubes P (2002) Overlapping roles of P-selectin and alpha 4 integrin to recruit leukocytes to the central nervous system in experimental autoimmune encephalomyelitis. *J Immunol* 169: 1000–1006

104 Vajkoczy P, Laschinger M, Engelhardt B (2001) Alpha4-integrin-VCAM-1 binding mediates G protein-independent capture of encephalitogenic T cell blasts to CNS white matter microvessels. *J Clin Invest* 108: 557–565

105 Kanwar JR, Harrison JE, Wang D, Leung E, Mueller W, Wagner N, Krissansen GW

(2000) Beta7 integrins contribute to demyelinating disease of the central nervous system. *J Neuroimmunol* 103: 146–152

106 Hamann A, Andrew DP, Jablonski-Westrich D, Holzmann B, Butcher EC (1994) Role of alpha 4-integrins in lymphocyte homing to mucosal tissues *in vivo*. *J Immunol* 152: 3282–3293

107 Gazitt Y, Akay C (2004) Mobilization of myeloma cells involves SDF–1/CXCR4 signaling and downregulation of VLA-4. *Stem Cells* 22: 65–73

108 Leger OJ, Yednock TA, Tanner L, Horner HC, Hines DK, Keen S, Saldanha J, Jones ST, Fritz LC, Bendig MM (1997) Humanization of a mouse antibody against human alpha-4 integrin: a potential therapeutic for the treatment of multiple sclerosis. *Hum Antibodies* 8: 3–16

109 Polman CH, O'Connor PW, Havrdova E, Hutchinson M, Kappos L, Miller DH, Phillips JT, Lublin FD, Giovannoni G, Wajgt A et al (2006) A randomized, placebo-controlled trial of natalizumab for relapsing multiple sclerosis. *N Engl J Med* 354: 899–910

110 Rudick RA, Stuart WH, Calabresi PA, Confavreux C, Galetta SL, Radue EW, Lublin FD, Weinstock-Guttman B, Wynn DR, Lynn F et al (2006) Natalizumab plus interferon beta–1a for relapsing multiple sclerosis. *N Engl J Med* 354: 911–923

111 Hesterberg PE, Winsor-Hines D, Briskin MJ, Soler-Ferran D, Merrill C, Mackay CR, Newman W, Ringler DJ (1996) Rapid resolution of chronic colitis in the cotton-top tamarin with an antibody to a gut-homing integrin alpha 4 beta 7. *Gastroenterology* 111: 1373–1380

112 Picarella D, Hurlbut P, Rottman J, Shi X, Butcher E, Ringler DJ (1997) Monoclonal antibodies specific for beta 7 integrin and mucosal addressin cell adhesion molecule-1 (MAdCAM-1) reduce inflammation in the colon of scid mice reconstituted with CD45RB^high CD4^+ T cells. *J Immunol* 158: 2099–2106

113 Macdonald JK, McDonald JW (2006) Natalizumab for induction of remission in Crohn's disease. *Cochrane Database Syst Rev* 3: CD006097

114 Langer-Gould A, Atlas SW, Bollen AW, Pelletier D (2005) Progressive multifocal leukoencephalopathy in a patient treated with natalizumab. *N Engl J Med* 353: 375–381

115 Kleinschmidt-Demasters BK, Tyler KL (2005) Progressive multifocal leukoencephalopathy complicating treatment with natalizumab and interferon beta-1a for multiple sclerosis. *N Engl J Med* 353: 369–374

116 Berger JR, Koralnik IJ (2005) Progressive multifocal leukoencephalopathy and natalizumab – Unforeseen consequences. *N Engl J Med* 353: 414–416

117 Drazen JM (2005) Patients at risk. *N Engl J Med* 353: 417

118 Yousry TA, Major EO, Ryschkewitsch C, Fahle G, Fischer S, Hou J, Curfman B, Miszkiel K, Mueller-Lenke N, Sanchez E et al (2006) Evaluation of patients treated with natalizumab for progressive multifocal leukoencephalopathy. *N Engl J Med* 354: 924–933

119 Stuve O, Marra CM, Bar-Or A, Niino M, Cravens PD, Cepok S, Frohman EM, Phillips JT, Arendt G (2006) Altered CD4^+/CD8^+ T-cell ratios in cerebrospinal fluid of natalizumab-treated patients with multiple sclerosis. *Arch Neurol* 63: 1383–1387

VCAM-1 and its functions in development and inflammatory diseases

Sharon J. Hyduk and Myron I. Cybulsky

Department of Laboratory Medicine and Pathobiology, University of Toronto, Toronto General Research Institute, Toronto, ON, M5G 2C4, Canada

Discovery of VCAM-1

VCAM-1 was identified in endothelium by monoclonal antibody and expression cloning approaches [1–3]. These approaches, as well as the discovery of other endothelial cell adhesion molecules, were dependent on the development of efficient and reproducible techniques for culturing human umbilical vein endothelial cells, and on observations that treatment with inflammatory cytokines, such as interleukin-1 (IL-1) and tumor necrosis factor-α (TNF), or with bacterial endotoxin resulted in endothelial activation [4]. Cytokine activation alters the phenotype of quiescent endothelial cells, resulting in a protein synthesis-dependent hyperadhesive state for leukocytes. The monoclonal antibody studies involved immunizing mice with activated endothelium and screening for monoclonal antibodies that recognized cytokine-inducible epitopes. The antibodies were then used to identify unique proteins by immunoprecipitation and leukocyte adhesion function was ascertained by antibody adhesion-blocking assays. Expression cloning used a subtracted cytokine-activated human umbilical vein endothelial cell library packaged in a eukaryotic expression vector that was transiently expressed in transfected COS cells, and cDNA from cells exhibiting increased leukocyte adhesion was extracted amplified, retransfected and rescreened [3].

VCAM-1 structure, genomic organization and alternative splicing

VCAM-1 is a type I transmembrane glycoprotein and member of the immunoglobulin (Ig) gene superfamily [3]. Initial expression cloning identified a form with six extracellular C2 or H-type Ig domains [3]; however, subsequently cDNAs were isolated from cytokine-activated endothelium that contained an additional Ig domain, designated domain 4 (the remaining domains were re-designated 5–7) [5–7] (Fig. 1). Expression of the seven-domain form of VCAM-1 was much more abundant in IL-1-stimulated human umbilical vein endothelial cells, and this was the only form detected on the cell surface by immunoprecipitation [5]. The low-abundance six-Ig domain VCAM-1 transcripts could be detected by PCR. Cloning of the human *vcam1* gene determined that six- and seven-Ig domain forms arise by alternative RNA splicing [8].

Genomic clones encoding the human and murine *vcam1* gene revealed that each of the extracellular Ig domains is encoded by a separate exon. All splice junctions occur after the first nucleotide of a codon (type 1) [8, 9]; therefore, any exon can be spliced in or out without disturbing the reading frame of the mRNA. The structure of the gene and homology between Ig domains 1–3 and 4–6 (55–75% amino acid identity, highest between domains 1 and 4) is consistent with the possibility that exon duplication played a role in the evolution of *vcam1*. This explains why the seven-Ig domain form has two ligand binding sites, in Ig domains 1 and 4 [10, 11] (Fig. 1).

A small fraction of murine VCAM-1 transcripts are alternatively spliced and are expressed as a three-Ig domain (domains 1–3) form (Fig. 1). This is because the *vcam1* gene of rodents contains a unique exon that is located between exons encoding Ig domains 3 and 4 [9, 12, 13]. This exon encodes a stop codon and polyadenylation signal, but not a transmembrane domain; therefore, alternatively spliced forms are attached to the cell membrane by a glycerophosphatidylinositol linkage. Rabbit VCAM-1 is expressed with seven or eight Ig domains. The additional Ig domain is located in position 8 and is most homologous to domains 3, 6, and 7.

VCAM-1 ligands

The primary ligand of VCAM-1 is $\alpha_4\beta_1$ integrin (VLA-4, CD49d/CD29) and it also binds weakly to $\alpha_4\beta_7$ integrin (LPAM-1) [14] (see the chapter by Britta Engelhardt). Several other integrins have also been identified as alternative VCAM-1 ligands, including $\alpha_9\beta_1$ and $\alpha_D\beta_2$ [15–17]. The α_9 subunit is homologous to α_4, and forms a heterodimer with the β_1 chain. The $\alpha_9\beta_1$ integrin also binds to the third fibronectin type III repeat of tenascin-C, an extracellular matrix protein. CHO cells transfected with $\alpha_9\beta_1$ adhere to immobilized recombinant VCAM-1 and TNF-α-stimulated endothelium [16]. The $\alpha_9\beta_1$ integrin is expressed on human peripheral blood neutrophils and to a lesser degree monocytes. In addition, $\alpha_9\beta_1$ integrin mediates neutrophil transmigration across recombinant VCAM-1-coated filters or TNF-α-activated endothelial monolayers. The $\alpha_D\beta_2$ integrin is the fourth member of the leukocyte β_2 integrin family. It is expressed by human eosinophils and supports the adhesion of eosinophils to endothelial VCAM-1 in static adhesion assays [15]. Expression of $\alpha_D\beta_2$ in an α_4 integrin-null Jurkat cell line enables these cells to adhere to VCAM-1 under flow conditions [17]. $\alpha_D\beta_2$ interacts with a region of VCAM-1 that overlaps with the binding site for α_4 integrin. As would be expected, the ligand binding site in $\alpha_D\beta_2$ is believed to be within the I domain [17], which is lacking in α_4 integrins.

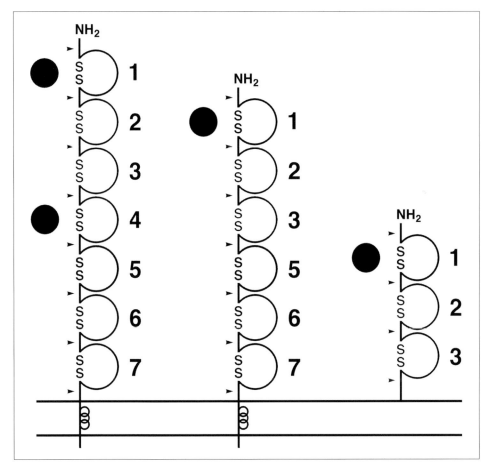

Figure 1
Schematic illustration of different forms of VCAM-1 generated by alternative RNA splicing. Binding sites of α4 integrins to Ig domains 1 and 4 are indicated by solid circles. Loops represent disulfide-linked Ig domains (numbered) and arrowheads point to exon splice junctions.

Recently, a phage display approach identified 'secreted protein acidic and rich in cysteine' (SPARC), a 32-kDa matricellular glycoprotein also known as osteonectin or BM-40, as a ligand of VCAM-1 [18]. *In vitro* experiments demonstrated that the binding of leukocyte-derived SPARC to VCAM-1 expressed by vascular endothelium was necessary for leukocyte transmigration through endothelial monolayers (diapedesis), but not for leukocyte adhesion. SPARC-deficient mice had abnormalities in leukocyte recruitment to the inflamed peritoneum. Previous studies demonstrated that SPARC regulates endothelial cell shape and barrier function and mediates focal adhesion disassembly [19, 20]. Consistent with this, SPARC binding to VCAM-1 induced rearrangement of the actin cytoskeletal in endothelial cells and the appearance of intercellular gaps, presumably through VCAM-1-mediated signal transduction.

The three-dimensional structure of $\alpha_4\beta_1$-binding fragment of VCAM-1 consisting of N-terminal domains 1 and 2 has been solved by X-ray crystallography [21, 22]. The integrin-binding motif (QIDSPL) is located in the loop between β-strands C and D in the lower portion of domain 1. This loop is highly exposed since it projects markedly from one face of the molecule near the contact between domains 1 and 2. The upper part of domain 2 also is involved in binding to $\alpha_4\beta_1$. In contrast, the $\alpha_L\beta_2$ (lymphocyte function-associated antigen-1/LFA-1) binding site on ICAM-1 and -2 is located on a flat surface in the upper part of domain 1, and is complementary to the flat surface of the LFA-1 I domain [23].

Regulation of VCAM-1 expression

VCAM-1 expression is low or absent on unactivated cultured human umbilical vein endothelium. Cell surface expression is induced rapidly by IL-1, TNF-α or bacterial lipopolysaccharide (LPS), reaches maximal levels at 10–24 h, and then it gradually declines [1, 2]. These stimuli activate NF-κB signal transduction (reviewed in [24, 25]), which is essential for induced VCAM-1 expression. The human and murine *vcam1* promoters contain two adjacent consensus elements for binding of NF-κB transcription factor family members [8, 9]. The functional significance of these *cis* elements was revealed through *in vitro* studies (reviewed in [26]). Transient transfection experiments with segments of human *vcam1* 5' flanking sequence coupled to a reporter defined a 258-bp region capable of directing full cytokine-induced expression in endothelial cells. Mutational analysis and further deletion experiments revealed that the integrity of both NF-κB *cis* elements (–73 and –58 bp) was necessary but not sufficient for full cytokine-mediated transcription activation [27, 28]. In TNF-activated endothelial cells, the *vcam1* promoter NF-κB *cis* elements bound primarily p50/p65 heterodimers, and NF-κB interacted with other transcription factors, including interferon regulatory factor (IRF)-1 and Sp1 in transactivating VCAM-1 expression [29, 30]. In fact, interferon (IFN)-α and -γ enhance TNF-induced VCAM-1 mRNA transcription and protein expression in human endothelial cells by an IRF-1-dependent pathway [31], and hyperosmotic stimuli inhibit VCAM-1 expression in cultured endothelial cells *via* effects on IRF-1 expression and activity [32]. Members of the GATA family of transcription factors function to enhance both constitutive and inducible *vcam1* transcription [27, 28]. TNF induces expression of c-Fos and c-Jun, and binding of c-Fos/c-Jun to an activating protein (AP)-1 consensus element also enhances *vcam1* transcription [33]. Ultimately, promoter-bound NF-κB and associated co-activators of transcription, such as cyclic-

AMP responsive element binding (CREB)-binding protein (CBP) and p300, regulate local chromatin remodeling by modulating the extent of histone acetylation. Inhibition of histone deacetylation with trichostatin A suppresses TNF-induced VCAM-1 expression [34].

Oct-1 is a transcription factor that represses the transcription of proinflammatory genes. Octamer binding sites act as silencers, and Oct-1 represses the expression of VCAM-1, E-selectin and von Willebrand factor in unstimulated endothelial cells [35, 36]. TNF overcomes the negative effects of octamers and activates the *vcam1* promoter through NF-κB. Oct-1 interacts with p65, which implicates it as a potential regulator of NF-κB transactivator function. The expression of NF-κB-dependent genes is more pronounced in Oct-1-deficient murine embryonic fibroblasts (relative to wild type), and reintroduction of human Oct-1 abolishes these differences [35]. Oct-1 expression is induced by IL-6, and this may be a mechanism by which NF-κB-dependent gene expression can be selectively reverted to quiescent levels.

IL-4 induces modest levels of VCAM-1 without inducing expression of ICAM-1 or E-selectin, and, in combination with TNF, enhances VCAM-1 and partly suppresses ICAM-1 and E-selectin [37, 38]. The combination of TNF and IL-4 induces a synergistic increase and prolongation of VCAM-1 expression on the cell surface. This results from a combination of transcriptional activation by TNF and stabilization of transcripts by IL-4 [39]. IL-4 stimulation of endothelium induces *vcam1* transcription by an oxidative stress-mediated activation of Sp-1, but not NF-κB, AP-1 or IRF-1 [40].

In addition to cytokines, many other stimuli can induce endothelial cell expression of VCAM-1. These include: binding of ligands to receptor for advanced glycation end products (RAGE) [41, 42]; binding of CD154, the 39-kDa CD40 ligand on activated T cells, to endothelial CD40 [43–45]; and exposure of endothelium to lysophosphatidylcholine (a component of oxidized low density lipoprotein that up-regulates VCAM-1 expression in cultured arterial but not venous endothelium [46]) and apolipoprotein CIII [47], which is enriched in very low-density lipoproteins (VLDL) and is associated with coronary heart disease. Homocysteine is an important risk factor for atherothrombosis, and it up-regulates VCAM-1 expression in cultured human aortic endothelial cells through a cyclooxygenase-dependent mechanism [48]. Similarly, C-reactive protein (CRP) may actively amplify the inflammatory response in atherosclerosis by directly activating endothelial cells. A conversion from a cyclic pentameric structure to monomeric CRP is a prerequisite for endothelial cell activation and expression of chemokines and adhesion molecules, including VCAM-1 [49].

Thrombin also induces VCAM-1 expression, through a time-dependent coordinate binding of p65/RelA and NFATc to a tandem NF-κB element in the *vcam1* promoter [50]. Binding of GATA-2 to a tandem GATA motif contributes to induction of maximal expression [51, 52]. Thrombin and vascular endothelial growth factor induce the expression of Down's syndrome critical region (DSCR)-1 gene, and this negative feedback regulator of calcineurin-NFAT signaling inhibits VCAM-1 expression.

In the last two decades, numerous studies identified pharmacological agents, dietary compounds and physiological pathways that modulate VCAM-1 expression. These studies are far too numerous to review comprehensively, and only highlights are included. IL-1 and TNF-inducible expression of VCAM-1 in endothelium can be inhibited by a variety of compounds that affect NF-κB signal transduction. These include various antioxidants [53, 54] and β-oxa polyunsaturated fatty acids [55], which inhibit the activity of IκB kinases [56], gallates [57], proteasome inhibitors [58], which prevent degradation of IκBα (an inhibitory component of NF-κB) and nitric oxide [59, 60], which augments IκBα expression [56]. Inhibition of the polyol pathway enzyme aldose reductase blocks TNF-induced NF-κB activation and VCAM-1 expression [61]. In contrast, inhibition of flavin-binding proteins and superoxide production or redox-sensitive genes by AGI-1067, a probucol derivative and potent anti-oxidant, results in NF-κB-independent suppression of endothelial VCAM-1 expression [62, 63]. Hormones including estrogen, estrogen metabolite 17-epiestriol and glucocorticoid [64, 65] and exogenous as well as endogenous agonists/ligands of peroxisome proliferator-activated receptor (PPAR)-γ and -α [66–69] also inhibit VCAM-1 expression.

Epidemiological studies suggest that Mediterranean diets and diets rich in certain fish oils are associated with reduced risk of cardiovascular disease. Antioxidant polyphenols in olive oil and red wine, which are abundant in the Mediterranean diet, and docosahexaenoate, an omega-3 fatty acid, reduce cytokine-induced endothelial cell expression of VCAM-1 and other adhesion molecules [64, 70].

A recently identified derivative of a fungus-derived cyclopeptolide acts as a selective inhibitor of VCAM-1 synthesis in endothelial cells [71]. The specificity of this compound (CAM741) is conferred by the signal peptide of VCAM-1. CAM741 represses the VCAM-1 protein biosynthesis by blocking the process of cotranslational translocation from the endoplasmic reticulum (ER) and, instead of translocating to the luminal side of the ER, the VCAM-1 precursor protein is synthesized towards the cytosolic compartment, where it is degraded.

Abundant studies suggest that hemodynamic forces have profound effects on endothelial cell biology and different shear stress profiles can induce unique repertoires of gene expression [72–75]. The molecular mechanisms of how endothelial cells sense shear stress and of how shear modulates endothelial gene expression remains an area of active investigation. Shear stress alters the expression of adhesion molecules in cultured endothelial cells [76, 77] and uniform laminar shear reduces cytokine-induced expression of certain adhesion molecules [78]. Uniform laminar shear inhibits TNF-α-mediated signal transduction in endothelial cells, in part, by stimulating mitogen-activated protein kinases (MAPKs) that phosphorylate transcription factors. Shear inhibits c-Jun N-terminal kinase (JNK) activation through multiple mechanisms. These include stimulation of counter-regulatory MAPKs, such as extracellular signal regulated kinases (ERK) 1/2 and ERK5, and inhibition of apoptosis signal-regulated kinase (ASK) 1, an activator of JNK and p38 [79]. Laminar shear stress decreases thioredoxin-interacting protein (TXNIP) expression in endothelial cells [80]. TXNIP is a stress-responsive protein that inhibits thioredoxin activity, which in turn can inhibit ASK1. Thus, reduction of TXNIP expression results in increased thioredoxin activity and decreased activity of ASK1, JNK and p38.

Uniform laminar shear stress induces the expression of a transcription factor called Kruppel-like factor 2 (KLF2) *via* a MAPK kinase 5 (MEK5)/ERK5/myocyte enhancer factor-2 (MEF2) signaling pathway [81]. KLF2 in turn up-regulates endothelial nitric oxide synthase expression and negatively regulates IL-1β, VCAM-1 and E-selectin expression, possibly through its interactions with cofactors CBP/p300 [82].

In addition to inducible expression on vascular endothelium, VCAM-1 can be expressed constitutively on a variety of cell types. In normal adult tissues, endothelial cell VCAM-1 expression is generally absent, although occasional blood vessels show positive immunohistochemical staining. VCAM-1 is constitutively expressed on some epithelial and monocyte-derived cells, including dendritic cells in lymphoid tissues and skin and Kupffer cells in the liver [83]. Cultured bone marrow stromal cells and skeletal muscle cells during myogenesis also express VCAM-1 [84–86]. In contrast to endothelium, there is relatively high basal expression of VCAM-1 in developing skeletal muscle, and expression is independent of NF-κB activation and cytokine stimulation. The absence of *vcam1* promoter NF-κB *cis* elements does not affect constitutive VCAM-1 expression in C2C12 mouse myoblasts, and a position-specific enhancer located between the TATA box and the transcriptional start site (bp –21 to –5) overrides the effect of other promoter elements and regulates constitutive VCAM-1 expression [87]. In muscle cells, this element binds IRF-2 [88], a member of the IRF family. IRF-2 is not dependent on cytokines for expression or activity, and can act as a repressor in other non-muscle cells. The expression of IRF-2 parallels that of VCAM-1 during mouse skeletal myogenesis.

VCAM-1 and α_4 integrin are expressed by vascular smooth muscle cells during development of the human fetal aorta [89]. VCAM-1 and α_4 integrin were strongly expressed in smooth muscle cells in 10-week-old fetuses, and their expression was dramatically reduced within the 24th week of gestation and disappeared in the adult aortic media. In the adult, VCAM-1 is expressed in a variety of pathological conditions, including atherosclerosis, vascular injury and transplant arteriopathy [90–92]. In cultured smooth muscle cells, differentiation by serum depravation induced VCAM-1 expression [93]. In other *in vitro* studies, a number of cytokines, particularly TNF, could induce VCAM-1 expression; however, there appears to be some variability depending on the study and source of cells [90, 94–96]. It appears that NF-κB signaling and nuclear translocation are not required for induced VCAM-1 expression in smooth muscle cells [97]. The production of TGF-β inhibits expression [98].

VCAM-1 in embryological development and hematopoiesis

The temporal and spatial expression patterns suggest that VCAM-1 and α_4 integrin participate in a variety of physiological and developmental processes. Immunostaining of wild-type mouse embryos at 8.0–8.5 days revealed expression of VCAM-1 on the tip of the allantois and the α_4 integrin on the inner surface of the chorionic plate and targeted disruption of these genes in mouse embryonic stem cells defined a critical developmental role in the formation of the umbilical cord and placenta [99–101]. Mice heterozygous for the disrupted *vcam1* allele developed normally; however, homozygous mice (VCAM-1-knockout mice) die during mid-gestation, indicative of a recessive lethal phenotype [99, 100]. Timed pregnancies revealed an alteration in the expected Mendelian genetic distribution at 10.5 days and at 9.5 days, a distinctive altered phenotype was present in VCAM-1-knockout embryos consisting of failure in fusion of the allantois to the chorionic plate, with resulting hydropic expansion of the allantois, and absence of blood vessel development in the placenta [99]. After fusion of the allantois and chorion in wild-type mice, these structures form the umbilical cord and fetal placenta. In VCAM-1-knockout mice the resulting malformations result in embryonic death and resorption within 1–3 days of virtually all embryos. Less than 3% of VCAM-1$^{-/-}$ embryos survived development, presumably by circumventing the placentation defects, and as adults these VCAM-1$^{-/-}$ mice were healthy, fertile and had organs with normal histological features [99]. Mice deficient in α_4 integrins displayed a morphologically similar chorioallantoic fusion defect [101].

Other roles of VCAM-1 and α_4 integrins in embryological development are not fully defined and controversial. Embryos deficient in α_4 integrin and VCAM-1 exhibited dissolution of the forming epicardium and coronary vessels, leading to pericardial edema and hemorrhage. This defect temporally followed the chorioallantoic fusion defect, and thus, may be a secondary event, at least in VCAM-1-deficient mice, because occasional VCAM-1$^{-/-}$ mice that survived development and VCAM-1 domain 4-deficient mice with hypomorphic VCAM-1 expression (<8% of wild type) had normal pericardium and coronary arteries [99, 102].

The role of VCAM-1 and α_4 integrins in myogenesis is controversial. This is a biphasic process, in which primary myoblasts fuse to form primary myotubes, and then secondary myoblasts align along the primary myotubes and form secondary myotubes that comprise most of the muscle in adult mammals. *In vivo* immunolocalization demonstrated expression of α_4 integrins on primary and secondary myotubes and VCAM-1 on secondary myoblasts. Antibodies to either molecule inhibited myotube formation in culture [85]. However, subsequent studies using α_4 integrin null cells derived from chimeric mice, demonstrated normal myogenesis [103].

VCAM-1 and α_4 integrins participate in hematopoiesis and lymphocyte homing to the bone marrow. Immunolocalization studies demonstrated VCAM-1 expression in the bone marrow stroma reticular and sinusoidal endothelial cells [86, 104–106].

Conditional Cre recombinase–LoxP-mediated deletion of VCAM-1 in mice was accomplished simultaneously by two groups. One employed the IFN-α-induced Cre and the other used Tie2-promoter-driven Cre to delete VCAM-1 efficiently in most tissues except the brain or in endothelial and hematopoietic cells, respectively [107, 108]. These studies revealed reduced immature B cells in the bone marrow, a mild leukocytosis, with elevated immature B cells in the blood, and impaired lymphocyte homing to the bone marrow, but normal composition of lymphoid organs. In addition, a humoral immune response to a T cell-dependent antigen was impaired [107].

In vitro antibody blocking studies suggest that VCAM-1 and α4 integrins mediate adherence of lymphoid precursors and CD34+ hematopoietic progenitor cells to stromal cells, and support proliferation of lymphoid cells in long-term bone marrow cultures [84, 86, 104]. *In vivo* injection of antibodies stimulates release of progenitor cells into the circulation in mice and primates [109–111]. Mice with markedly reduced VCAM-1 expression [102] did not have hematopoietic insufficiencies in myeloid or lymphoid compartments [112]. In Dexter-type long-term bone marrow cultures, VCAM-1-deficient stromal cells supported normal myeloid differentiation and proliferation [112]. In contrast, in mice with a conditional ablation of VCAM-1 through a Tie2-driven Cre transgene, there is an increase in circulating progenitors as a consequence of their ongoing release from bone marrow, a process that was enhanced by splenectomy [113]. A redundancy of adhesion mechanisms, such as binding of stem cells *via* β_1 and β_2 integrins to stromal cell fibronectin and ICAM-1 [114], may account for a mild phenotype during VCAM-1 deficiency. It appears that α_4 integrins have a more important role in hematopoiesis, perhaps because they bind multiple ligands in addition to VCAM-1, such as the CS-1 fragment of fibronectin and osteopontin. IFN-induced deletion of α_4 integrins revealed their role in homeostasis, regeneration, and homing of adult hematopoietic cells [115]. Deletion of α_4 integrin by Tie2-driven Cre resulted in sustained alterations in the biodistribution of hematopoietic progenitor cells and these changes were seen only when α_4-deficient donor cells were transplanted [116]. Studies with α_4 integrin chimeric mice revealed that T cell development was dependent on α4 integrins after birth, but not in the fetus [117]. Both B and T cell precursors, but not monocytes or natural killer cells, require α_4 integrins for normal development in the bone marrow. In peripheral tissues, α_4 integrins participate in T cell homing to Peyer's patches, but not other secondary organs [117, 118].

VCAM-1 and α_4 integrin-mediated leukocyte-endothelial interactions during emigration

The process of leukocyte emigration from blood into tissues can be subdivided into discrete stages, which include tethering, rolling, chemokine-triggered arrest, adhesion strengthening and transendothelial migration or diapedesis [119–122]. VCAM-1 and α_4 integrin, its principal leukocyte ligand, participate in all stages of the adhesion cascade. The first adhesive interaction is tethering. It slows the velocity of leukocytes that contact inflamed endothelium. Multiple sequential tethering events constitute leukocyte rolling. Tethering and rolling are mediated primarily by binding of E-, P- and L-selectins to carbohydrate moieties on proteins such as PSGL-1. Selectin bonds have rapid binding and release rates (high k_{on} and k_{off}) [123, 124], and these transient adhesive interactions are required for leukocyte rolling. VCAM-1 interactions with $\alpha4$ integrins can also promote tethering and rolling of lymphocytes and monocytes at low shear rates [125–127]. In these studies, arrest of rolling cells on VCAM-1 occasionally occurred spontaneously or was triggered by activation of integrins with manganese, phorbol ester and monoclonal antibody TS2/16 [125, 128]. T lymphocytes with spontaneously active α_4 integrins do not roll, but undergo immediate arrest both *in vitro* and *in vivo* [129, 130]. In addition to binding VCAM-1, leukocyte α_4 integrins can engage CD44, which improves lymphocyte homing [131]. In a pathophysiological context, $\alpha_4\beta_1$-dependent monocyte and lymphocyte rolling appears to be relevant to atherosclerosis and inflammation [132–134].

Adhesion mediated by VCAM-1, other immunoglobulin superfamily adhesion molecules and selectins is directly dependent on the cell surface expression levels of these molecules. In contrast, integrin-mediated adhesion is more complex. Circulating leukocytes express abundant integrins on their surface, yet they have relatively low ligand binding capability. To mediate stable adhesion, integrins must be "activated" by appropriate "inside-out" intracellular signals. The overall strength of integrin-mediated cell adhesion is referred to as "avidity", and is dependent on the "affinity" or properties of individual integrin bonds and the "valency" or number of bonds. Affinity is determined by the conformation of integrin molecules. In response to inside-out signaling, integrins undergo conformation changes that include conversion from bent to extended conformations and opening of the ligand-binding pocket, which increase the intrinsic affinity for ligand and result in stable or persistent bonds [135, 136]. Integrin affinity can be assessed by monomeric soluble ligand binding assays and with antibodies that recognize unique conformation-dependent epitopes. The valency of ligand binding depends on integrin expression levels, lateral mobility in the plasma membrane and clustering. Inside-out signaling can up-regulate both affinity and valency of ligand binding, but signaling steps that determine each are poorly understood. Finally, the ability of a relatively small number of integrins to mediate rapid leukocyte adhesive interactions such as rolling or arrest appears to be critically dependent on anchoring to the cortical actin cytoskeleton. Fluctuations or oscillations of individual integrin molecules between low and high affinity states likely accounts for the formation and dissolution of bonds, which is required for complex cellular phenomena such as migration.

Leukocytes rolling in close proximity to the endothelium are exposed to chemokines presented by proteoglycans on the endothelial cell surface [137, 138]. Chemokines including SDF-1, MIP3α and β, MCP-1, GROα and RANTES initiate inside-out signaling cascades *via* G protein-coupled receptors (GPCRs) that up-regulate $\alpha_4\beta_1$ and $\alpha_L\beta_2$ integrin binding to VCAM-1 and ICAM-1 [139–141]. This results in leukocyte arrest, adhesion strengthening and subsequently diapedesis. A rapid up-regulation of α_4 integrin affinity in response to stimulation by chemoattractants and the chemokine SDF-1α mediates the arrest of rolling monocytes and T lymphocytes [142, 143]. This is a sudden and persistent event that is likely mediated by relatively few molecular bonds that must form rapidly (high k_{on}) and persist (slow release rate or low k_{off}). These features are characteristic of high-affinity integrins. Rapid integrin clustering may contribute either to arrest or to stabilization of arrested cells [144], but this is still controversial, because clustering is difficult to assess directly in a sub-second time frame that is relevant to arrest. Furthermore, both α_4 and β_2 integrins are preclustered on the leukocyte surface [126, 145, 146]. Clusters of α_4 integrins are located on the tips of lymphocyte microvilli and monocyte microridges [126, 145].

The signaling cascade initiated by GPCRs that leads to activation of integrin affinity is not fully understood. Based on studies in leukocytes, platelets and transfected cell lines, the key signaling stages include activation of phospholipase C (PLC) and small GTPases, leading to induction of integrin conformational transition through association with actin-binding proteins. Activation of PLC leads to intracellular calcium flux and generation of diacylglycerol. In U937 cells and human monocytes, PLC/calcium signaling is required for induction of high α_4 affinity and arrest [147]. It is likely that different leukocyte types, species and integrins use different intermediate signaling steps, thus providing diverse specificities.

In addition to mediating adhesion, ligand-bound integrins produce intracellular signals, referred to as "outside-in" signaling. These signals regulate a variety of leukocyte functions including adhesion strengthening, spreading, motility, diapedesis and proliferation. For example, binding of α_4 integrins to VCAM-1 induces a signal transduction pathway in leukocytes that up-regulates β_2 integrin avidity and increases the strength of adhesion to ICAM-1 [148, 149]. Furthermore, a role for VCAM-1 and α_4 integrins in diapedesis has been documented by several laboratories using *in vitro* models [150, 151].

Ligand-induced integrin clustering and allosteric conformational changes contribute to the initiation of outside-in signaling. Since integrin cytoplasmic domains have no enzymatic functions, activation of signaling cascades requires interactions with cytoplasmic signaling adaptors such as talin, filamin and α actinin, usually *via* the β chain. The cytoplasmic domain of the α_4 integrin is unique because it specifically binds paxillin, a 68-kDa signaling adaptor molecule that contains LIM protein-protein interaction motifs and LD motifs that mediate protein-protein interactions [152, 153]. Paxillin binding to α_4 is regulated by the phosphorylation of the α_4 cytoplasmic tail and paxillin binds only when Ser988 is de-phosphorylated [154]. This occurs when leukocyte $\alpha_4\beta_1$ integrins are in a high-affinity conformation either constitutively, such as during lymphoid development in the thymus and bone marrow, or induced transiently during recruitment by inside-out signaling *via* GPCRs [155]. The implication of these observations is that stable binding of ligand to high-affinity $\alpha_4\beta_1$ integrins initiates outside-in signaling through adaptor proteins associated with the β chain, as well as paxillin associated with the α_4 tail. The role of α_4 integrin-paxillin in outside-in signaling is relevant to leukocyte migration [156]. Topographically specific integrin phosphorylation can control cell migration and polarization by spatial segregation of adaptor protein binding. Paxillin and phospho-α_4 were observed in distinct clusters at the leading edge of migrating cells, whereas unphosphorylated α_4 and paxillin colocalized along the lateral edges of those cells [157]. Through association with paxillin, unphosphorylated α_4 integrins can activate distinct signaling pathways. For example, paxillin recruits an ADP-ribosylation factor GTPase-activating protein (Arf-GAP) *via* its LD4 domain, which decreases Arf activity and inhibits Rac, thereby restricting Rac activation to the leading edge of migrating cells [158]. Recently, mice were developed bearing a Y991A mutation in the α_4 cytoplasmic tail, which blocks paxillin binding. Unlike α_4 null mice, Y991A α_4 mice were viable and had impaired recruitment of mononuclear leukocytes to inflammation in the peritoneal cavity [159]. This may be due to a defect in leukocyte migration following adhesion to endothelium or even reduced adhesion. The latter notion is based on a report that paxillin binding to α_4 integrins in Jurkat cells mediates association with the actin cytoskeleton that is required for adhesion strengthening under conditions of mechanical strain [160]. Unlike circulating T cells, Jurkat cells express α_4 integrins that are constitutively activated and associated with paxillin [155], and it remains to be determined whether the above observation applies to situations where integrin affinity is transiently activated by chemokine GPCR signaling.

Coordinated phosphorylation and dephosphorylation of α_4 integrin and the consequent dissociation and association of paxillin is critical for integrin function [161]. Protein kinase A (PKA) phosphorylates Ser988 in the α_4 integrin cytoplasmic tail *in vitro* at a PKA consensus site [154, 157]. Inhibition of PKA activity abrogates phosphorylation of α_4 integrin at the leading edge of migrating cells and inhibits migration [157]. Recently, α_4 integrin was identified as a novel A-kinase anchoring protein. Association of PKA with α_4 integrin at the plasma membrane was required for phosphorylation of α_4 integrin at the leading edge [162]. Paxillin association requires de-phosphorylation of Ser988 in the α_4 integrin cytoplasmic tail. We have recently demonstrated that inhibition of protein phosphatase 2B (calcineurin, PPP3), a Ca^{2+}-calmodulin-dependent serine/threonine kinase, prevents fMLP-induced association of paxillin with α_4 integrin in U937 cells, a monocytic cell line [163]. Calcineurin inhibition abrogates α_4 integrin/VCAM-1-dependent chemotaxis, but has minimal or no effect on arrest or adhesion strengthening on VCAM-1.

VCAM-1 expression and function in atherosclerosis

Hypercholesterolemia promotes the infiltration and retention of low-density lipo-proteins (LDL) in the artery wall through specific interaction with proteoglycans [164], and this is an important pathogenic stimulus for the initiation and progression of atherosclerosis. LDL retained in the intima undergoes aggregation, oxidation and/or enzymatic modification, which leads to release of bioactive phospholipids that can activate endothelial cells and incite an inflammatory response [165, 166]. Endothelial cell activation during atherogenesis was initially observed in hypercholesterolemic rabbit models [167]. VCAM-1 expression was selectively up-regulated on endothelial cells overlying early atherosclerotic lesions in the aorta and was not observed in regions devoid of atherosclerotic lesions. These data suggested that the endothelium that lines the arterial luminal surface actively regulates the adherence and recruitment of blood mononuclear leukocytes to the intima, which is one of the earliest events observed in atherogenesis and a necessary step for macrophage accumulation in the intima and transformation into foam cells.

The expression of VCAM-1, as well as ICAM-1 and E-selectin was investigated in hypercholesterolemic rabbits, and subsequently LDLR$^{-/-}$ and ApoE$^{-/-}$ murine models of atherosclerosis [168–171]. Northern blot analysis demonstrated increased VCAM-1 and ICAM-1, but not E-selectin steady state mRNA levels in hypercholesterolemic mouse and rabbit aortas, which correlated with the extent of atherosclerotic lesion formation, as was determined by staining intimal lipid deposits with oil red O [170]. In small lesions, VCAM-1 and ICAM-1 were expressed predominantly by endothelial cells, whereas in large foam cell-rich lesions many intimal cells expressed these molecules. It is likely that expression by intimal cells accounted for increased VCAM-1 and ICAM-1 steady state mRNA levels in Northern blots. VCAM-1 was also expressed by medial smooth muscle cells adjacent to lesions [90]. This phenotypic change may occur in activated smooth muscle cells or cells in the process of migration to the intima.

In humans, several groups detected the expression of inducible adhesion molecules in advanced atherosclerotic plaques obtained at autopsy or from hearts of transplant recipients [172–177]. Coronary artery plaques displayed focal VCAM-1 expression in luminal endothelial cells, usually in association with inflammatory infiltrates [177]. Focal endothelial VCAM-1 expression was also found in uninvolved vessels with diffuse intimal thickening. VCAM-1 was expressed by some smooth muscle cells and macrophages and by endothelial cells of neovasculature at the base of plaques.

During the last two decades, many lines of transgenic mice bearing a deficiency of an adhesion molecule or chemokine have been developed, and the contribution of these adhesion molecules in atherogenesis has been investigated (reviewed in [178]). The ApoE$^{-/-}$ or LDLR$^{-/-}$ models have been studied most frequently and the extent of atherosclerotic lesion formation was assessed by estimating the volume of lesions in the aortic root or determining the surface area of the aorta occupied by lesions. The histological features of lesions and their cellular composition were also determined. As was mentioned above, deficiency of VCAM-1 or α_4 integrin results in embryonic lethality. Thus, we produced VCAM-1 domain 4-deficient (VCAM-1 D4D) mice that express a mutant form of VCAM-1 at markedly reduced levels. Approximately 25% of VCAM-1 D4D mice were viable. VCAM-1$^{D4D/D4D}$ mice express a six-Ig domain form of VCAM-1 and lack Ig domain 4, which contains an α_4 integrin binding site (Fig. 1). Thus, the six-Ig domain form has only one ligand binding site unlike two found in wild-type mice. Also, the expression levels of D4D VCAM-1 are markedly reduced or hypomorphic (3–8% of wild type). VCAM-1 D4D mice were bred into the LDLR$^{-/-}$ background and were fed a 1.25% cholesterol-enriched diet for 8 weeks. *En face* analysis of oil red O-stained aortas revealed reduced lesion area compared to VCAM-1$^{+/+}$ mice [102]. In contrast, significant changes in lesion area were not found in parallel experiments using ICAM-1-deficient mice [102]. VCAM-1 D4D mice were also bred into the Apo E$^{-/-}$ background and lesion formation in the aortic root was quantified. These studies revealed a VCAM-1 gene dosage effect on aortic root atherosclerotic lesions at 16 weeks of age. The aortic root lesion area was reduced by 84% and 56% in VCAM-1$^{D4D/D4D}$ and VCAM-1$^{+/D4D}$ mice, respectively, and lesions in VCAM-1$^{D4D/D4D}$ mice were limited to very small nascent fatty streaks [179]. Together these studies suggest that VCAM-1 has a critical role in atherogenesis. Based on these data, one would expect that the α_4 integrin should have an important function in atherosclerosis. This has been difficult to test, since mice deficient in α_4 integrin are not viable; however, supporting data have been generated using infusion of α_4 integrin blocking peptide, which reduced lesion formation in mice [180]. VCAM-1 and α_4 integrin were also key mediators of U937 cell rolling and adhesion in an *ex vivo* perfusion model of the carotid artery bifurcation harvested from ApoE$^{-/-}$ mice [132].

VCAM-1 in inflammatory diseases

VCAM-1 expression is induced in many acute and chronic inflammatory conditions and other pathological processes. Expression levels are actively regulated by proinflammatory cytokines such as IL-1β, IFN-γ and TNF-α. The conditions associated with alterations in VCAM-1 expression are too numerous to review comprehensively and only a few examples are discussed.

Perhaps the most well-studied role for VCAM-1 and its leukocyte ligand α_4 integrin in human disease is in multiple sclerosis (MS). VCAM-1 is absent or expressed at very low levels on microvascular endothelium in the non-inflamed central nervous system (CNS). In MS, and its experimental animal model experimental allergic encephalomyelitis (EAE), VCAM-1 expression is up-regulated on CNS microvascular endothelium [181]. VCAM-1 is also seen on microglial cells and astrocytes in the inflamed, but not normal CNS. VCAM-1 expression in the microvasculature of the mouse spinal cord is modulated by TNF-α; anti-TNF-α therapy decreases CNS VCAM-1 expression and leukocyte infiltration in EAE [182]. Additionally, antibodies to VCAM-1 delay the onset of EAE [183]. Humanized antibody to α_4 integrin (natalizumab) has been approved for use in patients with relapsing-remitting MS. In clinical trials, patients on natalizumab therapy showed significant improvement in clinical parameters and disease activity measured by magnetic resonance imaging (MRI) (reviewed in chapter 5).

Migration to the inflamed rheumatoid synovium is dependent on α_4/VCAM-1 adhesive interactions. VCAM-1 expression is elevated in synovial tissue in rheumatoid arthritis compared to control or osteoarthritic synovia [184, 185]. This expression is seen on synoviocytes, stromal cells and cells of the vascular wall [184, 186]. Proinflammatory cytokines TNF-α, IFN-γ and IL-1 increase expression of VCAM-1 on synoviocytes *in vitro* [186] and anti-TNF-α treatment resulted in significant clinical improvement in patients with rheumatoid arthritis [187].

Ischemia/reperfusion injury and cellular rejection in solid organ transplantation are characterized by adhesion molecule up-regulation on the graft endothelium. Graft rejection is a significant consideration for allogenic organ transplantation. Endothelial expression of VCAM-1 correlates with the severity of cellular infiltration of grafts and, in animal models, anti-adhesion molecule therapy reduces acute rejection and prolongs graft survival. Immunosuppressive agents such as FK778 have been shown to reduce early up-regulation of adhesion molecules, abolish leukocyte infiltration and resulted in prolonged cardiac allograft survival [188].

Contact hypersensitivity to allergens such as nickel chloride and cobalt chloride is common in industrialized countries and is associated with induction of gene transcription of adhesion molecules including VCAM-1. These agents induce NF-κB activity that subsequently modulates transcription of cytokine and adhesion molecule genes and leukocyte infiltration [189, 190]. In mice, increased expression of VCAM-1 is observed at the onset of oxazolone contact hypersensitivity and is primarily due to immune-dependent local release of TNF-α [191].

VCAM-1 is also induced on vascular endothelium in asthma, colitis, autoimmune thyroiditis, acute appendicitis, acute diverticulitis, sarcoidosis, a variety of dermatoses, various vasculitides and many other disorders.

Soluble VCAM-1 a marker of disease activity

In addition to expression on the cell surface, VCAM-1 and indeed other adhesion molecules are found in soluble form in plasma and other body fluids. Soluble adhesion molecules most likely arise by proteolytic cleavage of the transmembrane form, and not *de novo* synthesis of a secreted molecule. The observed molecular weight of soluble VCAM-1 (sVCAM-1) is similar to that expected for the extracellular domain of the membrane bound form. Phorbol esters increase the levels of sVCAM-1 released from cultured cells, while decreasing expression of the transmembrane form [192, 193]. Shedding of the ectodomain of membrane proteins is often mediated by the Zn^{2+}-dependent protease superfamily, which includes matrix metalloproteases (MMPs) and disintegrin and metalloproteinases (ADAMs). MMP activity in cerebral microvascular endothelial cells correlates with VCAM-1 expression; inhibition of MMP activity leads to an increase in cell surface VCAM-1 and a decrease in sVCAM-1 [194]. ADAM-17 is a sheddase for a variety of proteins, including TNF and IL-1 receptors, L-selectin, fractalkine and VCAM-1 [193]. ADAM-17 proteolytic activity and cleavage of VCAM-1 is regulated by tissue inhibitor of metalloproteinase-3 (TIMP-3) [195]. ADAM-8 also mediates VCAM-1 shedding and has been implicated in lung disorders and asthma [196, 197]. Neutrophil elastase can also cleave VCAM-1 from the cell surface; however, the cleaved form produced in this manner is a 65-kDa not the 100-kDa ectodomain fragment that is found in serum and produced by ADAM activity [198].

Increased levels of soluble adhesion molecules, including VCAM-1, correlate with a variety of inflammatory disorders. The release of soluble forms of adhesion molecules may be a consequence of cell damage arising from cytokine production in inflammatory diseases. Cultured endothelial cells that have been activated by cytokines release soluble adhesion molecules into the culture media. Thus, soluble forms of adhesion molecules may have a pathological role, but are also useful as a marker for disease activity and endothelial cell injury. Elevated circulating adhesion molecules may also compete with membrane-bound forms and limit the adhesive interactions between leukocytes and endothelium.

In MS patients, serum levels of sVCAM-1 are elevated and correlate with clinical activity and the number of enhancing (active) lesions in contrast-enhanced MRI [199–202]. sVCAM-1 is also detected in the cerebrospinal fluid of patients with active MS [201–203]. sVCAM-1 is, therefore, a useful marker for inflammatory activity in CNS inflammatory diseases. Interestingly, levels of sVCAM-1 are modified by therapeutics; IFN-β reduces clinical exacerbations and disease activity while increasing sVCAM-1 in the serum, suggesting increased shedding of the membrane-associated VCAM-1 that is up-regulated in inflammatory lesions [204, 205].

Plasma sVCAM-1 is also elevated in patients with inflammatory bowel disease, including Crohn's disease and ulcerative colitis [206, 207], in individuals with rheumatoid arthritis [208] and patients with type 1 diabetes [209]. Both serum and urinary sVCAM-1 levels are elevated in patients with systemic lupus erythematosus and correlate with disease exacerbations [210, 211]. Endothelial VCAM-1 plays a significant role in the development of atherosclerosis as described above. Serum concentrations of sVCAM-1 also correlate with the extent of atherosclerosis (measured by angiography) and may be useful to determine the stage of atherosclerosis [212].

Graft rejection is the major obstacle to long-term organ transplantation and noninvasive methods to evaluate acute rejection will improve diagnosis and prevention of rejection. Elevated levels of sVCAM-1 have been detected in patients with cardiac transplant vasculopathy, acute renal allograft rejection and liver allograft rejection [213–215]. Furthermore, levels of sVCAM-1 (and sICAM-1) decrease following successful treatment for rejection [213]. Thus, measurement of circulating adhesion molecules may predict rejection episodes and prompt immediate treatment.

sVCAM-1 has prognostic significance in patients with carcinoma. sVCAM-1 is significantly elevated in patients with advanced breast cancer compared with controls. [216, 217]. In patients with less advanced disease (stage 2) elevated sVCAM-1 is predictive of decreased survival [216]. Elevated serum sVCAM-1 has also been described in gastric cancer [218], colorectal cancer [219], bladder cancer [220], leukemia [221] and advanced non-Hodgkins lymphoma [222]. Significant correlations between sVCAM-1 levels and tumor staging and development of metastases were also observed. Circulating sVCAM-1 levels may also have diagnostic potential for pre-clinical cancer. In a cohort database of patients followed for 14 years, elevated serum sVCAM-1 was observed in patients that developed cancer during the follow-up period [223].

Concluding remarks

In the two decades since its discovery and identification, significant strides have been made towards understanding the regulation of VCAM-1 expression and its function. VCAM-1 plays an important role in development, hematopoiesis, leukocyte recruitment and the pathogenesis of many inflammatory diseases. It is a potential therapeutic target as well as a marker of disease activity. Its main ligand, the α_4 integrin, must be activated and undergo changes in conformation, clustering and association with other proteins to mediate stable adhesion of leukocytes to VCAM-1. The signaling pathways that lead to integrin activation are not fully understood and this remains an important area of ongoing research.

References

1   Rice GE, Bevilacqua MP (1989) An inducible endothelial cell surface glycoprotein mediates melanoma adhesion. *Science* 246: 1303–1306
2   Carlos TM, Schwartz BR, Kovach NL, Yee E, Rosa M, Osborn L, Chi-Rosso G, Newman B, Lobb R, Harlan JM (1990) Vascular cell adhesion molecule-1 mediates lymphocyte adherence to cytokine-activated cultured human endothelial cells [published erratum appears in *Blood* (1990) 76: 2420]. *Blood* 76: 965–970
3   Osborn L, Hession C, Tizard R, Vassallo C, Luhowskyj S, Chi-Rosso G, Lobb R (1989) Direct expression cloning of vascular cell adhesion molecule 1, a cytokine-induced endothelial protein that binds to lymphocytes. *Cell* 59: 1203–1211

4   Pober JS, Cotran RS (1990) Cytokines and endothelial cell biology. *Physiol Rev* 70: 427–451

5   Cybulsky MI, Fries JW, Williams AJ, Sultan P, Davis VM, Gimbrone MA Jr, Collins T (1991) Alternative splicing of human VCAM-1 in activated vascular endothelium. *Am J Pathol* 138: 815–820

6   Hession C, Tizard R, Vassallo C, Schiffer SB, Goff D, Moy P, Chi-Rosso G, Luhowskyj S, Lobb R, Osborn L (1991) Cloning of an alternate form of vascular cell adhesion molecule-1 (VCAM1). *J Biol Chem* 266: 6682–6685

7   Polte T, Newman W, Raghunathan G, Gopal TV (1991) Structural and functional studies of full-length vascular cell adhesion molecule-1: internal duplication and homology to several adhesion proteins. *DNA Cell Biol* 10: 349–357

8   Cybulsky MI, Fries JW, Williams AJ, Sultan P, Eddy R, Byers M, Shows T, Gimbrone MA Jr, Collins T (1991) Gene structure, chromosomal location, and basis for alternative mRNA splicing of the human VCAM1 gene. *Proc Natl Acad Sci USA* 88: 7859–7863

9   Cybulsky MI, Allan-Motamed M, Collins T (1993) Structure of the murine VCAM1 gene. *Genomics* 18: 387–391

10   Osborn L, Vassallo C, Browning BG, Tizard R, Haskard DO, Benjamin CD, Dougas I, Kirchhausen T (1994) Arrangement of domains, and amino acid residues required for binding of vascular cell adhesion molecule-1 to its counter-receptor VLA-4 (alpha 4 beta 1). *J Cell Biol* 124: 601–608

11   Vonderheide RH, Tedder TF, Springer TA, Staunton DE (1994) Residues within a conserved amino acid motif of domains 1 and 4 of VCAM-1 are required for binding to VLA-4. *J Cell Biol* 125: 215–222

12   Moy P, Lobb R, Tizard R, Olson D, Hession C (1993) Cloning of an inflammation-specific phosphatidyl inositol-linked form of murine vascular cell adhesion molecule-1. *J Biol Chem* 268: 8835–8841

13   Terry RW, Kwee L, Levine JF, Labow MA (1993) Cytokine induction of an alternatively spliced murine vascular cell adhesion molecule (VCAM) mRNA encoding a glycosylphosphatidylinositol-anchored VCAM protein. *Proc Natl Acad Sci USA* 90: 5919–5923

14   Elices MJ, Osborn L, Takada Y, Crouse C, Luhowskyj S, Hemler ME, Lobb RR (1990) VCAM-1 on activated endothelium interacts with the leukocyte integrin VLA-4 at a site distinct from the VLA-4/fibronectin binding site. *Cell* 60: 577–584

15   Grayson MH, Van der Vieren M, Sterbinsky SA, Michael Gallatin W, Hoffman PA, Staunton DE, Bochner BS (1998) αdβ2 integrin is expressed on human eosinophils and functions as an alternative ligand for vascular cell adhesion molecule 1 (VCAM-1). *J Exp Med* 188: 2187–2191

16   Taooka Y, Chen J, Yednock T, Sheppard D (1999) The integrin α9β1 mediates adhesion to activated endothelial cells and transendothelial neutrophil migration through interaction with vascular cell adhesion molecule-1. *J Cell Biol* 145: 413–420

17   Van der Vieren M, Crowe DT, Hoekstra D, Vazeux R, Hoffman PA, Grayson MH, Bochner BS, Gallatin WM, Staunton DE (1999) The leukocyte integrin αdβ2 binds

VCAM-1: evidence for a binding interface between I domain and VCAM-1. *J Immunol* 163: 1984–1990

18 Kelly KA, Allport JR, Yu AM, Sinh S, Sage EH, Gerszten RE, Weissleder R (2007) SPARC is a VCAM-1 counter-ligand that mediates leukocyte transmigration. *J Leukoc Biol* 81: 748–756

19 Goldblum SE, Ding X, Funk SE, Sage EH (1994) SPARC (secreted protein acidic and rich in cysteine) regulates endothelial cell shape and barrier function. *Proc Natl Acad Sci USA* 91: 3448–3452

20 Murphy-Ullrich JE, Lane TF, Pallero MA, Sage EH (1995) SPARC mediates focal adhesion disassembly in endothelial cells through a follistatin-like region and the Ca(2+)-binding EF-hand. *J Cell Biochem* 57: 341–350

21 Jones EY, Harlos K, Bottomley MJ, Robinson RC, Driscoll PC, Edwards RM, Clements JM, Dudgeon TJ, Stuart DI (1995) Crystal structure of an integrin-binding fragment of vascular cell adhesion molecule-1 at 1.8 Å resolution. *Nature* 373: 539–544

22 Wang JH, Pepinsky RB, Stehle T, Liu JH, Karpusas M, Browning B, Osborn L (1995) The crystal structure of an N-terminal two-domain fragment of vascular cell adhesion molecule 1 (VCAM-1): a cyclic peptide based on the domain 1 C-D loop can inhibit VCAM-1-alpha 4 integrin interaction. *Proc Natl Acad Sci USA* 92: 5714–5718

23 Wang J, Springer TA (1998) Structural specializations of immunoglobulin superfamily members for adhesion to integrins and viruses. *Immunol Rev* 163: 197–215

24 Hoffmann A, Baltimore D (2006) Circuitry of nuclear factor kappaB signaling. *Immunol Rev* 210: 171–186

25 Perkins ND (2007) Integrating cell-signalling pathways with NF-kappaB and IKK function. *Nat Rev Mol Cell Biol* 8: 49–62

26 Collins T, Read MA, Neish AS, Whitley MZ, Thanos D, Maniatis T (1995) Transcriptional regulation of endothelial cell adhesion molecules: NF-kappa B and cytokine-inducible enhancers. *FASEB J* 9: 899–909

27 Iademarco MF, McQuillan JJ, Rosen GD, Dean DC (1992) Characterization of the promoter for vascular cell adhesion molecule-1 (VCAM-1). *J Biol Chem* 267: 16323–16329

28 Neish AS, Williams AJ, Palmer HJ, Whitley MZ, Collins T (1992) Functional analysis of the human vascular cell adhesion molecule 1 promoter. *J Exp Med* 176: 1583–1593

29 Neish AS, Khachigian LM, Park A, Baichwal VR, Collins T (1995) Sp1 is a component of the cytokine-inducible enhancer in the promoter of vascular cell adhesion molecule-1. *J Biol Chem* 270: 28903–28909

30 Neish AS, Read MA, Thanos D, Pine R, Maniatis T, Collins T (1995) Endothelial interferon regulatory factor 1 cooperates with NF-kappa B as a transcriptional activator of vascular cell adhesion molecule 1. *Mol Cell Biol* 15: 2558–2569

31 Lechleitner S, Gille J, Johnson DR, Petzelbauer P (1998) Interferon enhances tumor necrosis factor-induced vascular cell adhesion molecule 1 (CD106) expression in human endothelial cells by an interferon-related factor 1-dependent pathway. *J Exp Med* 187: 2023–2030

32   Ochi H, Masuda J, Gimbrone MA (2002) Hyperosmotic stimuli inhibit VCAM-1 expression in cultured endothelial cells *via* effects on interferon regulatory factor–1 expression and activity. *Eur J Immunol* 32: 1821–1831

33   Ahmad M, Theofanidis P, Medford RM (1998) Role of activating protein-1 in the regulation of the vascular cell adhesion molecule-1 gene expression by tumor necrosis factor-alpha. *J Biol Chem* 273: 4616–4621

34   Inoue K, Kobayashi M, Yano K, Miura M, Izumi A, Mataki C, Doi T, Hamakubo T, Reid PC, Hume DA et al (2006) Histone deacetylase inhibitor reduces monocyte adhesion to endothelium through the suppression of vascular cell adhesion molecule-1 expression. *Arterioscler Thromb Vasc Biol* 26: 2652–2659

35   dela Paz NG, Simeonidis S, Leo C, Rose DW, Collins T (2007) Regulation of NF-kappaB-dependent gene expression by the POU domain transcription factor Oct-1. *J Biol Chem* 282: 8424–8434

36   Schwachtgen JL, Remacle JE, Janel N, Brys R, Huylebroeck D, Meyer D, Kerbiriou-Nabias D (1998) Oct-1 is involved in the transcriptional repression of the von Willebrand factor gene promoter. *Blood* 92: 1247–1258

37   Masinovsky B, Urdal D, Gallatin WM (1990) IL-4 acts synergistically with IL-1 beta to promote lymphocyte adhesion to microvascular endothelium by induction of vascular cell adhesion molecule-1. *J Immunol* 145: 2886–2895

38   Thornhill MH, Haskard DO (1990) IL-4 regulates endothelial cell activation by IL-1, tumor necrosis factor, or IFN-gamma. *J Immunol* 145: 865–872

39   Iademarco MF, Barks JL, Dean DC (1995) Regulation of vascular cell adhesion molecule-1 expression by IL-4 and TNF-alpha in cultured endothelial cells. *J Clin Invest* 95: 264–271

40   Lee YW, Kuhn H, Hennig B, Neish AS, Toborek M (2001) IL-4-induced oxidative stress upregulates VCAM-1 gene expression in human endothelial cells. *J Mol Cell Cardiol* 33: 83–94

41   Schmidt AM, Hori O, Cao R, Yan SD, Brett J, Wautier JL, Ogawa S, Kuwabara K, Matsumoto M, Stern D (1996) RAGE: a novel cellular receptor for advanced glycation end products. *Diabetes* 45: S77–80

42   Cines DB, Pollak ES, Buck CA, Loscalzo J, Zimmerman GA, McEver RP, Pober JS, Wick TM, Konkle BA, Schwartz BS et al (1998) Endothelial cells in physiology and in the pathophysiology of vascular disorders. *Blood* 91: 3527–3561

43   Hollenbaugh D, Mischel-Petty N, Edwards CP, Simon JC, Denfeld RW, Kiener PA, Aruffo A 1995 Expression of functional CD40 by vascular endothelial cells. *J Exp Med* 182: 33–40

44   Karmann K, Hughes CC, Schechner J, Fanslow WC, Pober JS (1995) CD40 on human endothelial cells: inducibility by cytokines and functional regulation of adhesion molecule expression. *Proc Natl Acad Sci USA* 92: 4342–4346

45   Yellin MJ, Brett J, Baum D, Matsushima A, Szabolcs M, Stern D, Chess L (1995) Functional interactions of T cells with endothelial cells: the role of CD40L-CD40-mediated signals. *J Exp Med* 182: 1857–1864

46 Kume N, Cybulsky MI, Gimbrone MA Jr (1992) Lysophosphatidylcholine, a component of atherogenic lipoproteins, induces mononuclear leukocyte adhesion molecules in cultured human and rabbit arterial endothelial cells. *J Clin Invest* 90: 1138–1144

47 Kawakami A, Aikawa M, Alcaide P, Luscinskas FW, Libby P, Sacks FM (2006) Apolipoprotein CIII induces expression of vascular cell adhesion molecule-1 in vascular endothelial cells and increases adhesion of monocytic cells. *Circulation* 114: 681–687

48 Silverman MD, Tumuluri RJ, Davis M, Lopez G, Rosenbaum JT, Lelkes PI (2002) Homocysteine upregulates vascular cell adhesion molecule-1 expression in cultured human aortic endothelial cells and enhances monocyte adhesion. *Arterioscler Thromb Vasc Biol* 22: 587–592

49 Khreiss T, Jozsef L, Potempa LA, Filep JG (2004) Conformational rearrangement in C-reactive protein is required for proinflammatory actions on human endothelial cells. *Circulation* 109: 2016–2022

50 Minami T, Miura M, Aird WC, Kodama T (2006) Thrombin-induced autoinhibitory factor, Down syndrome critical region-1, attenuates NFAT-dependent vascular cell adhesion molecule-1 expression and inflammation in the endothelium. *J Biol Chem* 281: 20503–20520

51 Minami T, Aird WC (2001) Thrombin stimulation of the vascular cell adhesion molecule-1 promoter in endothelial cells is mediated by tandem nuclear factor-kappa B and GATA motifs. *J Biol Chem* 276: 47632–47641

52 Minami T, Abid MR, Zhang J, King G, Kodama T, Aird WC (2003) Thrombin stimulation of vascular adhesion molecule-1 in endothelial cells is mediated by protein kinase C (PKC)-delta-NF-kappa B and PKC-zeta-GATA signaling pathways. *J Biol Chem* 278: 6976–6984

53 Marui N, Offermann MK, Swerlick R, Kunsch C, Rosen CA, Ahmad M, Alexander RW, Medford RM (1993) Vascular cell adhesion molecule-1 (VCAM-1) gene transcription and expression are regulated through an antioxidant-sensitive mechanism in human vascular endothelial cells. *J Clin Invest* 92: 1866–1874

54 Weber C, Erl W, Pietsch A, Strobe M, Ziegler-Heitbrock HW, Weber PC (1994) Antioxidants inhibit monocyte adhesion by suppressing nuclear factor-kappa B mobilization and induction of vascular cell adhesion molecule-1 in endothelial cells stimulated to generate radicals. *Arterioscler Thromb* 14: 1665–1673

55 Ferrante A, Robinson BS, Singh H, Jersmann HP, Ferrante JV, Huang ZH, Trout NA, Pitt MJ, Rathjen DA, Easton CJ et al (2006) A novel beta-oxa polyunsaturated fatty acid downregulates the activation of the IkappaB kinase/nuclear factor kappaB pathway, inhibits expression of endothelial cell adhesion molecules, and depresses inflammation. *Circ Res* 99: 34–41

56 Spiecker M, Peng HB, Liao JK (1997) Inhibition of endothelial vascular cell adhesion molecule-1 expression by nitric oxide involves the induction and nuclear translocation of IkappaBalpha. *J Biol Chem* 272: 30969–30974

57 Murase T, Kume N, Hase T, Shibuya Y, Nishizawa Y, Tokimitsu I, Kita T (1999) Gallates inhibit cytokine-induced nuclear translocation of NF-kappaB and expression of

leukocyte adhesion molecules in vascular endothelial cells. *Arterioscler Thromb Vasc Biol* 19: 1412–1420

58 Read MA, Neish AS, Luscinskas FW, Palombella VJ, Maniatis T, Collins T (1995) The proteasome pathway is required for cytokine-induced endothelial-leukocyte adhesion molecule expression. *Immunity* 2: 493–506

59 De Caterina R, Libby P, Peng HB, Thannickal VJ, Rajavashisth TB, Gimbrone MA Jr, Shin WS, Liao JK (1995) Nitric oxide decreases cytokine-induced endothelial activation. Nitric oxide selectively reduces endothelial expression of adhesion molecules and proinflammatory cytokines. *J Clin Invest* 96: 60–68

60 Khan BV, Harrison DG, Olbrych MT, Alexander RW, Medford RM (1996) Nitric oxide regulates vascular cell adhesion molecule 1 gene expression and redox-sensitive transcriptional events in human vascular endothelial cells. *Proc Natl Acad Sci USA* 93: 9114–9119

61 Ramana KV, Bhatnagar A, Srivastava SK (2004) Inhibition of aldose reductase attenuates TNF-alpha-induced expression of adhesion molecules in endothelial cells. *FASEB J* 18: 1209–1218

62 Tummala PE, Chen XL, Medford RM (2000) NF- kappa B independent suppression of endothelial vascular cell adhesion molecule-1 and intercellular adhesion molecule-1 gene expression by inhibition of flavin binding proteins and superoxide production. *J Mol Cell Cardiol* 32: 1499–1508

63 Kunsch C, Luchoomun J, Grey JY, Olliff LK, Saint LB, Arrendale RF, Wasserman MA, Saxena U, Medford RM (2004) Selective inhibition of endothelial and monocyte redox-sensitive genes by AGI–1067: a novel antioxidant and anti-inflammatory agent. *J Pharmacol Exp Ther* 308: 820–829

64 Carluccio MA, Siculella L, Ancora MA, Massaro M, Scoditti E, Storelli C, Visioli F, Distante A, De Caterina R (2003) Olive oil and red wine antioxidant polyphenols inhibit endothelial activation: antiatherogenic properties of Mediterranean diet phytochemicals. *Arterioscler Thromb Vasc Biol* 23: 622–629

65 Mukherjee TK, Nathan L, Dinh H, Reddy ST, Chaudhuri G (2003) 17-epiestriol, an estrogen metabolite, is more potent than estradiol in inhibiting vascular cell adhesion molecule 1 (VCAM-1) mRNA expression. *J Biol Chem* 278: 11746–11752

66 Pasceri V, Wu HD, Willerson JT, Yeh ET (2000) Modulation of vascular inflammation *in vitro* and *in vivo* by peroxisome proliferator-activated receptor-gamma activators. *Circulation* 101: 235–238

67 Wang N, Verna L, Chen NG, Chen J, Li H, Forman BM, Stemerman MB (2002) Constitutive activation of peroxisome proliferator-activated receptor-gamma suppresses pro-inflammatory adhesion molecules in human vascular endothelial cells. *J Biol Chem* 277: 34176–34181

68 Ahmed W, Orasanu G, Nehra V, Asatryan L, Rader DJ, Ziouzenkova O, Plutzky J (2006) High-density lipoprotein hydrolysis by endothelial lipase activates PPARalpha: a candidate mechanism for high-density lipoprotein-mediated repression of leukocyte adhesion. *Circ Res* 98: 490–498

69   Ziouzenkova O, Perrey S, Asatryan L, Hwang J, MacNaul KL, Moller DE, Rader DJ, Sevanian A, Zechner R, Hoefler G, Plutzky J (2003) Lipolysis of triglyceride-rich lipoproteins generates PPAR ligands: evidence for an antiinflammatory role for lipoprotein lipase. *Proc Natl Acad Sci USA* 100: 2730–2735

70   De Caterina R, Cybulsky MI, Clinton SK, Gimbrone MA Jr, Libby P (1994) The omega–3 fatty acid docosahexaenoate reduces cytokine-induced expression of proatherogenic and proinflammatory proteins in human endothelial cells. *Arterioscler Thromb* 14: 1829–1836

71   Besemer J, Harant H, Wang S, Oberhauser B, Marquardt K, Foster CA, Schreiner EP, de Vries JE, Dascher-Nadel C, Lindley IJ (2005) Selective inhibition of cotranslational translocation of vascular cell adhesion molecule 1. *Nature* 436: 290–293

72   Gimbrone MA Jr, Nagel T, Topper JN (1997) Biomechanical activation: an emerging paradigm in endothelial adhesion biology. *J Clin Invest* 100: S61–65

73   Topper JN, Cai J, Falb D, Gimbrone MA Jr (1996) Identification of vascular endothelial genes differentially responsive to fluid mechanical stimuli: cyclooxygenase-2, manganese superoxide dismutase, and endothelial cell nitric oxide synthase are selectively up-regulated by steady laminar shear stress. *Proc Natl Acad Sci USA* 93: 10417–10422

74   Dai G, Kaazempur-Mofrad MR, Natarajan S, Zhang Y, Vaughn S, Blackman BR, Kamm RD, Garcia-Cardena G, Gimbrone MA Jr (2004) Distinct endothelial phenotypes evoked by arterial waveforms derived from atherosclerosis-susceptible and -resistant regions of human vasculature. *Proc Natl Acad Sci USA* 101: 14871–14876

75   Hajra L, Evans AI, Chen M, Hyduk SJ, Collins T, Cybulsky MI (2000) The NF-kappa B signal transduction pathway in aortic endothelial cells is primed for activation in regions predisposed to atherosclerotic lesion formation. *Proc Natl Acad Sci USA* 97: 9052–9057

76   Nagel T, Resnick N, Atkinson WJ, Dewey CF Jr, Gimbrone MA Jr (1994) Shear stress selectively upregulates intercellular adhesion molecule-1 expression in cultured human vascular endothelial cells. *J Clin Invest* 94: 885–891

77   Ando J, Tsuboi H, Korenaga R, Takada Y, Toyama-Sorimachi N, Miyasaka M, Kamiya A (1994) Shear stress inhibits adhesion of cultured mouse endothelial cells to lymphocytes by downregulating VCAM-1 expression. *Am J Physiol* 267: C679–687

78   Chiu JJ, Lee PL, Chen CN, Lee CI, Chang SF, Chen LJ, Lien SC, Ko YC, Usami S, Chien S (2004) Shear stress increases ICAM-1 and decreases VCAM-1 and E-selectin expressions induced by tumor necrosis factor-[alpha] in endothelial cells. *Arterioscler Thromb Vasc Biol* 24: 73–79

79   Berk BC, Min W, Yan C, Surapisitchat J, Liu Y, Hoefen R (2002) Atheroprotective mechanisms activated by fluid shear stress in endothelial cells. *Drug News Perspect* 15: 133–139

80   Yamawaki H, Pan S, Lee RT, Berk BC (2005) Fluid shear stress inhibits vascular inflammation by decreasing thioredoxin-interacting protein in endothelial cells. *J Clin Invest* 115: 733–738

81   Parmar KM, Larman HB, Dai G, Zhang Y, Wang ET, Moorthy SN, Kratz JR, Lin Z,

Jain MK, Gimbrone MA Jr, Garcia-Cardena G (2006) Integration of flow-dependent endothelial phenotypes by Kruppel-like factor 2. *J Clin Invest* 116: 49–58

82 SenBanerjee S, Lin Z, Atkins GB, Greif DM, Rao RM, Kumar A, Feinberg MW, Chen Z, Simon DI, Luscinskas FW et al (2004) KLF2 Is a novel transcriptional regulator of endothelial proinflammatory activation. *J Exp Med* 199: 1305–1315

83 Rice GE, Munro JM, Corless C, Bevilacqua MP (1991) Vascular and nonvascular expression of INCAM-110. A target for mononuclear leukocyte adhesion in normal and inflamed human tissues. *Am J Pathol* 138: 385–393

84 Miyake K, Medina K, Ishihara K, Kimoto M, Auerbach R, Kincade PW (1991) A VCAM-like adhesion molecule on murine bone marrow stromal cells mediates binding of lymphocyte precursors in culture. *J Cell Biol* 114: 557–565

85 Rosen GD, Sanes JR, LaChance R, Cunningham JM, Roman J, Dean DC (1992) Roles for the integrin VLA-4 and its counter receptor VCAM-1 in myogenesis. *Cell* 69: 1107–1119

86 Simmons PJ, Masinovsky B, Longenecker BM, Berenson R, Torok-Storb B, Gallatin WM (1992) Vascular cell adhesion molecule-1 expressed by bone marrow stromal cells mediates the binding of hematopoietic progenitor cells. *Blood* 80: 388–395

87 Iademarco MF, McQuillan JJ, Dean DC (1993) Vascular cell adhesion molecule 1: contrasting transcriptional control mechanisms in muscle and endothelium. *Proc Natl Acad Sci USA* 90: 3943–3947

88 Jesse TL, LaChance R, Iademarco MF, Dean DC (1998) Interferon regulatory factor–2 is a transcriptional activator in muscle where It regulates expression of vascular cell adhesion molecule-1. *J Cell Biol* 140: 1265–1276

89 Duplaa C, Couffinhal T, Dufourcq P, Llanas B, Moreau C, Bonnet J (1997) The integrin very late antigen-4 is expressed in human smooth muscle cell. Involvement of alpha 4 and vascular cell adhesion molecule-1 during smooth muscle cell differentiation. *Circ Res* 80: 159–169

90 Li H, Cybulsky MI, Gimbrone MA Jr, Libby P (1993) Inducible expression of vascular cell adhesion molecule-1 by vascular smooth muscle cells *in vitro* and within rabbit atheroma. *Am J Pathol* 143: 1551–1559

91 Ardehali A, Laks H, Drinkwater DC, Ziv E, Drake TA (1995) Vascular cell adhesion molecule-1 is induced on vascular endothelia and medial smooth muscle cells in experimental cardiac allograft vasculopathy. *Circulation* 92: 450–456

92 Landry DB, Couper LL, Bryant SR, Lindner V (1997) Activation of the NF-kappa B and I kappa B system in smooth muscle cells after rat arterial injury. Induction of vascular cell adhesion molecule-1 and monocyte chemoattractant protein-1. *Am J Pathol* 151: 1085–1095

93 Lavie J, Dandre F, Louis H, Lamaziere JM, Bonnet J (1999) Vascular cell adhesion molecule-1 gene expression during human smooth muscle cell differentiation is independent of NF-kappaB activation. *J Biol Chem* 274: 2308–2314

94 Couffinhal T, Duplaa C, Moreau C, Lamaziere JM, Bonnet J (1994) Regulation of vas-

cular cell adhesion molecule-1 and intercellular adhesion molecule-1 in human vascular smooth muscle cells. *Circ Res* 74: 225–234

95 Braun M, Pietsch P, Felix SB, Baumann G (1995) Modulation of intercellular adhesion molecule-1 and vascular cell adhesion molecule-1 on human coronary smooth muscle cells by cytokines. *J Mol Cell Cardiol* 27: 2571–2579

96 Barks JL, McQuillan JJ, Iademarco MF (1997) TNF-alpha and IL-4 synergistically increase vascular cell adhesion molecule-1 expression in cultured vascular smooth muscle cells. *J Immunol* 159: 4532–4538

97 Wright PS, Cooper JR, Kropp KE, Busch SJ (1999) Induction of vascular cell adhesion molecule-1 expression by IL-4 in human aortic smooth muscle cells is not associated with increased nuclear NF-kappaB levels. *J Cell Physiol* 180: 381–389

98 Gamble JR, Bradley S, Noack L, Vadas MA (1995) TGF-beta and endothelial cells inhibit VCAM-1 expression on human vascular smooth muscle cells. *Arterioscler Thromb Vasc Biol* 15: 949–955

99 Gurtner GC, Davis V, Li H, McCoy MJ, Sharpe A, Cybulsky MI (1995) Targeted disruption of the murine VCAM1 gene: essential role of VCAM-1 in chorioallantoic fusion and placentation. *Genes Dev* 9: 1–14

100 Kwee L, Baldwin HS, Shen HM, Stewart CL, Buck C, Buck CA, Labow MA (1995) Defective development of the embryonic and extraembryonic circulatory systems in vascular cell adhesion molecule (VCAM-1) deficient mice. *Development* 121: 489–503

101 Yang JT, Rayburn H, Hynes RO (1995) Cell adhesion events mediated by alpha 4 integrins are essential in placental and cardiac development. *Development* 121: 549–560

102 Cybulsky MI, Iiyama K, Li H, Zhu S, Chen M, Iiyama M, Davis V, Gutierrez-Ramos JC, Connelly PW, Milstone DS (2001) A major role for VCAM-1, but not ICAM-1, in early atherosclerosis. *J Clin Invest* 107: 1255–1262

103 Yang JT, Rando TA, Mohler WA, Rayburn H, Blau HM, Hynes RO (1996) Genetic analysis of alpha 4 integrin functions in the development of mouse skeletal muscle. *J Cell Biol* 135: 829–835

104 Miyake K, Weissman IL, Greenberger JS, Kincade PW (1991) Evidence for a role of the integrin VLA-4 in lympho-hemopoiesis. *J Exp Med* 173: 599–607

105 Jacobsen K, Kravitz J, Kincade PW, Osmond DG (1996) Adhesion receptors on bone marrow stromal cells: *in vivo* expression of vascular cell adhesion molecule-1 by reticular cells and sinusoidal endothelium in normal and gamma-irradiated mice. *Blood* 87: 73–82

106 Tada T, Widayati DT, Fukuta K (2006) Morphological study of the transition of haematopoietic sites in the developing mouse during the peri-natal period. *Anat Histol Embryol* 35: 235–240

107 Leuker CE, M Labow, W Muller, Wagner N (2001) Neonatally induced inactivation of the vascular cell adhesion molecule 1 gene impairs B cell localization and T cell-dependent humoral immune response. *J Exp Med* 193: 755–768

108 Koni PA, Joshi SK, Temann UA, Olson D, Burkly L, Flavell RA (2001) Conditional

vascular cell adhesion molecule 1 deletion in mice: impaired lymphocyte migration to bone marrow. *J Exp Med* 193: 741–754

109 Papayannopoulou T, Nakamoto B (1993) Peripheralization of hemopoietic progenitors in primates treated with anti-VLA4 integrin. *Proc Natl Acad Sci USA* 90: 9374–9378

110 Funk PE, Kincade PW, Witte PL (1994) Native associations of early hematopoietic stem cells and stromal cells isolated in bone marrow cell aggregates. *Blood* 83: 361–369

111 Craddock CF, Nakamoto B, Andrews RG, Priestley GV, Papayannopoulou T (1997) Antibodies to VLA4 integrin mobilize long-term repopulating cells and augment cytokine-induced mobilization in primates and mice. *Blood* 90: 4779–4788

112 Friedrich C, Cybulsky MI, Gutierrez-Ramos JC (1996) Vascular cell adhesion molecule-1 expression by hematopoiesis-supporting stromal cells is not essential for lymphoid or myeloid differentiation *in vivo* or *in vitro*. *Eur J Immunol* 26: 2773–2780

113 Ulyanova T, Scott LM, Priestley GV, Jiang Y, Nakamoto B, Koni PA, Papayannopoulou T (2005) VCAM-1 expression in adult hematopoietic and nonhematopoietic cells is controlled by tissue-inductive signals and reflects their developmental origin. *Blood* 106: 86–94

114 Teixido J, Hemler ME, Greenberger JS, Anklesaria P (1992) Role of beta 1 and beta 2 integrins in the adhesion of human CD34hi stem cells to bone marrow stroma. *J Clin Invest* 90: 358–367

115 Scott LM, Priestley GV, Papayannopoulou T (2003) Deletion of alpha4 integrins from adult hematopoietic cells reveals roles in homeostasis, regeneration, and homing. *Mol Cell Biol* 23: 9349–9360

116 Priestley GV, Ulyanova T, Papayannopoulou T (2007) Sustained alterations in biodistribution of stem/progenitor cells in Tie2Cre+ alpha4(f/f) mice are hematopoietic cell autonomous. *Blood* 109: 109–111

117 Arroyo AG, Yang JT, Rayburn H, Hynes RO (1996) Differential requirements for alpha4 integrins during fetal and adult hematopoiesis. *Cell* 85: 997–1008

118 Berlin-Rufenach C, Otto F, Mathies M, Westermann J, Owen MJ, Hamann A, Hogg N (1999) Lymphocyte migration in lymphocyte function-associated antigen (LFA)-1-deficient mice. *J Exp Med* 189: 1467–1478

119 Springer TA (1994) Traffic signals for lymphocyte recirculation and leukocyte emigration: the multistep paradigm. *Cell* 76: 301–314

120 Butcher EC, Picker LJ (1996) Lymphocyte homing and homeostasis. *Science* 272: 60–66

121 Campbell JJ, Butcher EC (2000) Chemokines in tissue-specific and microenvironment-specific lymphocyte homing. *Curr Opin Immunol* 12: 336–341

122 Ley K (2001) Pathways and bottlenecks in the web of inflammatory adhesion molecules and chemoattractants. *Immunol Res* 24: 87–95

123 Alon R, Hammer DA, Springer TA (1995) Lifetime of the P-selectin-carbohydrate bond and its response to tensile force in hydrodynamic flow. *Nature* 374: 539–542

124 Alon R, Chen S, Puri KD, Finger EB, Springer TA (1997) The kinetics of L-selectin tethers and the mechanics of selectin-mediated rolling. *J Cell Biol* 138: 1169–1180

125 Alon R, Kassner PD, Carr MW, Finger EB, Hemler ME, Springer TA (1995) The integrin VLA-4 supports tethering and rolling in flow on VCAM-1. *J Cell Biol* 128: 1243–1253

126 Berlin C, Bargatze RF, Campbell JJ, von Andrian UH, Szabo MC, Hasslen SR, Nelson RD, Berg EL, Erlandsen SL, Butcher EC (1995) alpha 4 integrins mediate lymphocyte attachment and rolling under physiologic flow. *Cell* 80: 413–422

127 Luscinskas FW, Ding H, Lichtman AH (1995) P-selectin and vascular cell adhesion molecule 1 mediate rolling and arrest, respectively, of CD4+ T lymphocytes on tumor necrosis factor alpha-activated vascular endothelium under flow. *J Exp Med* 181: 1179–1186

128 Lalor PF, Clements JM, Pigott R, Humphries MJ, Spragg JH, Nash GB (1997) Association between receptor density, cellular activation, and transformation of adhesive behavior of flowing lymphocytes binding to VCAM-1. *Eur J Immunol* 27: 1422–1426

129 Chen C, Mobley JL, Dwir O, Shimron F, Grabovsky V, Lobb RR, Shimizu Y, Alon R (1999) High affinity very late antigen-4 subsets expressed on T cells are mandatory for spontaneous adhesion strengthening but not for rolling on VCAM-1 in shear flow. *J Immunol* 162: 1084–1095

130 Vajkoczy P, Laschinger M, Engelhardt B (2001) alpha4-integrin-VCAM-1 binding mediates G protein-independent capture of encephalitogenic T cell blasts to CNS white matter microvessels. *J Clin Invest* 108: 557–565

131 Nandi A, Estess P, Siegelman M (2004) Bimolecular complex between rolling and firm adhesion receptors required for cell arrest; CD44 association with VLA-4 in T cell extravasation. *Immunity* 20: 455–465

132 Huo Y, Hafezi-Moghadam A, Ley K (2000) Role of vascular cell adhesion molecule-1 and fibronectin connecting segment–1 in monocyte rolling and adhesion on early atherosclerotic lesions. *Circ Res* 87: 153–159

133 Singbartl K, Thatte J, Smith ML, Wethmar K, Day K, Ley K (2001) A CD2-green fluorescence protein-transgenic mouse reveals very late antigen-4-dependent CD8+ lymphocyte rolling in inflamed venules. *J Immunol* 166: 7520–7526

134 Kerfoot SM, Kubes P (2002) Overlapping roles of P-selectin and alpha 4 integrin to recruit leukocytes to the central nervous system in experimental autoimmune encephalomyelitis. *J Immunol* 169: 1000–1006

135 Carman CV, Springer TA (2003) Integrin avidity regulation: are changes in affinity and conformation underemphasized? *Curr Opin Cell Biol* 15: 547–556

136 Luo BH, Springer TA (2006) Integrin structures and conformational signaling. *Curr Opin Cell Biol* 18: 579–586

137 Middleton J, Neil S, Wintle J, Clark-Lewis I, Moore H, Lam C, Auer M, Hub E, Rot A (1997) Transcytosis and surface presentation of IL-8 by venular endothelial cells. *Cell* 91: 385–395

138 Tanaka Y, Adams DH, Hubscher S, Hirano H, Siebenlist U, Shaw S (1993) T-cell adhesion induced by proteoglycan-immobilized cytokine MIP-1 beta. *Nature* 361: 79–82

139 Lloyd AR, Oppenheim JJ, Kelvin DJ, Taub DD (1996) Chemokines regulate T cell

adherence to recombinant adhesion molecules and extracellular matrix proteins. *J Immunol* 156: 932–938

140 Campbell JJ, Qin S, Bacon KB, Mackay CR, Butcher EC (1996) Biology of chemokine and classical chemoattractant receptors: differential requirements for adhesion-triggering *versus* chemotactic responses in lymphoid cells. *J Cell Biol* 134: 255–266

141 Campbell JJ, Hedrick J, Zlotnik A, Siani MA, Thompson DA, Butcher EC (1998) Chemokines and the arrest of lymphocytes rolling under flow conditions. *Science* 279: 381–384

142 Chan JR, Hyduk SJ, Cybulsky MI (2001) Chemoattractants induce a rapid and transient upregulation of monocyte alpha4 integrin affinity for vascular cell adhesion molecule 1 which mediates arrest: an early step in the process of emigration. *J Exp Med* 193: 1149–1158

143 DiVietro JA, Brown DC, Sklar LA, Larson RS, Lawrence MB (2007) Immobilized stromal cell-derived factor-1{alpha} triggers rapid VLA-4 affinity increases to stabilize lymphocyte tethers on VCAM-1 and subsequently initiate firm adhesion. *J Immunol* 178: 3903–3911

144 Grabovsky V, Feigelson S, Chen C, Bleijs DA, Peled A, Cinamon G, Baleux F, Arenzana-Seisdedos F, Lapidot T, van Kooyk Y et al (2000) Subsecond induction of alpha4 integrin clustering by immobilized chemokines stimulates leukocyte tethering and rolling on endothelial vascular cell adhesion molecule 1 under flow conditions. *J Exp Med* 192: 495–506

145 Abitorabi MA, Pachynski RK, Ferrando RE, Tidswell M, Erle DJ (1997) Presentation of integrins on leukocyte microvilli: a role for the extracellular domain in determining membrane localization. *J Cell Biol* 139: 563–571

146 Cambi A, Joosten B, Koopman M, de Lange F, Beeren I, Torensma R, Fransen JA, Garcia-Parajo M, van Leeuwen FN, Figdor CG (2006) Organization of the integrin LFA-1 in nanoclusters regulates its activity. *Mol Biol Cell* 17: 4270–4281

147 Hyduk SJ, Chan, Jason R, Duffy ST, Chen M, Peterson MD, Waddell TK, Digby GC, Szaszi K, Kapus A, Cybulsky MI (2007) Phospholipase C, calcium, and calmodulin are critical for {alpha}4{beta}1 integrin affinity up-regulation and monocyte arrest triggered by chemoattractants. *Blood* 109: 176–184

148 Chan JR, Hyduk SJ, Cybulsky MI (2000) Alpha 4 beta 1 integrin/VCAM-1 interaction activates alpha L beta 2 integrin-mediated adhesion to ICAM-1 in human T cells. *J Immunol* 164: 746–753

149 Rose DM, Liu S, Woodside DG, Han J, Schlaepfer DD, Ginsberg MH (2003) Paxillin binding to the alpha 4 integrin subunit stimulates LFA-1 (integrin alpha L beta 2)-dependent T cell migration by augmenting the activation of focal adhesion kinase/proline-rich tyrosine kinase-2. *J Immunol* 170: 5912–5918

150 Chuluyan HE, Osborn L, Lobb R, Issekutz AC (1995) Domains 1 and 4 of vascular cell adhesion molecule-1 (CD106) both support very late activation antigen-4 (CD49d/CD29)-dependent monocyte transendothelial migration. *J Immunol* 155: 3135–3134

151 Meerschaert J, Furie MB (1995) The adhesion molecules used by monocytes for

migration across endothelium include CD11a/CD18, CD11b/CD18, and VLA-4 on monocytes and ICAM-1, VCAM-1, and other ligands on endothelium. *J Immunol* 154: 4099–4112

152 Liu S, Thomas SM, Woodside DG, Rose DM, Kiosses WB, Pfaff M, Ginsberg MH (1999) Binding of paxillin to alpha4 integrins modifies integrin-dependent biological responses. *Nature* 402: 676–681

153 Liu S, Ginsberg MH (2000) Paxillin binding to a conserved sequence motif in the alpha 4 integrin cytoplasmic domain. *J Biol Chem* 275: 22736–22742

154 Han J, Liu S, Rose DM, Schlaepfer DD, McDonald H, Ginsberg MH (2001) Phosphorylation of the integrin alpha 4 cytoplasmic domain regulates paxillin binding. *J Biol Chem* 276: 40903–40909

155 Hyduk SJ, Oh J, Xiao H, Chen M, Cybulsky MI (2004) Paxillin selectively associates with constitutive and chemoattractant-induced high-affinity alpha4beta1 integrins: implications for integrin signaling. *Blood* 104: 2818–2824

156 Liu S, Kiosses WB, Rose DM, Slepak M, Salgia R, Griffin JD, Turner CE, Schwartz MA, Ginsberg MZH (2002) A fragment of paxillin binds the alpha 4 integrin cytoplasmic domain (tail) and selectively inhibits alpha 4-mediated cell migration. *J Biol Chem* 277: 20887–20894

157 Goldfinger LE, Han J, Kiosses WB, Howe AK, Ginsberg MH (2003) Spatial restriction of alpha4 integrin phosphorylation regulates lamellipodial stability and alpha4beta1-dependent cell migration. *J Cell Biol* 162: 731–741

158 Nishiya N, Kiosses WB, Han J, Ginsberg MH (2005) An alpha(4) integrin-paxillin-Arf-GAP complex restricts Rac activation to the leading edge of migrating cells. *Nat Cell Biol* 7: 343–352

159 Feral CC, Rose DM, Han J, Fox N, Silverman GJ, Kaushansky K, Ginsberg MH (2006) Blocking the alpha4 integrin paxillin interaction selectively impairs mononuclear leukocyte recruitment to an inflammatory site. *J Clin Invest* 116: 715–723

160 Alon R, Feigelson SW, Manevich E, Rose DM, Schmitz J, Overby DR, Winter E, Grabovsky V, Shinder V, Matthews BD et al (2005) {alpha}4{beta}1-dependent adhesion strengthening under mechanical strain is regulated by paxillin association with the {alpha}4-cytoplasmic domain. *J Cell Biol* 171: 1073–1084

161 Han J, Rose DM, Woodside DG, Goldfinger LE, Ginsberg MH (2003) Integrin alpha 4 beta 1-dependent T cell migration requires both phosphorylation and dephosphorylation of the alpha 4 cytoplasmic domain to regulate the reversible binding of paxillin. *J Biol Chem* 278: 34845–34853

162 Lim CJ, Han J, Yousefi N, Ma Y, Amieux PS, McKnight GS, Taylor SS, Ginsberg MH (2007) Alpha4 integrins are type I cAMP-dependent protein kinase-anchoring proteins. *Nat Cell Biol* 9: 415–421

163 Wong T, Hyduk SJ, Cybulsky MI (2007) Calcineurin, a serine/threonine phosphatase, is required for GPCR-triggered association of paxillin, a cytoplasmic adapter protein, with high affinity {alpha}4{beta}1 integrins in U937 cells. *FASEB J* 21: A126-a-.

164 Skalen K, Gustafsson M, Rydberg EK, Hulten LM, Wiklund O, Innerarity TL, Boren

J (2002) Subendothelial retention of atherogenic lipoproteins in early atherosclerosis. *Nature* 417: 750–754

165 Glas CK, Witztum JL (2001) Atherosclerosis. the road ahead. *Cell* 104: 503–516

166 Steinberg D (2002) Atherogenesis in perspective: hypercholesterolemia and inflammation as partners in crime. *Nat Med* 8: 1211–1217

167 Cybulsky MI, Gimbrone MA Jr (1991) Endothelial expression of a mononuclear leukocyte adhesion molecule during atherogenesis. *Science* 251: 788–791

168 Li H, Cybulsky MI, Gimbrone MA Jr, Libby P (1993) An atherogenic diet rapidly induces VCAM-1, a cytokine-regulatable mononuclear leukocyte adhesion molecule, in rabbit aortic endothelium. *Arterioscler Thromb* 13: 197–204

169 Sakai A, Kume N, Nishi E, Tanoue K, Miyasaka M, Kita T (1997) P-selectin and vascular cell adhesion molecule-1 are focally expressed in aortas of hypercholesterolemic rabbits before intimal accumulation of macrophages and T lymphocytes. *Arterioscler Thromb Vasc Biol* 17: 310–316

170 Iiyama K, Hajra L, Iiyama M, Li H, DiChiara M, Medoff BD, Cybulsky MI (1999) Patterns of vascular cell adhesion molecule-1 and intercellular adhesion molecule-1 expression in rabbit and mouse atherosclerotic lesions and at sites predisposed to lesion formation. *Circ Res* 85: 199–207

171 Nakashima Y, Raines EW, Plump AS, Breslow JL, Ross R (1998) Upregulation of VCAM-1 and ICAM-1 at atherosclerosis-prone sites on the endothelium in the ApoE-deficient mouse. *Arterioscler Thromb Vasc Biol* 18: 842–851

172 Poston RN, Haskard DO, Coucher JR, Gall NP, Johnson-Tidey RR (1992) Expression of intercellular adhesion molecule-1 in atherosclerotic plaques. *Am J Pathol* 140: 665–673

173 Printseva O, Peclo MM, Gown AM (1992) Various cell types in human atherosclerotic lesions express ICAM-1. Further immunocytochemical and immunochemical studies employing monoclonal antibody 10F3. *Am J Pathol* 140: 889–896

174 Wood KM, Cadogan MD, Ramshaw AL, Parums DV (1993) The distribution of adhesion molecules in human atherosclerosis. *Histopathology* 22: 437–444

175 van der Wal AC, Das PK, Tigges AJ, Becker AE (1992) Adhesion molecules on the endothelium and mononuclear cells in human atherosclerotic lesions. *Am J Pathol* 141: 1427–1433

176 Davies MJ, Gordon JL, Gearing AJ, Pigott R, Woolf N, Katz D, Kyriakopoulos A (1993) The expression of the adhesion molecules ICAM-1, VCAM-1, PECAM, and E-selectin in human atherosclerosis. *J Pathol* 171: 223–229

177 O'Brien KD, Allen MD, McDonald TO, Chait A, Harlan JM, Fishbein D, McCarty J, Ferguson M, Hudkins K, Benjamin CD et al (1993) Vascular cell adhesion molecule-1 is expressed in human coronary atherosclerotic plaques. Implications for the mode of progression of advanced coronary atherosclerosis. *J Clin Invest* 92: 945–951

178 Cybulsky MI, Charo IF (2005) Leukocytes, adhesion molecules and chemokines in atherothrombosis. In: V Fuster, EJ Topol, EG Nabel (eds): *Atherothrombosis and Coronary Artery Disease*, 2nd edition. Lippincott Williams & Wilkins, Philadelphia, 487–504

179 Dansky HM, Barlow CB, Lominska C, Sikes JL, Kao C, Weinsaft J, Cybulsky MI, Smith JD (2001) Adhesion of monocytes to arterial endothelium and initiation of atherosclerosis are critically dependent on vascular cell adhesion molecule-1 gene dosage. *Arterioscler Thromb Vasc Biol* 21: 1662–1667

180 Shih PT, Brennan ML, Vora DK, Territo MC, Strahl D, Elices MJ, Lusis AJ, Berliner JA (1999) Blocking very late antigen-4 integrin decreases leukocyte entry and fatty streak formation in mice fed an atherogenic diet. *Circ Res* 84: 345–351

181 Cannella B, Raine CS (1995) The adhesion molecule and cytokine profile of multiple sclerosis lesions. *Ann Neurol* 37: 424–435

182 Barten DM, Ruddle NH (1994) Vascular cell adhesion molecule-1 modulation by tumor necrosis factor in experimental allergic encephalomyelitis. *J Neuroimmunol* 51: 123–133

183 Engelhardt B, Laschinger M, Schulz M, Samulowitz U, Vestweber D, Hoch G (1998) The development of experimental autoimmune encephalomyelitis in the mouse requires alpha4-integrin but not alpha4beta7-integrin. *J Clin Invest* 102: 2096–2105

184 Wilkinson LS, Edwards JC, Poston RN, Haskard DO (1993) Expression of vascular cell adhesion molecule-1 in normal and inflamed synovium. *Lab Invest* 68: 82–88

185 Furuzawa-Carballeda J, Alcocer-Varela J (1999) Interleukin-8, interleukin-10, intercellular adhesion molecule-1 and vascular cell adhesion molecule-1 expression levels are higher in synovial tissue from patients with rheumatoid arthritis than in osteoarthritis. *Scand J Immunol* 50: 215–222

186 Morales-Ducret J, Wayner E, Elices MJ, Alvaro-Gracia JM, Zvaifler NJ, Firestein GS (1992) Alpha 4/beta 1 integrin (VLA-4) ligands in arthritis. Vascular cell adhesion molecule-1 expression in synovium and on fibroblast-like synoviocytes. *J Immunol* 149: 1424–1431

187 Elliott MJ, Maini RN, Feldmann M, Long-Fox A, Charles P, Katsikis P, Brennan FM, Walker J, Bijl H, Ghrayeb J et al (1993) Treatment of rheumatoid arthritis with chimeric monoclonal antibodies to tumor necrosis factor alpha. *Arthritis Rheum* 36: 1681–1690

188 Schrepfer S, Deuse T, Schafer H, Reichenspurner H (2005) FK778, a novel immunosuppressive agent, reduces early adhesion molecule up-regulation and prolongs cardiac allograft survival. *Transpl Int* 18: 215–220

189 Goebeler M, Roth J, Brocker EB, Sorg C, Schulze-Osthoff K (1995) Activation of nuclear factor-kappa B and gene expression in human endothelial cells by the common haptens nickel and cobalt. *J Immunol* 155: 2459–2467

190 Viemann D, Schmidt M, Tenbrock K, Schmid S, Muller V, Klimmek K, Ludwig S, Roth J, Goebeler M (2007) The contact allergen nickel triggers a unique inflammatory and proangiogenic gene expression pattern *via* activation of NF-kappaB and hypoxia-inducible factor-1alpha. *J Immunol* 178: 3198–3207

191 McHale JF, Harari OA, Marshall D, Haskard DO (1999) Vascular endothelial cell expression of ICAM-1 and VCAM-1 at the onset of eliciting contact hypersensitivity in mice: evidence for a dominant role of TNF-alpha. *J Immunol* 162: 1648–1655

192 Leca G, Mansur SE, Bensussan A (1995) Expression of VCAM-1 (CD106) by a subset of TCR gamma delta-bearing lymphocyte clones. Involvement of a metalloprotease in the specific hydrolytic release of the soluble isoform. *J Immunol* 154: 1069–1077

193 Garton KJ, Gough PJ, Philalay J, Wille PT, Blobel CP, Whitehead RH, Dempsey PJ, Raines EW (2003) Stimulated shedding of vascular cell adhesion molecule 1 (VCAM-1) is mediated by tumor necrosis factor-alpha-converting enzyme (ADAM 17). *J Biol Chem* 278: 37459–37464

194 Hummel V, Kallmann BA, Wagner S, Fuller T, Bayas A, Tonn JC, Benveniste EN, Toyka KV, Rieckmann P (2001) Production of MMPs in human cerebral endothelial cells and their role in shedding adhesion molecules. *J Neuropathol Exp Neurol* 60: 320–327

195 Singh RJ, Mason JC, Lidington EA, Edwards DR, Nuttall RK, Khokha R, Knauper V, Murphy G, Gavrilovic J (2005) Cytokine stimulated vascular cell adhesion molecule-1 (VCAM-1) ectodomain release is regulated by TIMP-3. *Cardiovasc Res* 67: 39–49

196 Matsuno O, Miyazaki E, Nureki S, Ueno T, Kumamoto T, Higuchi Y (2006) Role of ADAM8 in experimental asthma. *Immunol Lett* 102: 67–73

197 Matsuno O, Miyazaki E, Nureki S, Ueno T, Ando M, Ito K, Kumamoto T, Higuchi Y (2006) Elevated soluble ADAM8 in bronchoalveolar lavage fluid in patients with eosinophilic pneumonia. *Int Arch Allergy Immunol* 142: 285–290

198 Levesque JP, Takamatsu Y, Nilsson SK, Haylock DN, Simmons PJ (2001) Vascular cell adhesion molecule-1 (CD106) is cleaved by neutrophil proteases in the bone marrow following hematopoietic progenitor cell mobilization by granulocyte colony-stimulating factor. *Blood* 98: 1289–1297

199 Hartung HP, Reiners K, Archelos JJ, Michels M, Seeldrayers P, Heidenreich F, Pflughaupt KW, Toyka KV (1995) Circulating adhesion molecules and tumor necrosis factor receptor in multiple sclerosis: correlation with magnetic resonance imaging. *Ann Neurol* 38: 186–193

200 Mossner R, Fassbender K, Kuhnen J, Schwartz A, Hennerici M (1996) Vascular cell adhesion molecule – a new approach to detect endothelial cell activation in MS and encephalitis *in vivo*. *Acta Neurol Scand* 93: 118–122

201 Dore-Duffy P, Newman W, Balabanov R, Lisak RP, Mainolfi E, Rothlein R, Peterson M (1995) Circulating, soluble adhesion proteins in cerebrospinal fluid and serum of patients with multiple sclerosis: correlation with clinical activity. *Ann Neurol* 37: 55–62

202 Droogan AG, McMillan SA, Douglas JP, Hawkins SA (1996) Serum and cerebrospinal fluid levels of soluble adhesion molecules in multiple sclerosis: predominant intrathecal release of vascular cell adhesion molecule-1. *J Neuroimmunol* 64: 185–191

203 Matsuda M, Tsukada N, Miyagi K, Yanagisawa N (1995) Increased levels of soluble vascular cell adhesion molecule-1 (VCAM-1) in the cerebrospinal fluid and sera of patients with multiple sclerosis and human T lymphotropic virus type-1-associated myelopathy. *J Neuroimmunol* 59: 35–40

204 Calabresi PA, Tranquill LR, Dambrosia JM, Stone LA, Maloni H, Bash CN, Frank JA,

McFarland HF (1997) Increases in soluble VCAM-1 correlate with a decrease in MRI lesions in multiple sclerosis treated with interferon beta-1b. *Ann Neurol* 41: 669–674

205 Matusevicius D, Kivisakk P, Navikas VV, Tian W, Soderstrom M, Fredrikson S, Link H (1998) Influence of IFN-beta1b (Betaferon) on cytokine mRNA profiles in blood mononuclear cells and plasma levels of soluble VCAM-1 in multiple sclerosis. *Eur J Neurol* 5: 265–275

206 Jones SC, Banks RE, Haidar A, Gearing AJ, Hemingway IK, Ibbotson SH, Dixon MF, and Axon AT (1995) Adhesion molecules in inflammatory bowel disease. *Gut* 36: 724–730

207 Goke M, Hoffmann JC, Evers J, Kruger H, Manns MP (1997) Elevated serum concentrations of soluble selectin and immunoglobulin type adhesion molecules in patients with inflammatory bowel disease. *J Gastroenterol* 32: 480–486

208 Wellicome SM, Kapahi P, Mason JC, Lebranchu Y, Yarwood H, Haskard DO (1993) Detection of a circulating form of vascular cell adhesion molecule-1: raised levels in rheumatoid arthritis and systemic lupus erythematosus. *Clin Exp Immunol* 92: 412–418

209 Clausen P, Jacobsen P, Rossing K, Jensen JS, Parving HH, Feldt-Rasmussen B (2000) Plasma concentrations of VCAM-1 and ICAM-1 are elevated in patients with Type 1 diabetes mellitus with microalbuminuria and overt nephropathy. *Diabet Med* 17: 644–649

210 Ikeda Y, Fujimoto T, Ameno M, Shiiki H, Dohi K (1998) Relationship between lupus nephritis activity and the serum level of soluble VCAM-1. *Lupus* 7: 347–354

211 Molad Y, Miroshnik E, Sulkes J, Pitlik S, Weinberger A, Monselise Y (2002) Urinary soluble VCAM-1 in systemic lupus erythematosus: a clinical marker for monitoring disease activity and damage. *Clin Exp Rheumatol* 20: 403–406

212 Peter K, Nawroth P, Conradt C, Nordt T, Weiss T, Boehme M, Wunsch A, Allenberg J, Kubler W, Bode C (1997) Circulating vascular cell adhesion molecule-1 correlates with the extent of human atherosclerosis in contrast to circulating intercellular adhesion molecule-1, E-selectin, P-selectin, and thrombomodulin. *Arterioscler Thromb Vasc Biol* 17: 505–512

213 Lang T, Krams SM, Villanueva JC, Cox K, So S, Martinez OM (1995) Differential patterns of circulating intercellular adhesion molecule-1 (cICAM-1) and vascular cell adhesion molecule-1 (cVCAM-1) during liver allograft rejection. *Transplantation* 59: 584–589

214 Lederer SR, Friedrich N, Regenbogen C, Getto R, Toepfer M, Sitter T (2003) Non-invasive monitoring of renal transplant recipients: urinary excretion of soluble adhesion molecules and of the complement-split product C4d. *Nephron Clin Pract* 94: c19–26

215 Wu YW, Lee CM, Lee YT, Wang SS, Huang PJ (2003) Value of circulating adhesion molecules in assessing cardiac allograft vasculopathy. *J Heart Lung Transplant* 22: 1284–1287

216 O'Hanlon DM, Fitzsimons H, Lynch J, Tormey S, Malone C, Given HF (2002) Soluble

adhesion molecules (E-selectin, ICAM-1 and VCAM-1) in breast carcinoma. *Eur J Cancer* 38: 2252–2257

217 Silva HC, Garcao F, Coutinho EC, De Oliveira CF, Regateiro FJ (2006) Soluble VCAM-1 and E-selectin in breast cancer: relationship with staging and with the detection of circulating cancer cells. *Neoplasma* 53: 538–543

218 Alexiou D, Karayiannakis AJ, Syrigos KN, Zbar A, Sekara E, Michail P, Rosenberg T, Diamantis T (2003) Clinical significance of serum levels of E-selectin, intercellular adhesion molecule-1, and vascular cell adhesion molecule-1 in gastric cancer patients. *Am J Gastroenterol* 98: 478–485

219 Alexiou D, Karayiannakis AJ, Syrigos KN, Zbar A, Kremmyda A, Bramis I, Tsigris C (2001) Serum levels of E-selectin, ICAM-1 and VCAM-1 in colorectal cancer patients: correlations with clinicopathological features, patient survival and tumour surgery. *Eur J Cancer* 37: 2392–2397

220 Coskun U, Sancak B, Sen I, Bukan N, Tufan MA, Gulbahar O, Sozen S (2006) Serum P-selectin, soluble vascular cell adhesion molecule-I (s-VCAM-I) and soluble intercellular adhesion molecule-I (s-ICAM-I) levels in bladder carcinoma patients with different stages. *Int Immunopharmacol* 6: 672–677

221 Christiansen I, Sundstrom C, Totterman TH (1998) Elevated serum levels of soluble vascular cell adhesion molecule-1 (sVCAM-1) closely reflect tumour burden in chronic B-lymphocytic leukaemia. *Br J Haematol* 103: 1129–1137

222 Christiansen I, Sundstrom C, Kalkner KM, Bring J, Totterman TH (1998) Serum levels of soluble vascular cell adhesion molecule-1 (sVCAM-1) are elevated in advanced stages of non-Hodgkin's lymphomas. *Eur J Haematol* 61: 311–318

223 Kamezaki S, Kurozawa Y, Iwai N, Hosoda T, Okamoto M, Nose T (2005) Serum levels of soluble ICAM-1 and VCAM-1 predict pre-clinical cancer. *Eur J Cancer* 41: 2355–2359

Lymphocyte function-associated antigen-1 (LFA-1) and macrophage antigen-1 (Mac-1): Cooperative partners in leukocyte emigration and function

Alan R. Schenkel[1] and Minsoo Kim[2]

[1]Department of Microbiology, Immunology and Pathology, College of Veterinary Medicine and Biomedical Sciences, Colorado State University, Fort Collins, CO 80523, USA;
[2]Department of Surgery, Division of Surgical Research, Rhode Island Hospital and Brown University School of Medicine, Providence, RI, 02903, USA

Introduction and nomenclature

Lymphocyte function-associated antigen-1 (LFA-1; [1]) and macrophage antigen-1 (Mac-1; [2]) are some of the best-characterized integrins. Like other integrins, their diverse relationships and functions have lead to complex nomenclature. Integrins are heterodimeric receptors, consisting of type-I transmembrane glycoprotein α and β subunits. The α subunits of LFA-1 and Mac-1 were first named as α_L and α_M, respectively, in 1983 [3–5]. The Third International Workshop on Human Leukocyte Differentiation Antigens assigned α_L subunit as CD11a and α_M subunit as CD11b, and the common β_2 subunit as CD18 [6] The common β_2 subunit of LFA-1 ($\alpha_L\beta_2$) and Mac-1 ($\alpha_M\beta_2$) , forms a family with $\alpha_X\beta_2$ (p150,95, CD11c/CD18), and $\alpha_D\beta_2$ (CD11d/CD18). All four are expressed predominantly on leukocytes [1, 7]. Another name of Mac-1 is complement receptor type-3 (CR3; [8]). Mac-1 had been called Mo-1 and OKM-1 early on until the antibodies were shown to recognize the same proteins [3, 6, 7, 9].

Discovery

Although cell-cell contact and recognition through adhesive interactions have long been known to be important for a number of fundamental physiological processes, little was known about how such interactions are formed and how they exert diverse effects on the body. By the mid 1980s, investigators were able to isolate cell surface adhesion receptors and determine their amino acid sequences [10–15]. Monoclonal antibodies (mAbs) were also very important experimental tools for isolating integrins. Early strategy for integrin discovery involved screening of antibodies that are able to block cellular adhesion and identifying their specific antigens as the small subset of surface molecules important in cell recognition [16–18]. In

1987, Alan Horwitz and Richard Hynes [19] proposed the name "integrin" for their newly cloned protein, which later turned out to be the β_1 subunit, denoting an integral membrane complex involved in the transmembrane association between the cytoskeleton and the extracellular matrix (ECM) to which cells adhere. Later, it was revealed that all the cell adhesion receptors discovered by different groups belonged to a large family of structurally related molecules, one or more of which were expressed on virtually every cell type. "Integrin" has been coined as a generic term for the members of this protein superfamily.

Even before the name "integrin" was proposed, several leukocyte integrins, including LFA-1 and Mac-1, had been discovered. The first identification of LFA-1 in Timothy Springer's lab came from a rat mAb that had been raised against mouse cytotoxic T lymphocytes (CTLs) in attempts to inhibit CTL-mediated killing of tumor cells [1]. Among many mAbs raised, one mAb named M7/14 consistently inhibited CTL killing by an average of 90%. The target antigen of M7/14 mAb was named LFA-1. Further studies revealed that M7/14 prevents the Mg^{2+}-dependent cell-cell conjugation formation, rather than actual killing events. The data supported the idea that LFA-1 mediates the first recognition-adhesion step. Subsequently, human LFA-1 was identified together with LFA-2 (CD2) and LFA-3 (CD58) using the same assay on human CTLs [20].

Mac-1 was originally identified in 1979, independent of LFA-1, as a marker for myeloid cells [2]. Once LFA-1 was cloned, it became obvious that there was remarkable structural similarity between LFA-1 and Mac-1, which led to the hypothesis that Mac-1 also functions in adhesion. A function of Mac-1 was first discovered when Mac-1 mAb inhibited the Mg^{2+}-dependent binding of the C3bi (C3b inactivator-cleaved C3b) fragment of complement by mouse and human myeloid cells, thus defining Mac-1 as complement receptor type-3 (CR3). In the meantime, the laboratory of Gordon Ross built on their own discovery of CR3, using the rosetting assays with C3bi-coated sheep erythrocytes or C3bi-coated fluorescent microspheres (C3bi-ms) [21]. They found that C3bi-ms bound to neutrophils and monocytes. This indicated that these cells express C3bi receptors (CR3). Two mAbs, OKM-1 and Mo-1, were raised against a human monocyte antigen in the Stuart Schlossman lab [22, 23]. This antigen turned out to be identical to Mac-1.

Although defects in molecules that have immune or hematopoietic function are frequently lethal, a rare syndrome illustrates the importance of the molecules LFA-1 and Mac-1. Now known as leukocyte adhesion deficiency type I, afflicted patients exhibit an immunodeficiency especially in response to bacterial infections. Neutrophils from these patients were unable to spread, move, and were poorly phagocytic. The first cases reported a genetic defect in children with recurrent bacterial infections and the absence of surface glycoproteins of various molecular weights, now known to be CD11a (LFA-1 α subunit), CD11b (Mac-1 α subunit), and CD11c, as well as the common β_2 subunit CD18. The deficiency is inherited, and siblings and the parents often have lower surface expression of these integrins [16–18].

Table 1 - Ligands of LFA-1 and Mac-1

	Ligand	Function of integrin–ligand interaction	References
LFA-1	ICAM-1	Emigration, antigen recognition	[24, 127]
	ICAM-2	Emigration	[26]
	ICAM-3	Antigen recognition	[27]
	ICAM-4	Erythropoiesis?	[28]
	ICAM-5	Unknown	[29]
	JAM-A	Emigration	[34]
Mac-1	ICAM-1, -2	Emigration	[18, 30]
	ICAM-4	Erythropoiesis?	[28]
	JAM-C	Emigration	[35]
	C3bi	Phagocytosis	[8]
	LPS/LPG/APG	Phagocytosis	[8]
	Bacterial and fungal polysaccharides		
	Fibronectin, laminin, collagen, fibrinogen	ECM migration, wound adhesion	[8, 36, 37]
	DC-SIGN	Dendritic cell–neutrophil communication	[39]

CAMs, JAMs and other ligands of LFA-1 and Mac-1

The ligands for LFA-1 and Mac-1 are diverse (Tab. 1). LFA-1 currently has six known ligands. Mac-1 has been called promiscuous, which is an apt description [8]. Both molecules are able to recognize the members of the well-known intracellular adhesion molecules (ICAM) family, ICAMs-1–5 [24–29]. Since LFA-1 and Mac-1 have the ability to recognize many of the same molecules, this in turn leads to functionally redundant interactions. This has been best studied with ICAM-1 and ICAM-2 during the diapedesis process, and is discussed further below. ICAM-3 is involved with the antigen recognition *via* LFA-1, but Mac-1 does not bind ICAM-3 [30, 31]. ICAM-4 is one of the minor blood group antigens, is important for erythropoiesis and binds both LFA-1 and Mac-1. Although a physiological role for this interaction is not known, it is may be used during erythrocyte turnover by macrophages [28]. LFA-1 has been shown to bind ICAM-5 (telencephalin) raising the possibility that it has a role in leukocyte trafficking in the central nervous system [29].

The junctional adhesion molecules (JAMs) are a relatively newly discovered family of molecules, with five members: JAM-A, JAM-B, JAM-C, JAM-4/JCAM (mouse only), and JAM-L [32, 33]. It is important to note that the JAM family has had a name change to prevent some earlier nomenclature confusion [33]. LFA-1 recognizes JAM-A [34], whereas Mac-1 recognizes JAM-C [35].

Complement recognition by Mac-1 was one of the earliest known functions of this integrin. Its earliest known name is complement receptor 3 (CR3), and Mac-1 binds to C3bi [8]. Lymphocytes do not express Mac-1 but other cells like neutrophils, macrophages and dendritic cells (DCs) need to phagocytose and process complement and antibody-coated pathogens.

Extracellular proteins provide a three-dimensional lattice that the cells must navigate to reach site of tissue damage and infection. Again, cells that are involved in the primary immune response, like monocytes and neutrophils, use Mac-1 to recognize proteins like fibrinogen [36], fibronectin, laminin, collagen and many other proteins that are simply denatured [37].

Bacterial/fungal ligands recognized by Mac-1 are also quite diverse. These include lipopolysaccharide (LPS) from gram-negative bacteria species in cooperation with CD14, *Leishmania* lipophosphoglycan (LPG), *Klebsiella pneumoniae* acylpolygalactoside (APG), *Mycobacterium tuberculosis* polysaccharides, and various soluble and particulate saccharides that include β-glucan and zymosan [8]. Mac-1 also associates with FcγRIIIB (CD16) to assist in phagocytosis of IgG-coated targets [38].

Mac-1 has been reported to bind DC-specific ICAM-grabbing non-integrin (DC-SIGN) during DC-neutrophil interactions, which may play a role in neutrophil chemokine/cytokine secretion and survival [39]. DC-SIGN also alternates with LFA-1 for ICAM-2 and ICAM-3 during T cell–antigen-presenting cell (APC) interactions [9, 40–42].

Structural regulation of affinity, avidity, and signaling

Conformational regulation ("affinity" regulation)

I domain

Integrins are heterodimeric receptors, consisting of α and β subunits that together form a globular, ligand-binding head region with two legs that contain the transmembrane and cytoplasmic domains of each subunit (Fig. 1). Of the 18 α subunits, 9 contain a von Willebrand factor-type A domain of about 200 amino acids with a central hydrophobic β-sheet surrounded by α-helices, referred to as an inserted (I) domain (Fig. 1). The ligand-binding capacity of LFA-1 and Mac-1 is contained solely in the I domain, whereas other domains play regulatory roles. A metal ion-dependent adhesion site (MIDAS) is located on the "top" of the I domain. Divalent cations are universally required for integrins to bind ligands and the metal at the MIDAS directly coordinates to a Glu or Asp residue in the ligand. This metal-dependent interaction through the MIDAS plays a central role in ligand recognition. The

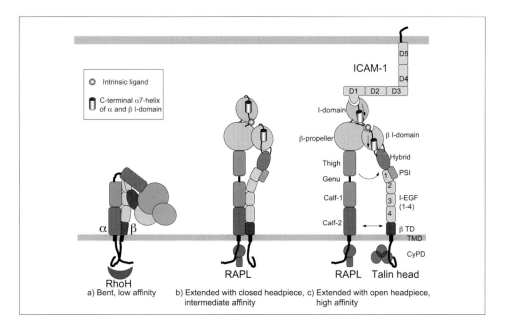

Figure 1
Model for LFA-1 activation.
Global and local integrin conformational changes are drawn based on the switchblade model
[50, 125]. Leukocyte integrins including LFA-1 are in an inactive, low-affinity state with a
bent conformation on resting cells (a). The maintenance of LFA-1 in the low-affinity state may
be an active process regulated by "default" inhibitory signals such as RhoH [59] (a). Chemo-
kine receptor or TCR stimulation induces activation and translocation of Rap-1 to the plasma
membrane. Activated Rap-1 interacts with a downstream effector, RAPL [57], which leads to
direct binding of RAPL to the juxtamembrane region of the αL cytoplasmic domain (b). The
binding of RAPL disrupts the cytoplasmic constraint between the α and β subunits, inducing
the extended conformation with intermediate affinity (b). The conformational change in the
cytoplasmic domain induced by RAPL binding unmasks the binding site for talin head domain
in the β2 cytoplasmic domain. Alternatively, GTP bound Rap-1–RAPL interaction recruits talin
to form the integrin activation complex [126]. As a direct consequence of the dual occupation
of RAPL and talin at the cytoplasmic domains, the extracellular interface between the α and β
subunits in the intermediate affinity LFA-1 become further destabilized, leading to the hybrid
domain swing out with respect to the β I-like domain, facilitating the downward movement
of the β I-like domain α7 helix (c). Next, the activated β I-like domain binds to the intrinsic
ligand in the linker between the I domain α7 helix and the β-propeller, thereby exerting a
downward pull on the α7 helix of the I domain, leading to the high-affinity conformation
(c). Binding of ligand (ICAM-1) may also induce or stabilize the high-affinity conformation.
I-EGF, integrin-epidermal growth factor domain; PSI, plexin/semaphorin/integrin; β TD, β tail
domain; TMD, transmembrane domain; CyPD, cytoplasmic domain.

ability of the I domain to bind ligand is controlled by conformational changes; the affinity of the I domain for its ligand is enhanced by a downward axial displacement of its C-terminal $\alpha7$ helix, which is conformationally linked to alterations in the MIDAS loops and Mg^{2+} coordination (Fig. 1) [43–45]. Compared to the default, low-affinity conformation, downward displacements by one and two turns of $\alpha7$ helix lead to intermediate- and high-affinity conformations with ~500 and 10000-fold increased affinity, respectively [45]. Conversely, binding of ligand to the MIDAS of the I domain also induces conformational changes, which stabilize the high-affinity conformation. Although structural studies demonstrated that conformational changes in the I domain regulate its affinity to ligands, it was controversial whether these changes are physiologically relevant.

Evidence to support the physiological relevance of the conformational changes seen in the I domain was provided by two mAbs. mAbs that bind selectively to the active conformation of the integrin have been a very useful alternative in studying integrin affinity regulation *in vivo*. The only such I domain mAbs reported for LFA-1 and Mac-1 are AL-57 [46], and CBRM1/5 [47], respectively. These mAbs show minimal binding to resting leukocytes, but upon cell stimulation by physiological chemoattractants, about 10% of total surface LFA-1 and Mac-1 molecules are recognized by their respective mAbs [46]. Therefore, the induction of the active I domain epitopes demonstrate the presence of conformational changes in the I domain on the cell surface.

β I-like domain
The β_2 subunit contains another von Willebrand factor-type A domain, an I-like domain, which is structurally similar to the α subunit I domain (Fig. 1). The function of the I-like domain is regulated by conformational changes similar to those observed in the α I domain, in which a downward movement of the C-terminal $\alpha7$-helix allosterically alters the geometry of the MIDAS and increases the affinity for ligand. The β I-like domain associates with the β-propeller domain of α subunit at the linkage to the I domain, forming a globular ligand binding integrin head (Fig. 1). It is proposed that the universally conserved glutamate residue in the I domain is an "intrinsic ligand" for β I-like domain and that binding of the activated β I-like domain to this intrinsic ligand pulls the C-terminal $\alpha7$ helix of the I domain downward, and activates high affinity for ligand (Fig. 1) [48].

The β I-like domain is attached to the hybrid domain through both its N terminus and its C-terminal $\alpha7$ helix (Fig. 1). Recently, high-resolution crystal structure of the $\alpha_{IIb}\beta_3$ headpiece that contains the β I-like, hybrid, and plexin/semaphorin/integrin (PSI) domains revealed that downward displacement of the β $\alpha7$-helix in the open, high-affinity conformation causes a 62° swing-out of the hybrid and PSI domains [49]. Therefore, the orientation of the hybrid domain in β subunit to the β I-like domain is critical for transmitting global conformational changes in the cytoplasmic and transmembrane domains into local conformational changes that directly regulate affinity to ligands (Fig. 1).

Cytoplasmic domains and conformational signal transmission

In the low-affinity conformation, the cytoplasmic tails of α and β subunits are associated with each other and constrain the extracellular head in folding over the legs owing to a bend at the knees ("Genu") (Fig. 1) [50], whereas in the high-affinity conformation, the cytoplasmic tails are dissociated and the extracellular domains are extended and upright (Fig. 1) [51]. Mutation of the highly conserved GFFKR sequence at the boundary between integrin α subunit transmembrane and cytoplasmic domains results in constitutive adhesiveness [52, 53] as well as increased exposure of extracellular domain activation epitopes [53], suggesting that the GFFKR sequence stabilizes the inactive integrin conformation by forming hydrophobic interactions and a salt bridge with the β subunit. During "inside-out" signaling, stimuli received by a variety of cell surface receptors, such as chemokine or T cell receptors (TCRs), initiate intracellular signals that impinge on integrin cytoplasmic domains and separate the cytoplasmic tails of α and β subunits. These events lead to disruption of the association between the integrin head and legs, and a switchblade-like extension (Fig. 1). This unbending event is linked to the outward swing of the hybrid domain and a pull-down of the C-terminal α helix of the β I-like domain, which induces the active high-affinity conformation. In integrins that lack I domains, the conformational changes of the β I-like domain are the last step for their ligand binding, where the active β I-like domain directly binds with its ligand. The N-terminal region of the integrin α-subunit contains seven repeats of ~60 amino acids, which fold into a seven-bladed β-propeller domain. The β-propeller domain directly participates in ligand recognition in those integrins that lack α-I domains. In the case of I domain-containing integrins such as LFA-1 and Mac-1, one more additional step is required; the active form of the β I-like domain interacts with the C-terminal I domain α7-helix and pull downward, which converts the I domain into the active form that binds extrinsic ligands such as ICAMs.

Ligand binding triggers conformational changes and clustering (multimerization), resulting in "outside-in" signals. "Outside-in" and "inside-out" signaling can affect a variety of cellular functions such as apoptosis, cytotoxicity, cell proliferation, cytokine production, antigen presentation, and gene activation [54, 55]. Several molecules mediate this signaling and recent studies have begun to link these events clearly.

Talin, a 250-kDa cytoskeletal protein that is composed of a 47-kDa N-terminal head domain and a 190-kDa C-terminal rod domain, was known for many years to associate with integrin cytoplasmic domains, and has now been recognized as an important modulator of integrin affinity. When freed from association with the rod domain by proteolysis or truncation, the talin head domain directly binds to β subunits promoting an unclasping and thus stabilizing of the high-affinity conformation [51]. Mutant talin, resistant to calpain proteolysis, revealed a reduction in the dynamics of adhesion turnover during cell migration in the absence of talin cleavage [56]. Although talin likely meets important functional requirements by linking LFA-1 to the cytoskeleton and at the same time regulating LFA-1 affinity, little is known about its precise role on LFA-1 and Mac-1 activation during leukocyte migration.

Rap-1, a member of Ras family of small GTPases, and its interacting molecule, RAPL, are required for TCR- and chemokine-induced adhesion of leukocyte to ICAM-1 and VCAM-1 [57]. Isolated B and T lymphocytes from RAPL knockout mice are much less adherent to ICAM-1 and fibronectin, impaired in migration, and do not show the characteristic cell polarization and clustering of integrins after stimulation with cytokines. In addition, lymphocytes and DCs from RAPL-deficient mice show a substantial impairment in their homing to the target organs, such as lymph nodes and spleen. RAPL binds to the α_L cytoplasmic domain residues Lys1097 and Lys1099 [58]. Therefore, it is likely that RAPL functions analogously to talin by destabilizing the α–β cytoplasmic interface.

The Rho family of small GTPases has been well characterized as key regulatory molecules in cell migration. Recently, it was shown that RhoH, a leukocyte-specific inhibitory Rho family member, is an important molecule that negatively regulates LFA-1 activation [59]. Unlike other Rho family GTPases, RhoH is GTPase deficient, remaining in the constitutively active, GTP-bound state [60]. Consequently, RhoH activity is not dynamically regulated by the cycle of GDP-GTP exchange. Therefore, RhoH might exert its "default" inhibitory signals for keeping LFA-1 in an inactive, nonadhesive state until an appropriate signal such as Rap-1 override the negative RhoH effect and triggers an adhesion-activating pathway. The concerted action of these two enzymes in regulation of LFA-1 activation during cell migration is not known.

Integrin clustering ("avidity" or "valency" regulation)

In addition to alterations in the conformation of individual integrin ab heterodimers ("affinity regulation"), the overall strength of cellular adhesive interactions is also regulated by the total number of bonds formed, *i.e.*, the "valency" of the interaction ("avidity regulation") [61]. Avidity (valency) regulation is mainly mediated by changes in cell surface receptor diffusivity or local clustering that alter the number of adhesive bonds that can form. However, the nature of avidity regulation or integrin "clusters", what drives their formation, and their role in dynamic adhesions has remained unclear until recently. Previously, it had been believed that release of cytoskeletal constraints by inside-out signals is rapidly followed by integrin clustering due to its intrinsic propensity to form cell surface integrin oligomers and thereby, enhancing cell adhesion. Recently, it was showed in living cells that LFA-1 clustering does not precede ligand binding but is a consequence of ligand binding. Instead, increased integrin diffusivity acts simply to facilitate ligand-dependent, mass action-driven accumulation of integrins into the site of contact with multivalent ligand substrates, resulting in "adhesion strengthening" [61]. Recently, it has been shown that both LFA-1 and Mac-1 signals uses molecules like Wiskott-Aldrich syndrome protein (WASp), Erk1/2, Syk, Pyk2, Vav-1,Cbl, Akt and PI3Kγ to enhance clustering and adhesion strengthening, as well transition to the next step in the diapedesis process, called locomotion [62].

The structural mechanism of integrin clustering is largely unknown. A recent study proposed homomeric transmembrane domain association (α-α or β-β) between neighboring integrins following heterodimeric transmembrane dissociation by integrin activation [63]. This model was based on the observation that peptides containing integrin α and β subunit transmembrane domains form homodimers and homotrimers in detergent micelles. However, in subsequent studies, it has been shown that this does not occur as a consequence of integrin activation by α and β subunit transmembrane separation, although it might occur after binding to multivalent ligands [64].

Cellular function of LFA-1 and Mac-1

Major cellular functions of both LFA-1 and Mac-1 occur during emigration (extravasation) of leukocytes and antigen recognition. Mac-1, as described above, also has the added ability to be used during phagocytosis and pathogen/foreign object recognition. We provide a synopsis of these functions here, although this field is quite large and some functions are still being explored or discovered.

Emigration

Emigration is the process by which leukocytes leave the blood and enter tissues. Many of the adhesion molecules discussed in this book are involved at many steps in the emigration process [4, 5]. Roles for LFA-1 and Mac-1 have been found at nearly every step in the process. There are several important things to note. First, most lymphocytes do not express Mac-1, and because there are so many subsets of leukocytes, some of the events during the emigration process may vary for each type of cell. Second, different tissues have different early events in the process to govern what types of cell enter and when. This may affect spatial and temporal functions of LFA-1 and Mac-1. Third, the expression of the ligands for LFA-1 and Mac-1 do overlap, and compensate for the blockade of a single ligand. However, there is evidence that the precise molecular steps may be distinctly different and sequential. Different systems used to study these events, such as antibody blockade and LFA-1- or Mac-1-deficient mice, has occasionally led to some potentially conflicting results in the literature, which require one to carefully compare and contrast methods and results.

Slow rolling

Rolling requires selectins, as discussed in earlier chapters. However, slow rolling and tethering of the leukocytes is aided by integrins. LFA-1-mediated rolling on ICAM-1 (and ICAM-3) was first shown on transfected cell lines *in vitro* [65]. Using LFA-1-, Mac-1- and CD18-deficient mice, it was shown that LFA-1 and Mac-1 contribute to leukocyte rolling cooperatively, but LFA-1 had the strongest effect at this step [66]. Further studies showed that the major binding partner for Mac-1 was ICAM-1, and the binding partner for LFA-1 was not ICAM-1, ICAM-2 or JAM-A [67]. Looking specifically at neutrophils, studies in LFA-1-deficient mice confirmed that LFA-1 and VLA-4 were important for rolling and firm adhesion of neutrophils, but found that Mac-1 was used for "emigration", which could have been either of the subsequent steps: locomotion or transendothelial migration [68]. Rolling of neutrophils *via* LFA-1 has also been shown using purified ICAM-1 and E-selectin [69]. As shown in the structural studies described above, the extended conformational state of LFA-1 is needed to mediate rolling using a low affinity I domain [70]. Earlier studies operated with the paradigm that rolling is mediated more by the selectin family members, as well VLA-4, rather than LFA-1 or Mac-1 [71–74], and it is clear that interactions between selectin family members, chemokines, and leukocyte integrins together are required for the rolling step to proceed normally.

Arrest (firm adhesion)

There is abundant evidence that both LFA-1 and Mac-1 play a critical role for the arrest of leukocytes on the endothelial cells, arguably one of the most critical functions of these integrins [71, 72, 75–78]. Using CD18-, CD11a- and CD11b-deficient mice, it has been shown that LFA-1 has the strongest effects at this stage, although Mac-1 seems to play a minor role as well [77]. Since lymphocytes lack Mac-1, LFA-1 functions in concert with α_4 integrins [71, 79, 80].

The counter ligands ICAM-1, ICAM-2, JAM-A, and JAM-C also appear to contribute to firm adhesion on endothelial cells [34, 35, 81–86]. ICAM-1 is expressed at sites of inflammation, whereas ICAM-2 is constitutively expressed [26, 87–89]. A recent study by Huang et al. [90] has shown that ICAM-2-deficient mice do not a have a defect in adhesion after inflammatory stimuli such as IL-1β or TNF-α are used. On resting endothelium, ICAM-2 has a peri-junctional pattern of expression [90–92]. Because of this expression pattern, it is reasonable to hypothesize that ICAM-2 may have a role for the constitutive surveillance of normal tissues. A similar function has been proposed for JAM-A and JAM-C. Both molecules localize to junctions and antibodies block adhesion [34, 35, 85, 86].

Post-arrest adhesion strengthening

A relatively recently described and distinct molecular event during emigration is called post-arrest adhesion strengthening. As described above, increased integrin diffusivity acts simply to facilitate ligand-dependent, mass action-driven accumulation of integrins into the site of contact with multivalent ligand substrates, resulting in this "adhesion strengthening" [61]. Further, it has been shown that WASp is involved in clustering and downstream Erk1/2 Syk, Pyk2, Vav-1,Cbl, Akt and PI3Kγ signaling. WASp thus plays a critical role at this step, which in turn affects subsequent cell function like locomotion [62, 93].

Locomotion/intraluminal crawling

Locomotion was widely overlooked as a distinct step in diapedesis for many years, despite descriptions of B lymphocyte and natural killer (NK) cells crawling on lipid bilayers that were embedded with purified ICAM-1 [94, 95]. More recently, locomotion of monocytes and neutrophils from the site of firm adhesion to an endothelial cell-endothelial cell junction has been shown by several groups [92, 96, 97]. In general, leukocytes adhere within a few microns of a junction just as a matter of physical space on the surface of the endothelial cells [96]. This still requires movement of the leukocyte to a junction. Using *in vitro* assays, there is also evidence that Mac-1 is the dominant integrin involved in chemotaxis of neutrophils to fMLP, and LFA-1 is the dominant integrin involved in chemotaxis to IL-8 [98]. This may be partially due to where each stimulus might found physiologically. fMLP is not likely to be found in the vascular bed like IL-8, but may be found in the ECM *in vivo*. This has been followed up to show that that LFA-1 and Mac-1 act sequentially on neutrophils, with LFA-1 upstream mediating firm adhesion and Mac-1 mediating locomotion [97].

It appears that ICAM-1 and ICAM-2 also have major roles at this stage [90–92]. Again, because ICAM-2 is constitutively expressed and localizes towards junctions on quiescent endothelium, it is possible that this forms a haptotactic gradient that is followed by the leukocytes, although preliminary studies were unable to confirm this [92]. JAM-A and JAM-C may also contribute to locomotion, but activity at this precise step has not been shown [34, 35, 84, 85].

Transendothelial migration

This part of the process has two possible routes, either through the endothelial cell-endothelial cell junctions (paracellular migration) or through the middle of a single endothelial cell (transcellular migration; reviewed in [99]). There are still a lot of open questions in this process. Unfortunately, many studies claiming to have blocked transendothelial migration have not identified the exact step in the process; upstream events like firm adhesion and locomotion obviously affect downstream events.

As previously discussed, the studies on locomotion hint that the interactions between LFA-1, Mac-1, ICAM-1 and ICAM-2 guide the leukocyte to a site where transendothelial migration can occur. Membranes enriched with ICAM-1 and VCAM-1 on endothelial cells and LFA-1 and Mac-1 on neutrophils, monocytes and lymphocytes have been seen as they cross through or around endothelial cells *in vitro* [100, 101]. This study was corroborated by another group, which showed that neutrophil migration was marked by LFA-1-ICAM-1 forming a ring around the neutrophil as they pass through or around endothelial cells [102]. One might consider this a "emigration synapse" as it is strikingly similar to the "immunological synapse" found during antigen presentation as discussed later in this chapter.

Again, putative roles for JAM-A and JAM-C have been placed during transendothelial migration [34, 35, 84, 85], although one study did not find a role for neutrophil migration *via* JAM-A *in vitro* [102].

Transepithelial migration

Similar to transendothelial migration, leukocytes must also cross epithelial layers to reach the sites of inflammation. This requires adhesion of the leukocytes to the basal surface of the epithelium, followed by extrusion through the tight epithelial cell junctions to reach the apical surface. Although roles for LFA-1 and Mac-1 have been shown in this process, the precise steps controlled by these integrins have not always been identified. Mac-1 and LFA-1 both have been shown to mediate transepithelial migration after indomethacin-induced injury of intestinal tissue *in vivo* [103]. Similar results have been shown *in vitro* with neutrophil migration across human airway cells [104] and lymphocyte migration across retinal pigment epithelial monolayers [105]. LFA-1 interactions with JAM-C appear to play a role in transepithelial migration of neutrophils [106].

Migration in the ECM

LFA-1 and Mac-1, in concert with other integrins, provides a clear mechanism for transport through the three-dimensional matrix at wounds and into sites of inflammation [107, 108]. Fibroblasts, especially in inflamed tissues have been shown to express ICAM-1 and ICAM-2 to support monocyte and neutrophil migration [109, 110]. Because lymphocytes lack Mac-1, other integrins and adhesion molecules must assume a primary role for movement [111]. Mac-1 has also been shown to closely associate with the serine protease urokinase type plasminogen activator receptor, which helps Mac-1 bind to fibrinogen [91]. This is also consistent with chemotaxis studies with neutrophils, demonstrating that Mac-1 is preferentially used by neutrophils in response to fMLP *in vitro* [98], a chemotactic gradient likely to exist from an infected wound *in vivo*.

Antigen presentation and signaling

Consistent with the ability of adhesion molecules to stabilize the interactions between two cells during diapedesis, this same stabilization is used between the contacts of T cells with APCs [112–114]. This may also function in contacts of neutrophils with DCs *via* Mac-1 and DC-SIGN [40]. CD18 integrins and DNAX accessory molecule-1 (DNAM-1/CD226) cooperate for signal function as well [115]. Finally, the elucidation of downstream signal pathways for many of the processes governed by LFA-1 and Mac-1 provides further insight into the "outside-in" and "inside-out" paradigm.

LFA-1, ICAM-1, and the immunological synapse
The role of LFA-1 in T cell interactions with APCs has been known for quite sometime [112], but there have been a number of recent advances in how this interaction works. The first studies to show the formation of a discreet immunological synapse remain landmark publications for several reasons. The distinct interactions between major histocompatibility complexes (MHC) surrounded by ICAM-1 on the APC and their respective cognate ligands TCRs are called supramolecular activation clusters (SMACs, [113]) or immunological synapses [114]. Fluorescent imaging was used to show the MHC surrounded by ICAM-1 on the APC and their respective cognate ligands TCR surrounded by LFA-1 [113, 114]. This dynamic complex confirms that clustering, aided by integrins, is a vital aspect of cell-cell signaling events. Two groups have also demonstrated a common signaling connection between the TCR and LFA-1 *via* the adapter molecule Fyb/Slap-130 [116, 117].

DC-SIGN signaling
DC-SIGN (CD209) was identified as a receptor on the specialized family of APCs, the DCs, for ICAM-2 and ICAM-3 [40]. LFA-1 alternates with DC-SIGN for ICAM-2 and ICAM-3 [42]. Neutrophils and DCs communicate partly *via* DC-SIGN and Mac-1 interactions, extending the life of the neutrophil and facilitating antigen recognition responses [39]. Interestingly, DC-SIGN interactions were also shown to mediate DC migration across both resting and activated endothelial cells *in vitro* [40].

DNAM signaling
DNAM-1 is expressed on most T lymphocytes, NK cells, and monocytes. LFA-1 constitutively associates with DNAM-1 in NK cells and on exogenously activated T cells. On activated T cells, cross-linking with anti-CD18 induced tyrosine phosphorylation of DNAM-1 *via* the Fyn protein tyrosine kinase [118]. The ligands for

DNAM-1 are the polio virus receptor PVR (CD155) and nectin-2 (PRR-2/CD112) [119]. It has recently been shown that DNAM-1 and PVR are also involved in cell adhesion and monocyte migration [102, 120].

Integrin signaling pathways

Several signal pathways have been described earlier in this review. There are also a growing number of comprehensive and excellent reviews on LFA-1- and Mac-1-mediated signaling. T cell signaling *via* LFA-1 has been reviewed by Hogg et al. [55]. LFA-1 signaling *via* molecules like the Rho GTPases is important during diapedesis [93, 121]. Cell signal pathways activated by Mac-1-mediated phagocytosis have been shown as well, acting though G proteins, especially Rho-GTPases, related members like Vav, and Fyb [122]. Neutrophil interactions like Mac-1 with ICAM members, ECM, and DC-SIGN have also been shown to activate proliferation or death signals [123]. Finally, tyrosine kinases, like Syk and Hck, mediate signaling for degranulation and the respiratory burst by neutrophils *in vitro* and *in vivo* [122, 124].

Summary

Here we have tried to summarize a vast amount of research that started in the early 1980s on LFA-1 and Mac-1 function. As diverse as any adhesion molecule family, the functions and ligands of these two integrins have a major impact on much of the immune response. Much progress has been made in elucidating the structural changes in these molecules and relating them to function. This is a major achievement considering the difficulties of analyzing the three-dimensional structure of cell surface proteins. Parallel and sometimes redundant functions have been shown in some specific parts of function, such as during emigration. In other cases the two molecules have distinct roles, such as the role of LFA-1 in TCR signaling and that of Mac-1 during phagocytosis. There is also much to be further explored as the molecular events that occur during adhesion molecule function reveal novel structural changes and cellular processes.

References

1   Davignon D, Martz E, Reynolds T, Kürzinger K, Springer TA (1981) Lymphocyte function-associated antigen 1 (LFA-1): A surface antigen distinct from Lyt-2,3 that participates in T lymphocyte-mediated killing. *Proc Natl Acad Sci USA* 78: 4535–4539
2   Springer TA, Galfre G, Secher DS, Milstein C (1979) Mac-1: a macrophage differentiation antigen identified by monoclonal antibody. *Eur J Immunol* 9: 301–306

3   Sanchez-Madrid F, Simon P, Thompson S, Springer TA (1983) Mapping of antigenic and functional epitopes on the α and β subunits of two related glycoproteins involved in cell interactions, LFA-1 and Mac-1. *J Exp Med* 158: 586–602

4   Springer TA (1994) Traffic signals for lymphocyte recirculation and leukocyte emigration: the multi-step paradigm. *Cell* 76: 301–314

5   Hynes RO (1992) Integrins: Versatility, modulation, and signaling in cell adhesion. *Cell* 69: 11–25

6   McMichael AJ (1987) *Leukocyte typing III: White cell differentiation antigens.* Oxford University Press, Oxford

7   Hynes RO (1987) Integrins: A family of cell surface receptors. *Cell* 48: 549–554

8   Ehlers MR (2000) CR3: a general purpose adhesion-recognition receptor essential for innate immunity. *Microbes Infect* 2: 289–294

9   Todd RF III, van Agthoven A, Schlossman SF, Terhorst C (1982) Structural analysis of differentiation antigens Mo1 and Mo2 on human monocytes. *Hybridoma* 1: 329–337

10   Hatta K, Takeichi M (1986) Expression of N-cadherin adhesion molecules associated with early morphogenetic events in chick development. *Nature* 320: 447–449

11   Nagafuchi A, Shirayoshi Y, Okazaki K, Yasuda K, Takeichi M (1987) Transformation of cell adhesion properties by exogenously introduced E-cadherin cDNA. *Nature* 329: 341–343

12   Pytela R, Pierschbacher MD, Ruoslahti E (1985) Identification and isolation of a 140 kd cell surface glycoprotein with properties expected of a fibronectin receptor. *Cell* 40: 191–198

13   Pytela R, Pierschbacher MD, Ruoslahti E (1985) A 125/115-kDa cell surface receptor specific for vitronectin interacts with the arginine-glycine-aspartic acid adhesion sequence derived from fibronectin. *Proc Natl Acad Sci USA* 82: 5766–5770

14   Brackenbury R, Rutishauser U, Edelman GM (1981) Distinct calcium-independent and calcium-dependent adhesion systems of chicken embryo cells. *Proc Natl Acad Sci USA* 78: 387–391

15   Hoffman S, Sorkin BC, White PC, Brackenbury R, Mailhammer R, Rutishauser U, Cunningham BA, Edelman GM (1982) Chemical characterization of a neural cell adhesion molecule purified from embryonic brain membranes. *J Biol Chem* 257: 7720–7729

16   Crowley CA, Curnutte JT, Rosin RE, Andre-Schwartz J, Gallin JI, Klempner M, Snyderman R, Southwick FS, Stossel TP, Babior BM (1980) An inherited abnormality of neutrophil adhesion. Its genetic transmission and its association with a missing protein. *N Engl J Med* 302: 1163–1168

17   Harlan JM, Killen PD, Senecal FM, Schwartz BR, Yee EK, Taylor RF, Beatty PG, Price TH, Ochs HD (1985) The role of neutrophil membrane glycoprotein GP–150 in neutrophil adherence to endothelium *in vitro*. *Blood* 66: 167–178

18   Anderson DC, Kishimoto TK, Smith CW (1995) Leukocyte adhesion deficiency and other disorders of leukocyte adherence and motility. In: CR Scriver, AL Beaudet, WS Sly, D Valle (eds): *The metabolic and molecular bases of inherited disease.* McGraw-Hill, New York, 3955–3994

19   Tamkun JW, DeSimone DW, Fonda D, Patel RS, Buck C, Horwitz AF, Hynes RO (1986) Structure of integrin, a glycoprotein involved in the transmembrane linkage between fibronectin and actin. *Cell* 46: 271–282

20   Sanchez-Madrid F, Krensky AM, Ware CF, Robbins E, Strominger JL, Burakoff SJ, Springer TA (1982) Three distinct antigens associated with human T lymphocyte-mediated cytolysis: LFA-1, LFA-2, and LFA-3. *Proc Natl Acad Sci USA* 79: 7489–7493

21   Ross GD, Lambris JD (1982) Identification of a C3bi-specific membrane complement receptor that is expressed on lymphocytes, monocytes, neutrophils, and erythrocytes. *J Exp Med* 155: 96–110

22   Breard J, Reinherz EL, Kung PC, Goldstein G, Schlossman SF (1980) A monoclonal antibody reactive with human peripheral blood monocytes. *J Immunol* 124: 1943–1948

23   Todd RF III, Nadler LM, Schlossman SF (1981) Antigens on human monocytes identified by monoclonal antibodies. *J Immunol* 126: 1435–1442

24   Marlin SD, Springer TA (1987) Purified intercellular adhesion molecule-1 (ICAM-1) is a ligand for lymphocyte function-associated antigen 1 (LFA-1). *Cell* 51: 813–819

25   Diamond MS, Staunton DE, de Fougerolles AR, Stacker SA, Garcia-Aguila J, Hibbs ML, Springer TA (1990) ICAM-1 (CD54): a counter-receptor for Mac-1 (CD11b/CD18). *J Cell Biol* 111: 3129–3139

26   de Fougerolles AR, Stacker SA, Schwarting R, Springer TA (1991) Characterization of ICAM-2 and evidence for a third counter-receptor for LFA-1. *J Exp Med* 174: 253–267

27   de Fougerolles AR, Qin X, Springer TA (1994) Characterization of the function of intercellular adhesion molecule (ICAM)–3 and comparison with ICAM-1 and ICAM-2 in immune responses. *J Exp Med* 179: 619–629

28   Ihanus E, Uotila L, Toivanen A, Stefanidakis M, Bailly P, Cartron JP, Gahmberg CG (2003) Characterization of ICAM-4 binding to the I domains of the CD11a/CD18 and CD11b/CD18 leukocyte integrins. *Eur J Biochem* 270: 1710–1723

29   Tian L, Kilgannon P, Yoshihara Y, Mori K, Gallatin WM, Carpen O, Gahmberg CG (2000) Binding of T lymphocytes to hippocampal neurons through ICAM-5 (telencephalin) and characterization of its interaction with the leukocyte integrin CD11a/CD18. *Eur J Immunol* 30: 810–818

30   de Fougerolles AR, Diamond MS, Springer TA (1995) Heterogenous glycosylation of ICAM-3 and lack of interaction with Mac-1 and p150,95. *Eur J Immunol* 25: 1008–1012

31   Neelamegham S, Taylor AD, Shankaran H, Smith CW, Simon SI (2000) Shear and time-dependent changes in Mac-1, LFA-1, and ICAM-3 binding regulate neutrophil homotypic adhesion. *J Immunol* 164: 3798–3805

32   Mandell KJ, Parkos CA (2005) The JAM family of proteins. *Adv Drug Deliv Rev* 57: 857–867

33   Muller WA (2003) Leukocyte-endothelial-cell interactions in leukocyte transmigration and the inflammatory response. *Trends Immunol* 24: 327–334

34   Ostermann G, Weber KS, Zernecke A, Schroder A, Weber C (2002) JAM-1 is a ligand

of the beta(2) integrin LFA-1 involved in transendothelial migration of leukocytes. *Nat Immunol* 3: 151–158

35 Santoso S, Sachs UJ, Kroll H, Linder M, Ruf A, Preissner KT, Chavakis T (2002) The junctional adhesion molecule 3 (JAM–3) on human platelets is a counterreceptor for the leukocyte integrin Mac-1. *J Exp Med* 196: 679–691

36 Trezzini C, Jungi TW, Kuhnert P, Peterhans E (1988) Fibrinogen association with human monocytes: evidence for constitutive expression of fibrinogen receptors and for involvement of Mac-1 (CD18, CR3) in the binding. *Biochem Biophys Res Commun* 156: 477–484

37 Davis GE (1992) The Mac-1 and p150,95 beta 2 integrins bind denatured proteins to mediate leukocyte cell-substrate adhesion. *Exp Cell Res* 200: 242–252

38 Todd RF 3rd, Petty HR (1997) Beta 2 (CD11/CD18) integrins can serve as signaling partners for other leukocyte receptors. *J Lab Clin Med* 129: 492–498

39 van Gisbergen KP, Sanchez-Hernandez M, Geijtenbeek TB, van Kooyk Y (2005) Neutrophils mediate immune modulation of dendritic cells through glycosylation-dependent interactions between Mac-1 and DC-SIGN. *J Exp Med* 201: 1281–1292

40 Geijtenbeek TB, Krooshoop DJ, Bleijs DA, van Vliet SJ, van Duijnhoven GC, Grabovsky V, Alon R, Figdor CG, van Kooyk Y (2000) DC-SIGN-ICAM-2 interaction mediates dendritic cell trafficking. *Nat Immunol* 1: 353–357

41 Geijtenbeek TB, Torensma R, van Vliet SJ, van Duijnhoven GC, Adema GJ, van Kooyk Y, Figdor CG (2000) Identification of DC-SIGN, a novel dendritic cell-specific ICAM-3 receptor that supports primary immune responses. *Cell* 100: 575–585

42 Bleijs DA, Geijtenbeek TB, Figdor CG, van Kooyk Y (2001) DC-SIGN and LFA-1: a battle for ligand. *Trends Immunol* 22: 457–463

43 Shimaoka M, Lu C, Palframan R, von Andrian UH, Takagi J, Springer TA (2001) Reversibly locking a protein fold in an active conformation with a disulfide bond: integrin αL I domains with high affinity and antagonist activity *in vivo*. *Proc Natl Acad Sci USA* 98: 6009–6014

44 Huth JR, Olejniczak ET, Mendoza R, Liang H, Harris EA, Lupher ML Jr, Wilson AE, Fesik SW, Staunton DE (2000) NMR and mutagenesis evidence for an I domain allosteric site that regulates lymphocyte function-associated antigen 1 ligand binding. *Proc Natl Acad Sci USA* 97: 5231–5236

45 Shimaoka M, Xiao T, Liu J-H, Yang Y, Dong Y, Jun C-D, McCormack A, Zhang R, Joachimiak A, Takagi J et al (2003) Structures of the αL I domain and its complex with ICAM-1 reveal a shape-shifting pathway for integrin regulation. *Cell* 112: 99–111

46 Shimaoka M, Kim M, Cohen EH, Yang W, Astrof N, Peer D, Salas A, Ferrand A, Springer TA (2006) AL-57, a ligand-mimetic antibody to integrin LFA-1, reveals chemokine-induced affinity up-regulation in lymphocytes. *Proc Natl Acad Sci USA* 103: 13991–13996

47 Diamond MS, Garcia-Aguilar J, Bickford JK, Corbi AL, Springer TA (1993) The I domain is a major recognition site on the leukocyte integrin Mac-1 (CD11b/CD18) for four distinct adhesion ligands. *J Cell Biol* 120: 1031–1043

48   Yang W, Shimaoka M, Salas A, Takagi J, Springer TA (2004) Inter-subunit signal trans-
     mission in integrins by a receptor-like interaction with a pull spring. *Proc Natl Acad Sci
     USA* 101: 2906–2911

49   Xiao T, Takagi J, Coller BS, Wang JH, Springer TA (2004) Structural basis for allostery
     in integrins and binding to fibrinogen-mimetic therapeutics. *Nature* 432: 59–67

50   Nishida N, Xie C, Shimaoka M, Cheng Y, Walz T, Springer TA (2006) Activation of leu-
     kocyte beta(2) integrins by conversion from bent to extended conformations. *Immunity*
     25: 583–594

51   Kim M, Carman CV, Springer TA (2003) Bidirectional transmembrane signaling by
     cytoplasmic domain separation in integrins. *Science* 301: 1720–1725

52   Hughes PE, Diaz-Gonzalez F, Leong L, Wu C, McDonald JA, Shattil SJ, Ginsberg MH
     (1996) Breaking the integrin hinge. A defined structural constraint regulates integrin
     signaling. *J Biol Chem* 271: 6571–6574

53   Lu CF, Springer TA (1997) The alpha subunit cytoplasmic domain regulates the assem-
     bly and adhesiveness of integrin lymphocyte function-associated antigen-1. *J Immunol*
     159: 268–278

54   Hogg N, Harvey J, Cabanas C, Landis RC (1993) Control of leukocyte integrin activa-
     tion. *Am Rev Respir Dis* 148: S55–59

55   Hogg N, Laschinger M, Giles K, McDowall A (2003) T-cell integrins: more than just
     sticking points. *J Cell Sci* 116: 4695–4705

56   Franco SJ, Rodgers MA, Perrin BJ, Han J, Bennin DA, Critchley DR, Huttenlocher A
     (2004) Calpain-mediated proteolysis of talin regulates adhesion dynamics. *Nat Cell Biol*
     6: 977–983

57   Kinashi T, Katagiri K (2004) Regulation of lymphocyte adhesion and migration by the
     small GTPase Rap1 and its effector molecule, RAPL. *Immunol Lett* 93: 1–5

58   Tohyama Y, Katagiri K, Pardi R, Lu C, Springer TA, Kinashi T (2003) The critical cyto-
     plasmic regions of the αL/β2 integrin in Rap1-induced adhesion and migration. *Mol
     Biol Cell* 14: 2570–2582

59   Cherry LK, Li X, Schwab P, Lim B, Klickstein LB (2004) RhoH is required to maintain
     the integrin LFA-1 in a nonadhesive state on lymphocytes. *Nat Immunol* 5: 961–967

60   Sahai E, Marshall CJ (2002) RHO-GTPases and cancer. *Nat Rev* Cancer 2: 133–142

61   Kim M, Carman CV, Yang W, Salas A, Springer TA (2004) The primacy of affinity
     over clustering in regulation of adhesiveness of the integrin αLβ2. *J Cell Biol* 167:
     1241–1253

62   Zhang H, Schaff UY, Green CE, Chen H, Sarantos MR, Hu Y, Wara D, Simon SI, Lowell
     CA (2006) Impaired integrin-dependent function in Wiskott-Aldrich syndrome protein-
     deficient murine and human neutrophils. *Immunity* 25: 285–295

63   Li R, Mitra N, Gratkowski H, Vilaire G, Litvinov SV, Nagasami C, Weisel JW, Lear JD,
     DeGrado WF, Bennett JS (2003) Activation of integrin αIIbβ3 by modulation of trans-
     membrane helix associations. *Science* 300: 795–798

64   Luo BH, Carman CV, Takagi J, Springer TA (2005) Disrupting integrin transmembrane

domain heterodimerization increases ligand binding affinity, not valency or clustering. *Proc Natl Acad Sci USA* 102: 3679–3684

65   Knorr R, Dustin ML (1997) The lymphocyte function-associated antigen 1 I domain is a transient binding module for intercellular adhesion molecule (ICAM)-1 and ICAM-3 in hydrodynamic flow. *J Exp Med* 186: 719–730

66   Dunne JL, Ballantyne CM, Beaudet AL, Ley K (2002) Control of leukocyte rolling velocity in TNF-alpha-induced inflammation by LFA-1 and Mac-1. *Blood* 99: 336–341

67   Dunne JL, Collins RG, Beaudet AL, Ballantyne CM, Ley K (2003) Mac-1, but not LFA-1, uses intercellular adhesion molecule-1 to mediate slow leukocyte rolling in TNF-alpha-induced inflammation. *J Immunol* 171: 6105–6111

68   Henderson RB, Lim LHK, Tessier PA, Gavins FNE, Mathies M, Perretti M, Hogg N (2001) The use of lymphocyte function-associated antigen (LFA)-1-deficient mice to determine the role of LFA-1, Mac-1, and α4 integrin in the inflammatory response of neutrophils. *J Exp Med* 194: 219–226

69   Chesnutt BC, Smith DF, Raffler NA, Smith ML, White EJ, Ley K (2006) Induction of LFA-1-dependent neutrophil rolling on ICAM-1 by engagement of E-selectin. *Microcirculation* 13: 99–109

70   Salas A, Shimaoka M, Phan U, Kim M, Springer TA (2006) Transition from rolling to firm adhesion can be mimicked by extension of integrin alphaLbeta2 in an intermediate affinity state. *J Biol Chem* 281: 10876–10882

71   Lawrence MB, Springer TA (1991) Leukocytes roll on a selectin at physiologic flow rates: Distinction from and prerequisite for adhesion through integrins. *Cell* 65: 859–873

72   Luscinskas FW, Kansas GS, Ding H, Pizcueta P, Schleiffenbaum BE, Tedder TF, Gimbrone MA (1994) Monocyte rolling, arrest and spreading on IL-4-activated vascular endothelium under flow is mediated *via* sequential action of L-selectin, β1-integrins and β2-integrins. *J Cell Biol* 125: 1417–1427

73   Berlin C, Bargatze RF, Campbell JJ, von Andrian UH, Szabo MC, Hasslen SR, Nelson RD, Berg EL, Erlandsen SL, Butcher EC (1995) Alpha 4 integrins mediate lymphocyte attachment and rolling under physiologic flow. *Cell* 80: 413–422

74   Warnock RA, Askari S, Butcher EC, von Andrian UH (1998) Molecular mechanisms of lymphocyte homing to peripheral lymph nodes. *J Exp Med* 187: 205–216

75   von Andrian UH, Chambers JD, McEvoy LM, Bargatze RF, Arfors K-E, Butcher EC (1991) Two-step model of leukocyte-endothelial cell interaction in inflammation: Distinct roles for LECAM-1 and the leukocyte beta 2 integrins *in vivo*. *Proc Natl Acad Sci USA* 88: 7538–7542

76   Diacovo TG, Roth SJ, Buccola JM, Bainton DF, Springer TA (1996) Neutrophil rolling, arrest, and transmigration across activated, surface-adherent platelets *via* sequential action of P-selectin and the beta 2-integrin CD11b/CD18. *Blood* 88: 146–157

77   Ding ZM, Babensee JE, Simon SI, Lu H, Perrard JL, Bullard DC, Dai XY, Bromley SK, Dustin ML, Entman ML et al (1999) Relative contribution of LFA-1 and Mac-1 to neutrophil adhesion and migration. *J Immunol* 163: 5029–5038

78   Gopalan PK, Burns AR, Simon SI, Sparks S, McIntyre LV, Smith CW (2000) Preferential sites for stationary adhesion of neutrophils to cytokine-stimulated HUVEC under flow conditions. *J Leukoc Biol* 68: 47–57

79   Shimizu Y, Newman W, Venkat Gopal T, Horgan KJ, Graber N, Beall LD, Van Seventer GA, Shaw S (1991) Four molecular pathways of T cell adhesion to endothelial cells: roles of LFA-1, VCAM-1 and ELAM-1 and changes in pathway heirarchy under different activation conditions. *J Cell Biol* 113: 1203–1212

80   Laschinger M, Engelhardt B (2000) Interaction of alpha4-integrin with VCAM-1 is involved in adhesion of encephalitogenic T cell blasts to brain endothelium but not in their transendothelial migration *in vitro*. *J Neuroimmunol* 102: 32–43

81   Dustin ML, Springer TA (1988) Lymphocyte function associated antigen-1 (LFA-1) interaction with intercellular molecule-1 (ICAM-1) is one of at least three mechanisms for lymphocyte adhesion to cultured endothelial cells. *J Cell Biol* 107: 321

82   Diamond MS, Springer TA (1993) A subpopulation of Mac-1 (CD11b/CD18) molecules mediates neutrophil adhesion to ICAM-1 and fibrinogen. *J Cell Biol* 120: 545–556

83   Issekutz AC, Rowter D, Springer TA (1999) Role of ICAM-1 and ICAM-2 and alternate CD11/CD18 ligands in neutrophil transendothelial migration. *J Leukoc Biol* 65: 117–126

84   Johnson-Leger CA, Aurrand-Lions M, Beltraminelli N, Fasel N, Imhof BA (2002) Junctional adhesion molecule-2 (JAM–2) promotes lymphocyte transendothelial migration. *Blood* 100: 2479–2486

85   Aurrand-Lions M, Lamagna C, Dangerfield JP, Wang S, Herrera P, Nourshargh S, Imhof BA (2005) Junctional adhesion molecule-C regulates the early influx of leukocytes into tissues during inflammation. *J Immunol* 174: 6406–6415

86   Chavakis T, Keiper T, Matz-Westphal R, Hersemeyer K, Sachs UJ, Nawroth PP, Preissner KT, Santoso S (2004) The junctional adhesion molecule-C promotes neutrophil transendothelial migration *in vitro* and *in vivo*. *J Biol Chem* 279: 55602–55608

87   Dustin ML, Rothlein R, Bhan AK, Dinarello CA, Springer TA (1986) Induction by IL-1 and interferon-gamma: tissue distribution, biochemistry and function of a natural adherence molecule (ICAM-1). *J Immunol* 137: 245–254

88   Rothlein R, Dustin ML, Marlin SD, Springer TA (1986) A human intercellular adhesion molecule (ICAM-1) distinct from LFA-1. *J Immunol* 137: 1270–1274

89   Staunton DE, Dustin ML, Springer TA (1989) Functional cloning of ICAM-2, a cell adhesion ligand for LFA-1 homologous to ICAM-1. *Nature* 339: 61–64

90   Huang MT, Larbi KY, Scheiermann C, Woodfin A, Gerwin N, Haskard DO, Nourshargh S (2006) ICAM-2 mediates neutrophil transmigration *in vivo*: evidence for stimulus specificity and a role in PECAM-1-independent transmigration. *Blood* 107: 4721–4727

91   McLaughlin F, Hayes BP, Horgan CM, Beesley JE, Campbell CJ, Randi AM (1998) Tumor necrosis factor (TNF)-alpha and interleukin (IL)-1beta down-regulate intercellular adhesion molecule (ICAM)–2 expression on the endothelium. *Cell Adhes Commun* 6: 381–400

92   Schenkel AR, Mamdouh Z, Muller WA (2004) Locomotion of monocytes on endothelium is a critical step during extravasation. *Nat Immunol* 5: 393–400

93   Cernuda-Morollon E, Ridley AJ (2006) Rho GTPases and leukocyte adhesion receptor expression and function in endothelial cells. *Circ Res* 98: 757–767

94   Dustin ML, Carpen O, Springer TA (1992) Regulation of locomotion and cell-cell contact area by the LFA-1 and ICAM-1 adhesion receptors. *J Immunol* 148: 2654–2663

95   Carpen O, Pallai P, Staunton DE, Springer TA (1992) Association of intercellular adhesion molecule-1 (ICAM-1) with actin-containing cytoskeleton and alpha-actinin. *J Cell Biol* 118: 1223–1234

96   Wojciechowski JC, Sarelius IH (2005) Preferential binding of leukocytes to the endothelial junction region in venules in situ. *Microcirculation* 12: 349–359

97   Phillipson M, Heit B, Colarusso P, Liu L, Ballantyne CM, Kubes P (2006) Intraluminal crawling of neutrophils to emigration sites: a molecularly distinct process from adhesion in the recruitment cascade. *J Exp Med* 203: 2569–2575

98   Heit B, Colarusso P, Kubes P (2005) Fundamentally different roles for LFA-1, Mac-1 and alpha4-integrin in neutrophil chemotaxis. *J Cell Sci* 118: 5205–5220

99   Engelhardt B, Wolburg H (2004) Mini-review: Transendothelial migration of leukocytes: through the front door or around the side of the house? *Eur J Immunol* 34: 2955–2963

100   Barreiro O, Yanez-Mo M, Serrador JM, Montoya MC, Vicente-Manzanares M, Tejedor R, Furthmayr H, Sanchez-Madrid F (2002) Dynamic interaction of VCAM-1 and ICAM-1 with moesin and ezrin in a novel endothelial docking structure for adherent leukocytes. *J Cell Biol* 157: 1233–1245

101   Carman CV, Springer TA (2004) A transmigratory cup in leukocyte diapedesis both through individual vascular endothelial cells and between them. *J Cell Biol* 167: 377–388

102   Fuchs A, Cella M, Giurisato E, Shaw AS, Colonna M (2004) Cutting edge: CD96 (tactile) promotes NK cell-target cell adhesion by interacting with the poliovirus receptor (CD155). *J Immunol* 172: 3994–3998

103   Stadnyk AW, Dollard C, Issekutz TB, Issekutz AC (2002) Neutrophil migration into indomethacin induced rat small intestinal injury is CD11a/CD18 and CD11b/CD18 co-dependent. *Gut* 50: 629–635

104   Kidney JC, Proud D (2000) Neutrophil transmigration across human airway epithelial monolayers: mechanisms and dependence on electrical resistance. *Am J Respir Cell Mol Biol* 23: 389–395

105   Devine L, Lightman SL, Greenwood J (1996) Role of LFA-1, ICAM-1, VLA-4 and VCAM-1 in lymphocyte migration across retinal pigment epithelial monolayers *in vitro*. *Immunology* 88: 456–462

106   Zen K, Babbin BA, Liu Y, Whelan JB, Nusrat A, Parkos CA (2004) JAM-C is a component of desmosomes and a ligand for CD11b/CD18-mediated neutrophil transepithelial migration. *Mol Biol Cell* 15: 3926–3937

107   Walzog B, Schuppan D, Heimpel C, Hafezi-Moghadam A, Gaehtgens P, Ley K (1995)

The leukocyte integrin Mac-1 (CD11b/CD18) contributes to binding of human granulocytes to collagen. *Exp Cell Res* 218: 28–38

108 van den Berg JM, Mul FP, Schippers E, Weening JJ, Roos D, Kuijpers TW (2001) Beta1 integrin activation on human neutrophils promotes beta2 integrin-mediated adhesion to fibronectin. *Eur J Immunol* 31: 276–284

109 Shang XZ, Issekutz AC (1998) Contribution of CD11a/CD18, CD11b/CD18, ICAM-1 (CD54) and -2 (CD102) to human monocyte migration through endothelium and connective tissue fibroblast barriers. *Eur J Immunol* 28: 1970–1979

110 Issekutz AC (1998) Adhesion molecules mediating neutrophil migration to arthritis *in vivo* and across endothelium and connective tissue barriers *in vitro*. *Inflamm Res* 47: 123–132

111 Friedl P, Brocker EB (2000) T cell migration in three-dimensional extracellular matrix: guidance by polarity and sensations. *Dev Immunol* 7: 249–266

112 Davignon D, Martz E, Reynolds T, Kurzinger K, Springer TA (1981) Lymphocyte function-associated antigen 1 (LFA-1): a surface antigen distinct from Lyt-2,3 that participates in T lymphocyte-mediated killing. *Proc Natl Acad Sci USA* 78: 4535–4539

113 Monks CR, Freiberg BA, Kupfer H, Sciaky N, Kupfer A (1998) Three-dimensional segregation of supramolecular activation clusters in T cells. *Nature* 395: 82–86

114 Grakoui A, Bromley SK, Sumen C, Davis MM, Shaw AS, Allen PM, Dustin ML (1999) The immunological synapse: a molecular machine controlling T cell activation. *Science* 285: 221–227

115 Shibuya K, Shirakawa J, Kameyama T, Honda S, Tahara-Hanaoka S, Miyamoto A, Onodera M, Sumida T, Nakauchi H, Miyoshi H et al (2003) CD226 (DNAM-1) is involved in lymphocyte function-associated antigen 1 costimulatory signal for naive T cell differentiation and proliferation. *J Exp Med* 198: 1829–1839

116 Griffiths EK, Krawczyk C, Kong YY, Raab M, Hyduk SJ, Bouchard D, Chan VS, Kozieradzki I, Oliveira-Dos-Santos AJ, Wakeham A et al (2001) Positive regulation of T cell activation and integrin adhesion by the adapter Fyb/Slap. *Science* 293: 2260–2263

117 Peterson EJ, Woods ML, Dmowski SA, Derimanov G, Jordan MS, Wu JN, Myung PS, Liu QH, Pribila JT, Freedman BD et al (2001) Coupling of the TCR to integrin activation by Slap-130/Fyb. *Science* 293: 2263–2265

118 Shibuya K, Lanier LL, Phillips JH, Ochs HD, Shimizu K, Nakayama E, Nakauchi H, Shibuya A (1999) Physical and functional association of LFA-1 with DNAM-1 adhesion molecule. *Immunity* 11: 615–623

119 Tahara-Hanaoka S, Shibuya K, Onoda Y, Zhang H, Yamazaki S, Miyamoto A, Honda S, Lanier LL, Shibuya A (2004) Functional characterization of DNAM-1 (CD226) interaction with its ligands PVR (CD155) and nectin-2 (PRR-2/CD112). *Int Immunol* 16: 533–538

120 Reymond N, Imbert AM, Devilard E, Fabre S, Chabannon C, Xerri L, Farnarier C, Cantoni C, Bottino C, Moretta A et al (2004) DNAM-1 and PVR regulate monocyte migration through endothelial junctions. *J Exp Med* 199: 1331–1341

121 Laudanna C, Alon R (2006) Right on the spot. Chemokine triggering of integrin-mediated arrest of rolling leukocytes. *Thromb Haemost* 95: 5–11

122 Hirahashi J, Mekala D, Van Ziffle J, Xiao L, Saffaripour S, Wagner DD, Shapiro SD, Lowell C, Mayadas TN (2006) Mac-1 signaling *via* Src-family and Syk kinases results in elastase-dependent thrombohemorrhagic vasculopathy. *Immunity* 25: 271–283

123 Mayadas TN, Cullere X (2005) Neutrophil beta2 integrins: moderators of life or death decisions. *Trends Immunol* 26: 388–395

124 Mocsai A, Zhou M, Meng F, Tybulewicz VL, Lowell CA (2002) Syk is required for integrin signaling in neutrophils. *Immunity* 16: 547–558

125 Luo BH, Springer TA (2006) Integrin structures and conformational signaling. *Curr Opin Cell Biol* 18: 579–586

126 Han J, Lim CJ, Watanabe N, Soriani A, Ratnikov B, Calderwood DA, Puzon-McLaughlin W, Lafuente EM, Boussiotis VA, Shattil SJ et al (2006) Reconstructing and deconstructing agonist-induced activation of integrin alphaIIbbeta3. *Curr Biol* 16: 1796–1806

127 Van Seventer GA, Shimizu Y, Horgan KJ, Shaw S (1990) The LFA-1 ligand ICAM-1 provides an important costimulatory signal for T cell receptor-mediated activation of resting T cells. *J Immunol* 144: 4579–4586

Transendothelial migration

PECAM: Regulating the start of diapedesis

William A. Muller

Department of Pathology and Laboratory Medicine and the Graduate Program in Immunology and Microbial Pathogenesis, Weill Medical College of Cornell University, 1300 York Avenue, New York, NY 10021, USA

Introduction: Diapedesis

The molecules involved in the capturing, rolling, adhesion, and locomotion steps of leukocyte emigration have been described in earlier chapters in this book. PECAM is primarily involved in diapedesis – the step in which leukocytes squeeze in ameboid fashion between the tightly apposed endothelial cells that line the blood vessels at a site of inflammation. It is also the "point of no return" for the leukocyte. *In vivo* as viewed by intravital microscopy, the preceding steps of capturing, rolling, locomotion, and adhesion are reversible. Most leukocytes that enter a postcapillary venule at the site of inflammation are not captured; many of the leukocytes that do adhere to the endothelium release back into the circulation. Once the leukocyte commits to diapedesis, however, it continues to migrate into the inflamed tissue and does not return to the circulation – at least not as the same cell.

The subsequent physiology of the leukocyte is changed following diapedesis: Neutrophils receive signals that turn off the apoptotic program that began when they left the bone marrow so they can stay alive to function in the inflammatory environment [1]. Monocytes are triggered to differentiate into macrophages, which remain in the tissue for many years as sentinels, or dendritic cells, which pick up antigen for presentation to T cells in the draining lymph nodes (leaving the site of inflammation by migrating across afferent lymphatic endothelium in the abluminal-to-luminal direction) [2].

There are qualitative changes in the nature of leukocyte-endothelial cell interactions when diapedesis begins. Preceding steps involve heterophilic interactions between an adhesion molecule on the leukocyte and a different adhesion molecule (or counter receptor) on the endothelial surface. These interactions occur essentially on a two-dimensional plane. In contrast, diapedesis largely involves homophilic interactions between PECAM-1 on the leukocyte and PECAM-1 at the endothelial cell border and subsequently homophilic interactions between CD99 on the leukocyte and CD99 at the endothelial cell borders. The portion of the leukocyte that is in the process of diapedesis interacts in three dimensions with the endothelial cells surrounding it. Since the leukocyte diameter is 10–12 μm, while endothelial cells are less than 0.5 μm thick at their borders, only a portion of the leukocyte is actually surrounded by the endothelial cells at any given moment during its passage.

PECAM-1

PECAM-1 (hereafter referred to in this review as PECAM, since no PECAM-2 has been found in the genome) is a 130-kDa type I transmembrane protein whose expression is restricted to endothelial cells, platelets, leukocytes, and their progenitors. It is concentrated at the borders of all continuous endothelia and expressed diffusely on the surfaces of platelets and most leukocyte types to varying degrees. N-linked glycosylation accounts for about 40% of its apparent molecular weight on SDS-PAGE [3]. It is a member of the immunoglobulin gene superfamily with six C-2 type immunoglobulin domains and a cytoplasmic tail containing two immunoreceptor tyrosine inhibitory motifs (ITIMs) [4]. Due to its high expression on continuous endothelia in all tissues [5], PECAM has been used as a marker for vasculature in histopathology. Perhaps due to its abundance in cells and its active ITIM domains, PECAM has been shown to function in processes as diverse as angiogenesis, platelet activation, T cell receptor modulation, and apoptosis. These roles of PECAM are not discussed here, but the interested reader is referred to some recent reviews on the subject [6–9].

PECAM is capable of homophilic adhesion, *i.e.*, binding to a molecule of PECAM on an apposing cell [5, 10] – as well as heterophilic adhesion [11]. Homophilic interaction between PECAM on leukocytes and PECAM at endothelial cell borders is the critical interaction for diapedesis, the subject of this review. Physiologically relevant heterophilic ligands for PECAM are not well defined. The integrin αvβ3 was thought to be a ligand [12], but this view has been modified [13]. Similarly, although heparin and heparan sulfate glycosaminoglycans can inhibit heterophilic PECAM interactions [14, 15], they do not serve as direct ligands for PECAM [16]. CD38 has been reported as a ligand for PECAM heterophilic interaction on B and T cells [17].

While the "P" in PECAM-1 stands for platelet, PECAM does not appear to contribute significantly to platelet adhesion. In fact, PECAM-defective mice have been reported to show defective thrombosis [18, 19] and defective "outside-in" signaling through integrin αIIb/β3 [19]. PECAM on endothelial cells has been shown to transmit mechanical signals from one cell to the other [20] and to collaborate with vascular endothelial cadherin (VE-cadherin, CD144) and vascular endothelial growth factor receptor 2 (VEGFR2) as a mechanosensory complex for arterial level fluid shear stress [21].

Studies *in vitro*

Arguably the best-studied and documented role for PECAM is in the process of diapedesis. (It also has a role in migration across the basement membrane and possibly an indirect role in leukocyte adhesion. These are discussed later.) This role has been documented both *in vitro* and *in vivo* for neutrophils, monocytes, and natural killer (NK) cells.

We first reported that Fab fragments of monoclonal antibody (mAb) against PECAM or soluble recombinant PECAM would selectively block transmigration of primary human monocytes or neutrophils across resting or cytokine-activated human umbilical vein endothelial cell (HUVEC) monolayers by 75–80% [22]. The block was selective for diapedesis, as there was no decrease in *adhesion* of leukocytes, which remained tightly adherent to the endothelial apical surface in the face of blocking reagents. Blocking PECAM did not affect the ability of leukocytes to move *per se*, as there was no effect of optimal concentrations of antibody on chemotaxis [22] or on locomotion of leukocytes on the apical surface of endothelial cells [23]. Furthermore, the block was reversible, as transmigration resumed and achieved control levels within hours of washing away the blocking reagents.

Homophilic PECAM-PECAM interactions were involved since blocking PECAM only on the leukocyte or only on the endothelial cell was just as effective as blocking both sides simultaneously. Furthermore, soluble recombinant PECAM blocked either side effectively [22, 24]. Subsequent studies demonstrated that only mAb that bound to the N-terminal domains of PECAM blocked transmigration [15, 25, 26] and that soluble recombinant PECAM-Fc molecules containing domain 1 (including domain 1-Fc alone), but not one missing domains 1 and 2, were effective blockers of transmigration [24]. Direct binding studies localized domain 1 as the homophilic interaction site of PECAM [10, 27].

Blockade of homophilic PECAM-PECAM interactions arrests leukocytes on the apical surface of the endothelial cell monolayer overlying the junctions [22, 28]. Ultrastructural evaluation demonstrates that they remain well-spread, sending filopodia to probe the junctions, but are unable to enter [22]. Live imaging of such co-cultures demonstrated that the blocked leukocytes often move over the surface of the endothelial monolayer along the junctions between adjacent endothelial cells probing for a way to get in, but are unable to do so [23].

Antibodies against PECAM were shown to block transendothelial migration (TEM) in other *in vitro* models, including migration of monocytes across endothelial cells in the presence of fluid shear [29]. It is important to point out again that while *in vitro* studies document a role for PECAM in the diapedesis of neutrophils, monocytes, and NK cells [30], migration of T cells across endothelium *in vitro* could not be blocked with anti-PECAM antibodies [31]. Admittedly, T cells do not transmigrate endothelial cells robustly in standard *in vitro* transmigration systems the way that myeloid cells do.

Studies *in vivo*: The importance of the mouse strain

The role for PECAM in diapedesis has been documented in many different animal models of inflammation (Tab. 1). Mouse PECAM is 79% homologous to human PECAM at the amino acid level and preserves the same six-domain structure [32]. mAbs against domain 1 of mouse PECAM block homophilic PECAM binding [32] *in vitro* and block diapedesis of neutrophils and monocytes into sites of inflammation *in vivo* [33]. This has been documented in many mouse models of inflammation including thioglycolate peritonitis [24, 34], acute dermatitis [35], and collagen-induced arthritis (CIA) [36]. In addition, blocking antibodies against PECAM limit inflammation in rat and cat models including immune complex pneumonitis [37] and ischemia-reperfusion injury [38, 39]. In contrast to *in vitro* models in which PECAM does not seem to play a role in T cell emigration, there is at least one report that migration of antigen-specific activated T cells into the central nervous system is PECAM dependent [40] (Tab. 1).

Furthermore, interfering with homophilic interaction of PECAM by administering murine PECAM-Fc (comprised of the full extracellular region of PECAM or domain 1 only) chimeric protein blocked leukocyte recruitment in these models [24, 34]. Quantitative analysis of histological sections from the sites of inflammation demonstrated that leukocyte recruitment to the vasculature supplying the inflamed region was not affected. However, the recruited leukocytes were impaired in their ability to leave the microvasculature. They were found accumulating in the vessel lumen, many of them apparently adherent to the apical surface of endothelial cells lining the blood vessels in these areas, similar to the appearance of leukocytes blocked by anti-PECAM reagents *in vitro* [24, 33].

Neutrophil migration into the peritoneal cavity in response to thioglycolate broth was blocked by up to 80% and monocyte migration was nearly totally abolished in response to anti-PECAM mAb or PECAM-Fc chimeras [24, 33]. Transgenic mice constitutively expressing PECAM-Fc at therapeutic levels showed a similar impairment in their ability to respond to acute inflammatory challenge [34]. However, even in these transgenic mice neutrophils were still able to get to a site of inflammation, albeit to only 15–20% of the wild-type levels. This strongly suggested the presence of a PECAM-independent pathway(s) for diapedesis. Therefore, it initially came as no surprise when PECAM-deficient mice created by homologous recombination (PECAM-knockout mice) had no phenotype in several models of acute inflammation [41]. It was assumed that the mechanisms responsible for the 15–20% of inflammation that was normally PECAM independent were expanded to support wild-type levels of leukocyte emigration.

Remarkably, when the PECAM knockout was bred out of the original C57BL/6 background into the FVB/n strain, the PECAM-deficient mice had a severe defect in their inflammatory response [42]. In fact, some of the homozygous PECAM knockouts in the FVB/n strain would succumb to a chronic inflammatory pulmonary disease (presumably due to normal respiratory tract flora) in our animal facility, while the knockouts in the C57BL/6 background housed in adjacent cages would remain perfectly healthy [43]. A review of the literature revealed that in none of the many reports in which PECAM antibodies or soluble chimeric protein were used to block inflammation had the C57BL/6 strain been used! Direct comparison of wild-type C57BL/6 mice with many other strains showed that in the peritonitis and dermatitis models, C57BL/6 were uniquely insensitive to blockade of inflammation by anti-PECAM reagents. That is, anti-PECAM mAb and/or PECAM-Fc blocked inflammation in every strain tested except C57BL/6, including outbred Swiss mice [43] and the closely related C57BL/10 strain (Seidman and Muller, unpublished). Unfortunately, all of the published studies on PECAM-knockout mice to this point had employed the original PECAM-deficient mice in the C57BL/6 strain background.

To be sure, inflammatory deficiencies in the C57BL/6 PECAM-knockout mice have been reported; these depend on the vascular bed and/or the eliciting stimulus. Thompson et al. [44] reported a defect in leukocyte recruitment in the cremaster muscle bed of PECAM-deficient C57BL/6 mice when inflammation was elicited with IL-1β but not when elicited with TNF-α. This was in line with their previous findings in rats using a cross-reacting polyclonal antibody [45]. In the lungs, where leukocytes cross capillaries rather than postcapillary venules to reach the alveoli, inflammation is often independent of adhesion molecules required for emigration elsewhere in the body. Nonetheless, here too, the C57BL/6 PECAM-knockout mice showed a deficient leukocyte response to intratracheal instillation of IL-1β but not TNF-α [46]. Absence of PECAM did not affect inflammatory response to several other stimuli (Tab. 1). One could easily assume that this PECAM-independent inflammation was seen because of the mouse strain. However, inflammatory response to superfusion of IL-1β but not formyl-methionyl leucyl phenylalanine (fMLF) could be blocked by anti-PECAM antibody in rats [47]. This strongly supports the idea that the degree of dependence on any particular adhesion molecule will vary with the vascular bed, the type of inflammatory stimulus, and point in the time course of inflammation. Inflammatory agents that strongly and directly activate leukocytes (e.g., TNF-α, fMLF) appear to stimulate PECAM-independent transmigration (Tab. 1) [44, 47].

The molecule(s) mediating PECAM-independent transmigration, i.e., the molecules that take the place of PECAM when it is not present, are not known. CD99 does not substitute for PECAM; inhibiting both PECAM and CD99 has an additive effect on blocking transmigration in vitro [28, 48], but CD99 functions at a later step than PECAM. In our hands, blocking JAM-A with either an mAb or polyclonal antibody did not inhibit the TEM of monocytes in the presence or absence of PECAM blockade [23], so this is unlikely to be a candidate for PECAM-independent transmigration, at least in this assay system. Nourshargh and colleagues [49] found that anti-ICAM-2 blocked neutrophil extravasation in PECAM-deficient mice in the IL-1β peritonitis model. This demonstrates that ICAM-2 can work independently

Table 1 - Blockade or deficiency of PECAM blocks inflammation in vivo. *The table lists, by species, published reports using various* in vivo *models of acute inflammation. The list is not complete, but representative; apologies are extended to any authors unintentionally omitted. Models in which xenogeneic systems were used (e.g., murine leukocytes interacting with human blood vessels in NOD/SCID mice) are not listed.*

Experimental model	Species (strain)	PECAM manipulation	Outcome	Refs.
Cats				
Glycogen-induced peritonitis	Cat	Rabbit anti-PECAM IgG[a]	↓ leukocyte emigration	[39]
Cardiac ischemia/reperfusion	Cat	Rabbit anti-PECAM IgG	↓ cardiac necrosis	[39]
Rats				
Immune complex pneumonitis	Rat (Long Evans)	Rabbit anti-PECAM IgG	↓ PMN infiltration	[37]
Glycogen-induced peritonitis	Rat (Long Evans)	Rabbit anti-PECAM IgG	↓ PMN infiltration	[37]
Mesenteric stimulation by IL-1β	Rat (Sprague-Dawley)	Rabbit anti-PECAM IgG	Arrest of leukocytes on endothelium and between endothelium and basal lamina	[47]
Mesenteric stimulation by fMLF	Rat (Sprague-Dawley)	Rabbit anti-PECAM IgG	No effect	[47]
Cardiac ischemia/reperfusion	Rat (Wistar)	Rabbit anti-PECAM IgG	↓ cardiac necrosis	[38]
Mesenteric microvasculature superfusion with H_2O_2 or L-NAME	Rat (Sprague-Dawley)	Rabbit anti-PECAM IgG	↓ leukocyte extravasation	[64]
Mesenteric microvasculature superfusion with thrombin	Rat (Sprague-Dawley)	Rabbit anti-PECAM IgG	No effect on extravasation	[64]
Muscle ischemia/reperfusion	Rat (Sprague-Dawley)	Rabbit anti-PECAM IgG	↓ leukocyte extravasation	[65]
Mesenteric microcirculation superfusion with IL-1β	Rat (Sprague-Dawley)	Cross-reacting mAb[b]	↓ leukocyte extravasation	[45]
Mesenteric microcirculation superfusion with fMLF	Rat (Sprague-Dawley)	Cross-reacting mAb	No effect	[45]

Experimental model	Species (strain)	PECAM manipulation	Outcome	Refs.
Intratracheal instillation of *S. pneumoniae* or *E. coli*	Rats (Lewis)	Rabbit anti-PECAM IgG	No effect	[66]
Mice	**Non-C57BL/6**			
Thioglycolate peritonitis	Mouse (CD2F1, AKR/J)	Anti-PECAM mAb	↓ PMN and mono-cytc infiltration	[33]
Thioglycolate peritonitis	Mouse (CD2F1)	PECAM-Fc chi-mera	↓ PMN and mono-cyte infiltration	[24]
Glycogen-induced peritonitis	Mouse (C3Heb/FeJ)	Anti-PECAM mAb	↓ PMN infiltration	[67]
Thioglycolate peritonitis	Mouse (FVB/n)	PECAM-Fc transgenic	↓ PMN and mono-cyte emigration	[34]
Croton oil dermatitis	Mouse (FVB/n)	PECAM-Fc transgenic	↓ PMN extravasation into dermis	[35]
Antigen-specific T cell trafficking into CNS	Mouse (AND)	Anti-PECAM mAb or PECAM-Fc	Blockade of T cell infiltration	[40]
OVA-induced asthma	Mouse (BALB/c)	Anti-PECAM mAb	No effect on eosinophils	[63]
Collagen-induced arthritis (but see text)	Mouse (DBA/1J)	Anti-PECAM mAb	↓ inflammation; ↓ synovial infiltration	[36]
Thioglycolate peritonitis	Mouse (FVB/n)	PECAM knock-out	↓ PMN and mono-cyte transmigration	[42]
Croton oil dermatitis	Mouse (FVB/n)	PECAM knock-out	↓ PMN emigration	[42]
Thioglycolate peritonitis	Mouse (FVB/n, outbred Swiss Webster, SJL)	Anti-PECAM mAb and/or PECAM-Fc	↓ PMN and Mo emigration	[42]
Mice	**C57BL/6**			
Thioglycolate peritonitis	Mouse (C57BL/6)	PECAM knock-out	No difference from wild type	[41]
IL-1β peritonitis	Mouse (C57BL/6)	PECAM knock-out	No effect on leuko-cyte extravasation; temporary arrest between endothelium and basal lamina	[41, 44]

Table 1 - continued

Experimental model	Species (strain)	PECAM manipulation	Outcome	Refs.
S. aureus peritonitis	Mouse (C57BL/6)	PECAM knock-out	No difference from wild type	[41]
Intratracheal instillation of *S. pneumoniae* or *E. coli*	Mice (C57BL/6)	Anti-PECAM mAb	No effect	[66]
Intratracheal instillation of IL-1β or immune complexes	Mouse (C57BL/6)	PECAM knock-out	↓ leukocyte emigration	[46]
Intratracheal instillation of acid, adenovirus, or TNF-α	Mouse (C57BL/6)	PECAM knock-out	No difference from wild type	[46]
Thioglycolate peritonitis	Mouse (C57BL/6)	PECAM knock-out	No difference from wild type	[42]
Croton oil dermatitis	Mouse (C57BL/6)	PECAM knock-out	No difference from wild type	[42]
Thioglycolate peritonitis	Mouse (C57BL/6)	Anti-PECAM mAb and/or PECAM-Fc	No effect on emigration	[42]

[a]*Rabbit anti-PECAM IgG. These experiments used a rabbit antibody generated against human PECAM that cross-reacts with rat and cat PECAM*
[b]*Cross-reacting mAb. These experiments used a mouse anti-human PECAM mAb that cross-reacts with rat PECAM.*

of PECAM, but it does not demonstrate that ICAM-2 takes over for PECAM when the latter is absent or blocked. The molecules that do this and the pathways taken are under active investigation.

The role of PECAM in multifaceted models of inflammatory disease is more complicated. In ongoing CIA in DBA mice, administration of anti-PECAM mAb blocked progression of disease [36]. Similarly, administration of murine PECAM-Fc after the onset of symptoms to DBA/1J mice that had received an arthritogenic anti-collagen II antibody cocktail arrested the progression of symptoms and reversed the inflammatory score (Dasgupta and Muller, unpublished). These were instances in which the effect of PECAM was measured on the effector phase of arthritis. In contrast, when CIA was induced in PECAM-deficient mice backcrossed six generations from C57BL/6 into the DBA/1 strain, they showed an accelerated onset of arthritis that was transient; the wild-type mice caught up after about 1 week [50]. The authors concluded that the accelerated disease was due to deficiency of

PECAM on T cells during the stimulation phase. This was reminiscent of a study in which PECAM-deficient mice in the C57BL/6 background had an earlier onset of experimental allergic encephalitis than wild type, which was ascribed to increased vascular permeability, not aberrant leukocyte emigration, in the knockout mice [51]. Similarly, when CIA was induced in PECAM-deficient mice in the C57BL/6 background an accelerated onset as well as more severe disease compared to wild type was reported [52]. In contrast, these authors found that in the K/BxN model of arthritis, in which arthritis is directly induced by deposition of reactive antibodies, the C57BL/6 knockouts behaved as the wild type [52]. This suggests again that loss of T cell PECAM, which can act through its ITIM domains as an inhibitory modulator of the T cell receptor [53], allows more active induction of disease. However, in models of arthritis in which T cell stimulation is not a factor, and outcome is dependent on myeloid cell trafficking into synovium, disease is ameliorated by blockade of PECAM [36] except in the C57BL/6 background [52].

The C57BL/6 strain is a favorite for immunology research because they are susceptible to so many inflammatory and immune-related diseases. However, it clearly is not at all representative of the murine requirement for PECAM, at least in these acute inflammatory models. Even though phenotypes were seen in some experimental models, and it is reasonable to expect that there are vascular beds and inflammatory conditions that are truly PECAM independent, it will be important to reinvestigate the role of PECAM, where possible, in these models using strains of mice other than C57BL/6. These studies suggest that blockade of PECAM is indeed a viable therapeutic strategy to combat unwanted leukocyte emigration.

PECAM-dependent mechanisms of transmigration

At least 90% of the total PECAM on the endothelial cell surface is concentrated at the intercellular borders when cells make confluent monolayers [5]. However, a compartment containing almost half as much PECAM as at the cell borders is present inside the cell in a reticulum of 50-nm membrane vesicles connected to each other and to the cell border. This network of PECAM-containing membrane encircles the cell just internal to the cell borders [54]. Anti-PECAM antibodies cannot access this compartment at 4°C, but surface PECAM is internalized into it at higher temperatures. We were able to study this compartment by labeling it and tracking it with Fab fragments of an anti-PECAM domain 5 antibody that does not inhibit any known PECAM function [15]. This compartment was described in HUVEC, but an ultrastructurally similar compartment exists in postcapillary venules *in vivo* (Mamdouh and Muller, unpublished).

Under resting tissue culture conditions, membrane cycles back and forth between this compartment and the plasma membrane at the junction with a half time of about 10–15 min. When cells are chilled on ice, the cycling stops, and this compartment is no longer accessible to small proteins. However, it is not completely shut off from the external surface, since protons can still enter. Thus, the compartment is not composed of true vesicles that completely pinch off the plasma membrane. At least some of the interconnected vesicle structures are always attached to the cell surface. The vesicles that make up this reticulum superficially resemble caveolae, but do not bear caveolin 1 nor do they have the biochemical characteristics of lipid rafts [54]. They appear to be a unique membrane compartment in endothelial cells. The role of this compartment and the constitutive recycling is not known. However, this compartment is critical for diapedesis.

When leukocytes transmigrate the membrane from this compartment is redirected. It no longer recycles evenly along the plasma membrane, but is directed to the zone of the endothelial border where the leukocyte makes contact. Membrane from this compartment surrounds the leukocyte for the entire process of diapedesis [54]. If homophilic interaction between leukocyte PECAM and endothelial cell PECAM is prevented, this "targeted recycling" does not occur. The membrane recycles evenly along the cell border as if the leukocyte were not there. The biology of this compartment explains the old observation that anti-PECAM mAb [22] or PECAM-Fc chimeras [24] added to endothelial cell monolayers blocked TEM when they were added at 37°C but not at 4°C. At 37°C, the antibodies had access to the membrane in the recycling compartment; at 4°C they only labeled the surface. Thus, blocking PECAM on the membrane in the recycling compartment is critical for blocking transmigration. Furthermore, directly impeding "targeted recycling" of this compartment independent of blocking PECAM also blocks diapedesis resulting in the same characteristic arrest of leukocytes over the intercellular borders as if PECAM itself had been blocked (Mamdouh and Muller, unpublished).

We hypothesize that homophilic interaction of PECAM on leukocytes with PECAM on endothelial cells initiates the targeted recycling of membrane from this compartment that is crucial for diapedesis (see Fig. 1). Targeted recycling brings a large surface area of membrane to surround the leukocyte. The surface area of the endothelial border necessarily increases when a leukocyte passes through. The extra surface area contributed by targeted recycling could be necessary to accommodate the leukocyte without the need for endothelial cell retraction. Alternatively or additionally, this compartment may provide a pool of unligated PECAM for interaction with leukocyte PECAM to propagate the signal for continued targeted recycling during diapedesis so that the existing PECAM-PECAM interactions between adjacent endothelial cells do not have to unzip [54].

Consequences of TEM

Nuclear accumulation of NF-κB in cytokine-activated endothelial cells is decreased following PMN transmigration *in vitro* [55]. This appeared to be at least partly due to PECAM-PECAM interactions, since cross-linking PECAM with mAbs produced a similar effect, and migration of mouse PMN across PECAM-deficient endothelial cells in Transwells did not affect nuclear NF-κB levels [55]. The authors postulated that even as homophilic PECAM interactions are necessary for bringing neutrophils into a site of inflammation, their entrance begins the process of turning off the inflammatory response. Presumably, whether inflammation continued would be dictated by the balance of pro- and anti-inflammatory stimuli impinging on the endothelial cells.

Additional roles of PECAM in leukocyte emigration

It has long been known that cross-linking PECAM on the surfaces of leukocytes activates integrins of the β_1 [56], β_2 [30, 57], and β_3 [58] families. This "inside-out" activation of integrins by PECAM involves activation of the GTPase Rap1 and requires intact ITIM domains on the cytoplasmic tail of PECAM [59]. Ligands for these integrins are present on the apical surface of endothelial cells. One could postulate that homophilic PECAM engagement between leukocyte and endothelial cell could stimulate the adhesion step of leukocyte emigration. However, endothelial cell PECAM is localized almost exclusively to the endothelial cell borders [5] so it is difficult to envisage how this direct interaction could occur until after the integrin-dependent adhesion and locomotion steps had already been completed [23]. Indeed, blocking PECAM does not reduce adhesion of leukocytes to the apical surface of endothelial cells [22]. On the other hand, when a leukocyte initiates diapedesis, most of it is still on the apical surface of the endothelial cell, and the trailing uropod remains in contact with the apical surface as the leading edge migrates through the junction. One could speculate that inside-out signals from PECAM-PECAM interactions in the junction activate integrins and contribute to stabilizing integrin-mediated adhesion of the remainder of the leukocyte to the apical surface (for traction?) as the more forward portions transmigrate.

A proven role for integrin activation by homophilic PECAM engagement during diapedesis is for subsequent interactions with basal lamina components following diapedesis. Indeed, homophilic engagement of PECAM induces exteriorization of the laminin binding integrin α6β1 on neutrophils [60], which is required for efficient migration of PMN across the subendothelial basal lamina. An extensive discussion of this appears in the chapter by Sussan Nourshargh in this volume. PECAM-deficient mice, even in the C57BL/6 background, have a slight delay in the migration of neutrophils across the subendothelial basement membrane in the mesenteric circulation in response to IL-1β, although normal numbers of neutrophils are recruited into the peritoneal cavity within 4 h of inciting intraperitoneal inflammation [41, 44].

Migration of leukocytes across the subendothelial basement membrane also may involve heterophilic interaction of leukocyte PECAM with some component(s) of the basal lamina. The membrane-proximal domain 6 of PECAM has been defined as the portion of the molecule responsible for heterophilic interactions [15]. In our *in vitro*

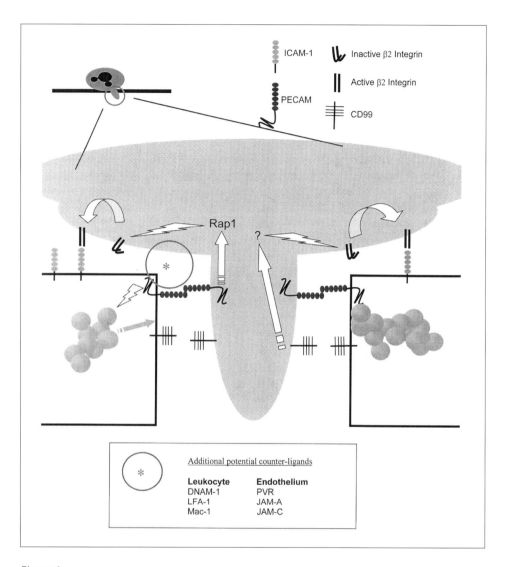

Figure 1

Molecular regulation of the initiation of diapedesis.

(Upper left) As a leukocyte traverses the endothelial border, it insinuates a pseudopod between the tightly apposed endothelial cells. The center image depicts an enlargement of the region circled. Homophilic engagement between PECAM on the leukocyte pseudopod and PECAM at the endothelial cell border is critical for transmigration to proceed. Signals from this interaction (lightning bolt within endothelial cell) stimulate the recruitment of membrane from a subjunctional compartment in endothelium to the site of diapedesis (shaded arrow) [54]. This provides a large amount of membrane surface area as well as addi-

model, all mAb against domain 6 of human PECAM, when bound to monocytes, had no effect on diapedesis, but selectively blocked their migration across the subendothelial basement membrane [15, 25]. Monocytes were observed trapped between the basal surface of the endothelial cell and the subendothelial basement membrane [15] in a manner that was predictive of the appearance of later studies in which a similar phenotype was observed *in vivo* using polyclonal antibody [47] or PECAM-deficient mice [41, 44]. The *in vivo* relevance of heterophilic PECAM interactions in migration of leukocytes across the basal lamina awaits the development of a blocking antibody selective for mouse PECAM domain 6.

Other molecules involved in diapedesis

This chapter has focused on PECAM-1, the best-studied molecule involved in diapedesis. However, it is clearly not the only molecule involved in the process. There are molecules that work in parallel (*i.e.*, control the same step as PECAM in a redundant fashion) and molecules that work in series at steps beyond (and perhaps before) PECAM.

Mice of the C57BL/6 strain obviously have a way to circumvent the absence of PECAM, but even in all of the other strains, which are dependent on PECAM, the blockade or genetic absence of PECAM does not totally abolish TEM. Therefore, molecules that function at the same step as (*i.e.*, in parallel with) PECAM must exist. These may have the same function as PECAM or send the leukocyte on a molecularly and physiologically distinct emigration pathway. However, the identity of such a molecule(s) is not known.

---

tional PECAM molecules to accommodate the leukocyte as it traverses the cell border. On the leukocyte side, PECAM engagement signals through Rap1 leading to inside-out activation of leukocyte integrins to their high-affinity conformation (lightning bolt; curved arrow) for the stable binding of ligands such as ICAM-1. This may help stabilize the leukocyte on the surface and provide traction as it pushes through the cell border. Other molecules on the surface of leukocytes are also capable of activating leukocyte integrins; this function is not unique to molecules involved in transmigration. ICAM-1 is recruited with β2 integrins to the site of transmigration. At a later step in diapedesis, homophilic interaction of CD99 is required to complete transendothelial migration. The consequences of CD99 engagement on the endothelial cell are not clear. On several leukocyte types, cross-linking of CD99 has been shown to activate integrins through an unknown signaling mechanism (question mark). The circled junctional space with an asterisk denotes that in various in vitro systems DNAM-1/poliovirus receptor, LFA-1/JAM-A, and Mac-1/JAM-C have been shown to play a role in leukocyte transmigration. The relationship of these interactions spatially and temporally to PECAM and CD99 is not clear, but is under investigation.

213

The step of diapedesis itself consists of at least two steps mediated in series by PECAM and CD99 [28]. CD99 is a highly glycosylated molecule expressed on the surfaces of leukocytes and concentrated at the borders of endothelial cells. Like PECAM, homophilic interaction of leukocyte CD99 with endothelial CD99 is necessary for diapedesis. However, while blockade of PECAM arrests leukocytes on the apical surface of the endothelial cell at the border, tightly adherent but unable to enter, blockade of CD99 arrests the leukocyte partway across the endothelial cell with the leading edge below the endothelium and the trailing uropod still on the apical surface [28, 48]. Like cells whose PECAM is selectively blocked, these cells can move laterally within the junction, but cannot proceed farther. Sequential blockade studies showed that CD99 functions after PECAM: If PECAM were first blocked, then the block removed, transmigration could still be arrested by blocking CD99. However, if cells were first blocked at the CD99-dependent step, they could no longer be blocked by anti-PECAM antibody once the CD99 blockade was removed [28]. Interfering with homophilic CD99 interactions has been more effective at blocking TEM *in vitro* than blocking PECAM [28, 48], and blocking both PECAM and CD99 produce additive effects. However, even this is not 100% effective, suggesting that there are CD99-independent pathways as well.

Recently, the interaction of DNAM-1 (CD226) on monocytes and poliovirus receptor (CD155) on endothelial cells was shown to play a role in diapedesis *in vitro* [61]. Blockade of either molecule arrested monocytes on the apical surface of endothelial cells, resembling, at least at the level of confocal microscopy, the blockade of PECAM. Sequential blockade experiments [28] are necessary to determine if these truly block at the same step as PECAM, but this is one molecular pair that is a candidate to mediate the same step in diapedesis as PECAM.

Other molecules reported to play a role in diapedesis include JAM-A and JAM-C (reviewed in [62]). The inflammatory response is a stereotyped response to tissue damage of any kind. However, it is already apparent that different adhesion molecules play dominant roles depending on the inflammatory stimulus, the vascular bed, and the point in the time course of the inflammatory response. For example, pretreatment of mice with mAb against PECAM did not block eosinophil recruitment into the lungs in response to allergen challenge (although the antibody was given 24 h prior to challenge [63]).

A major focus of research in our laboratory is to determine whether other molecules that regulate diapedesis (particularly those that can substitute for PECAM) can trigger targeted recycling from the lateral border recycling membrane pool. There may be a common mechanism for diapedesis that is triggered by various different surface receptors with different efficacies depending on the inflammatory stimulus, vascular bed, and state of the inflammatory response.

Acknowledgements

The author would like to thank the National Institutes of Health for support of the studies leading to this work: R01 HL046849 and R37 HL064774.

References

1   Coxon A, Tang T, Mayadas TN (1999) Cytokine-activated endothelial cells delay neutrophil apoptosis *in vitro* and *in vivo*. A role for granulocyte/macrophage colony-stimulating factor. *J Exp Med* 190: 923–934

2   Randolph GJ, Inaba K, Robbiani DF, Steinman RM, Muller WA (1999) Differentiation of phagocytic monocytes into lymph node dendritic cells *in vivo*. *Immunity* 11: 753–761

3   Newman PJ, Berndt MC, Gorski J, White GC II, Lyman S, Paddock C, Muller WA (1990) PECAM-1 (CD31) cloning and relation to adhesion molecules of the immunoglobulin gene superfamily. *Science* 247: 1219–1222

4   Newman PJ (1999) Switched at birth: A new family for PECAM-1. *J Clin Invest* 103: 5–9

5   Muller WA, Ratti CM, McDonnell SL, Cohn ZA (1989) A human endothelial cell-restricted, externally disposed plasmalemmal protein enriched in intercellular junctions. *J Exp Med* 170: 399–414

6   Newman PJ, Newman DK (2003) Signal transduction pathways mediated by PECAM-1: New roles for an old molecule in platelet and vascular cell biology. *Arterioscler Thromb Vasc Biol* 23: 953–964

7   Ilan N, Madri JA (2003) PECAM-1: old friend, new partners. *Curr Opin Cell Biol* 15: 515–524

8   Zocchi MR, Poggi A (2004) PECAM-1, apoptosis and CD34$^+$ precursors. *Leuk Lymphoma* 45: 2205–2213

9   Wong MX, Jacson DE (2004) Regulation of B cell activation by PECAM-1: implications for the development of autoimmune disease. *Curr Pharm Des* 10: 155–161

10 Sun Q-H, Delisser HM, Zukowski MM, Paddock C, Albelda SM, Newman PJ (1996) Individually distinct Ig homology domains in PECAM-1 regulate homophilic binding and modulate receptor affinity. *J Biol Chem* 271: 11090–11098

11 Muller WA, Berman ME, Newman PJ, DeLisser HM, Albelda SM (1992) A heterophilic adhesion mechanism for platelet/endothelial cell adhesion molecule 1 (CD31). *J Exp Med* 175: 1401–1404

12 Piali L, Hammel P, Uherek C, Bachmann F, Gisler R, Dunon D, Imhof B (1995) CD31/PECAM-1 is a ligand for alpha v beta 3 integrin involved in adhesion of leukocytes to endothelium. *J Cell Biol* 130: 451–460

13 Wong CW, Wiedle G, Ballestrem C, Wehrle-Haller B, Etteldorf S, Bruckner M, Engelhardt B, Gisler RH, Imhof BA (2000) PECAM-1/CD31 trans-homophilic binding at the

intercellular junctions is independent of its cytoplasmic domain; evidence for hetero-philic interaction with integrin $\alpha_v\beta_3$ in cis. *Mol Biol Cell* 11: 3109–3121

14 Delisser HM, Yan HC, Newman PJ, Muller WA, Buck CA, Albelda SM (1993) Platelet/endothelial cell adhesion molecule-1 (CD31)-mediated cellular aggregation involves cell surface glycosaminoglycans. *J Biol Chem* 268: 16037–16046

15 Liao F, Huynh HK, Eiroa A, Greene T, Polizzi E, Muller WA (1995) Migration of mono-cytes across endothelium and passage through extracellular matrix involve separate molecular domains of PECAM-1. *J Exp Med* 182: 1337–1343

16 Sun Q-H, Paddock C, Visentin GP, Zukowski MM, Muller WA, Newman PJ (1998) Cell surface glycosaminoglycans do not serve as ligands for PECAM-1. PECAM-1 is not a heparin-binding protein. *J Biol Chem* 273: 11483–11490

17 Deaglio S, Morra M, Mallone R, Ausiello CM, Prager E, Garbarino G, Dianzani U, Stockinger H, Malavasi F (1998) Human CD38 (ADP-ribosyl cyclase) is a counter-receptor of CD31, an Ig superfamily member. *J Immunol* 160: 395–402

18 Falati S, Patil S, Gross PL, Stapleton M, Merrill-Skoloff G, Barrett NE, Pixton KL, Weiler H, Cooley B, Newman DK et al (2006) Platelet PECAM-1 inhibits thrombus formation *in vivo*. *Blood* 107: 535–541

19 Wee JL, Jackson DE (2005) The Ig-ITIM superfamily member PECAM-1 regulates the "outside-in" signaling properties of integrin alpha(IIb)beta3 in platelets. *Blood* 106: 3816–3823

20 Osawa M, Masuda M, Kusano K-I, Fujiwara K (2002) Evidence for a role of platelet endothelial cell adhesion molecule-1 in endothelial cell mechanosignal transduction: is it a mechanoresponsive molecule? *J Cell Biol* 158: 773–785

21 Tzima E, Irani-Tehrani M, Kiosses WB, Dejana E, Schultz DA, Engelhardt B, Cao G, DeLisser H, Schwartz MA (2005) A mechanosensory complex that mediates the endo-thelial cell response to fluid shear stress. *Nature* 437: 426–431

22 Muller WA, Weigl SA, Deng X, Phillips DM (1993) PECAM-1 is required for transen-dothelial migration of leukocytes. *J Exp Med* 178: 449–460

23 Schenkel AR, Mamdouh Z, Muller WA (2004) Locomotion of monocytes on endothe-lium is a critical step during extravasation. *Nat Immunol* 5: 393–400

24 Liao F, Ali J, Greene T, Muller WA (1997) Soluble domain 1 of platelet-endothelial cell adhesion molecule (PECAM) is sufficient to block transendothelial migration *in vitro* and *in vivo*. *J Exp Med* 185: 1349–1357

25 Muller WA, Greene T, Liao F (1997) Transendothelial migration and interstitial migration of monocytes are mediated by separate domains of monocyte CD31. In: T Kishimoto (ed): *Leukocyte Typing VI. Proceedings of the VIth International Leukocyte Differentiation Antigen Workshop, Kobe, Japan, 1996.* Garland Publishers, London, 370–372

26 Nakada MT, Amin K, Christofidou-Solomidou M, O'Brien CD, Sun J, Gurubhagavatula I, Heavner GA, Taylor AH, Paddock C, Sun QH et al (2000) Antibodies against the first Ig-like domain of human platelet endothelial cell adhesion molecule-1 (PECAM-1) that

inhibit PECAM-1-dependent homophilic adhesion block *in vivo* neutrophil recruitment. *J Immunol* 164: 452–462

27   Newton JP, Buckley CD, Jones EY, Simmons DL (1997) Residues on both faces of the first immunoglobulin fold contribute to homophilic binding sites of PECAM-1/CD31. *J Biol Chem* 272: 20555–20563

28   Schenkel AR, Mamdouh Z, Chen X, Liebman RM, Muller WA (2002) CD99 plays a major role in the migration of monocytes through endothelial junctions. *Nat Immunol* 3: 143–150

29   Allport JR, Muller WA, Luscinskas FW (2000) Monocytes induce reversible focal changes in vascular endothelial cadherin complex during transendothelial migration under flow. *J Cell Biol* 148: 203–216

30   Berman ME, Xie Y, Muller WA (1996) Roles of platelet/endothelial cell adhesion molecule-1 (PECAM-1, CD31) in natural killer cell transendothelial migration and beta 2 integrin activation. *J Immunol* 156: 1515–1524

31   Bird IN, Spragg JH, Ager AH, Matthews N (1993) Studies of lymphocyte transendothelial migration: analysis of migrated cell phenotypes with regard to CD31 (PECAM-1), CD45RA and CD45RO. *Immunology* 80: 553–560

32   Xie Y, Muller WA (1993) Molecular cloning and adhesive properties of murine platelet/endothelial cell adhesion molecule-1. *Proc Natl Acad Sci USA* 90: 5569–5573

33   Bogen S, Pak J, Garifallou M, Deng X, Muller WA (1994) Monoclonal antibody to murine PECAM-1 (CD31) blocks acute inflammation *in vivo*. *J Exp Med* 179: 1059–1064

34   Liao F, Schenkel AR, Muller WA (1999) Transgenic mice expressing different levels of soluble platelet/endothelial cell adhesion molecule-IgG display distinct inflammatory phenotypes. *J Immunol* 163: 5640–5648

35   Muller WA, Randolph GJ (1999) Migration of leukocytes across endothelium and beyond: Molecules involved in the transmigration and fate of monocytes. *J Leukoc Biol* 66: 698–704

36   Ishikawa J, Okada Y, Bird IN, Jasani B, Spragg JH, Yamada T (2002) Use of anti-platelet-endothelial cell adhesion molecule-1 antibody in the control of disease progression in established collagen-induced arthritis in DBA/1J mice. *Jpn J Pharmacol* 88: 332–340

37   Vaporciyan AA, Delisser HM, Yan H-C, Mendiguren II, Thom SR, Jones ML, Ward PA, Albelda SM (1993) Involvement of platelet-endothelial cell adhesion molecule-1 in neutrophil recruitment *in vivo*. *Science* 262: 1580–1582

38   Gumina RJ, Schultz JE, Yao Z, Kenny D, Warltier DC, Newman PJ, Gross GJ (1996) Antibody to platelet/endothelial cell adhesion molecule-1 reduces myocardial infarct size in a rat model of ischemia-reperfusion injury. *Circulation* 94: 3327–3333

39   Murohara T, Delyani JA, Albelda SM, Lefer AM (1996) Blockade of platelet endothelial cell adhesion molecule-1 protects against myocardial ischemia and reperfusion injury in cats. *J Immunol* 156: 3550–3557

40   Qing Z, Sandor M, Radvany Z, Sewell D, Falus A, Potthoff D, Muller WA, Fabry Z

(2001) Inhibition of antigen-specific T cell trafficking into the central nervous system *via* blocking PECAM1/CD31 molecule. *J Neuropathol Exp Neurol* 60: 798–807

41 Duncan GS, Andrew DP, Takimoto H, Kaufman SA, Yoshida H, Spellberg J, Luis de la Pompa J, Elia A, Wakeham A, Karan-Tamir B et al (1999) Genetic evidence for functional redundancy of platelet/endothelial cell adhesion molecule-1 (PECAM-1): CD31-deficient mice reveal PECAM-1-dependent and PECAM-1-independent functions. *J Immunol* 162: 3022–3030

42 Schenkel AR, Chew TW, Muller WA (2004) Platelet endothelial cell adhesion molecule deficiency or blockade significantly reduces leukocyte emigration in a majority of mouse strains. *J Immunol* 173: 6403–6408

43 Schenkel AR, Chew TW, Chlipala E, Harbord MW, Muller WA (2006) Different susceptibilities of PECAM-deficient mouse strains to spontaneous idiopathic pneumonitis. *Exp Mol Pathol* 81: 23–30

44 Thompson RD, Noble KE, Larbi KY, Dewar A, Duncan GS, Mak TW, Nourshargh S (2001) Platelet-endothelial cell adhesion molecule-1 (PECAM-1)-deficient mice demonstrate a transient and cytokine-specific role for PECAM-1 in leukocyte migration through the perivascular basement membrane. *Blood* 97: 1854–1860

45 Thompson RD, Wakelin MW, Larbi KY, Dewar A, Asimakopoulous G, Horton MA, Nakada MT, Nourshargh S (2000) Divergent effects of platelet-endothelial cell adhesion molecule-1 and β_3 integrin blockade on leukocyte transmigration *in vivo*. *J Immunol* 165: 426–434

46 Albelda SM, Lau KC, Chien P, Huang Z, Arguiris E, Bohen A, Sun J, Billet JA, Christofidou-Solomidou M, Indik ZK et al (2004) Role for platelet-endothelial cell adhesion molecule-1 in macrophage Fcγ receptor function. *Am J Respir Cell Mol Biol* 31: 246–255

47 Wakelin MW, Sanz M-J, Dewar A, Albelda SM, Larkin SW, Boughton-Smith N, Williams TJ, Nourshargh S (1996) An anti-platelet/endothelial cell adhesion molecule-1 antibody inhibits leukocyte extravasation from mesenteric microvessels *in vivo* by blocking the passage through basement membrane. *J Exp Med* 184: 229–239

48 Lou O, Alcaide P, Luscinskas FW, Muller WA (2007) CD99 is a key mediator of the transendothelial migration of neutrophils. *J Immunol* 178: 1136–1143

49 Huang MT, Larbi KY, Scheiermann C, Woodfin A, Gerwin N, Haskard DO, Nourshargh S (2006) ICAM-2 mediates neutrophil transmigration *in vivo*: evidence for stimulus specificity and a role in PECAM-1-independent transmigration. *Blood* 107: 4721–4727

50 Tada Y, Koarada S, Morito F, Ushiyama O, Haruta Y, Kanegae F, Ohta A, Ho A, Mak TW, Nagasawa K (2003) Acceleration of the onset of collagen-induced arthritis by a deficiency of platelet endothelial cell adhesion molecule 1. *Arthritis Rheum* 48: 3280–3290

51 Graesser D, Solowiej A, Bruckner M, Osterweil E, Juedes A, Davis S, Ruddle NH, Engelhardt B, Madri JA (2002) Altered vascular permeability and early onset of experimental autoimmune encephalomyelitis in PECAM-1-deficient mice. *J Clin Invest* 109: 383–392

52 Wong MX, Hayball JD, Hogarth PM, Jackson DE (2005) The inhibitory co-receptor, PECAM-1 provides a protective effect in suppression of collagen-induced arthritis. *J Clin Immunol* 25: 19–28

53 Newman DK, Hamilton C, Newman PJ (2001) Inhibition of antigen-receptor signaling by platelet endothelial cell adhesion molecule-1 (CD31) requires functional ITIMs, SHP-2, and p56(lck). *Blood* 97: 2351–2357

54 Mamdouh Z, Chen X, Pierini LM, Maxfield FR, Muller WA (2003) Targeted recycling of PECAM from endothelial cell surface-connected compartments during diapedesis. *Nature* 421: 748–753

55 Cepinskas G, Savickiene J, Ionescu CV, Kvietys PR (2003) PMN transendothelial migration decreases nuclear NFkappaB in IL-1beta-activated endothelial cells: role of PECAM-1. *J Cell Biol* 161: 641–651

56 Tanaka Y, Albelda SM, Horgan KJ, Van Seventer GA, Shimizu Y, Newman W, Hallam J, Newman PJ, Buck CA, Shaw S (1992) CD31 expressed on distinctive T cell subsets is a preferential amplifier of beta1 integrin-mediated adhesion. *J Exp Med* 176: 245–253

57 Berman ME, Muller WA (1995) Ligation of platelet/endothelial cell adhesion molecule 1 (PECAM-1/CD31) on monocytes and neutrophils increases binding capacity of leukocyte CR3 (CD11b/CD18). *J Immunol* 154: 299–307

58 Chiba R, Nakagawa N, Kurasawa K, Tanaka Y, Saito Y, Iwamoto I (1999) Ligation of CD31 (PECAM-1) on endothelial cells increases adhesive function of $\beta_v\beta_3$ integrin and enhances $\beta_1$integrin-mediated adhesion of eosinophils to endothelial cells. *Blood* 94: 1319–1329

59 Reedquist KA, Ross E, Koop EA, Wolthuis RMF, Zwartkruis FJT, Kooyk Yv, Salmon M, Buckley CD, Bos JL (2000) The small GTPase, Rap1, mediates CD31-induced integrin adhesion. *J Cell Biol* 148: 1151–1158

60 Dangerfield J, Larbi KY, Huang MT, Dewar A, Nourshargh S (2002) PECAM-1 (CD31) homophilic interaction up-regulates alpha6beta1 on transmigrated neutrophils *in vivo* and plays a functional role in the ability of $\alpha6\beta1$ integrins to mediate leukocyte migration through the perivascular basement membrane. *J Exp Med* 196: 1201–1211

61 Reymond N, Imbert AM, Devilard E, Fabre S, Chabannon C, Xerri L, Farnarier C, Cantoni C, Bottino C, Moretta A et al (2004) DNAM-1 and PVR regulate monocyte migration through endothelial junctions. *J Exp Med* 199: 1331–1341

62 Muller WA (2003) Leukocyte-endothelial-cell interactions in leukocyte transmigration and the inflammatory response. *Trends Immunol* 24: 326–333

63 Miller M, Sung K-LP, Muller WA, Cho JY, Roman M, Castaneda D, Nayar J, Condon T, Kim J, Sriramarao P et al (2001) Eosinophil tissue recruitment to sites of allergic inflammation in the lung is platelet endothelial cell adhesion molecule independent. *J Immunol* 167: 2292–2297

64 Scalia R, Lefer AM (1998) *In vivo* regulation of PECAM-1 activity during acute endothelial dysfunction in the rat mesenteric microvasculature. *J Leukoc Biol* 64: 163–169

65 Turegun M, Gudemez E, Newman P, Zins J, Siemionow M (1999) Blockade of platelet

endothelial cell adhesion molecule-1 (PECAM-1) protects against ischemia-reperfusion injury in muscle flaps at microcirculatory level. *Plast Reconstr Surg* 104: 1033–1040

66 Tasaka S, Qin L, Saijo A, Albelda SM, DeLisser HM, Doerschuk CM (2003) Platelet endothelial cell adhesion molecule-1 in neutrophil emigration during acute bacterial pneumonia in mice and rats. *Am J Respir Crit Care Med* 167: 164–170

67 Chosay JG, Fisher MA, Farhood A, Ready KA, Dunn CJ, Jaeschke H (1998) Role of PECAM-1 (CD31) in neutrophil transmigration in murine models of liver and peritoneal inflammation. *Am J Physiol* 274: G776–782

Role of $\alpha_6\beta_1$ integrin in leukocyte adhesion and transmigration

Mathieu-Benoit Voisin and Sussan Nourshargh

Centre for Microvascular Research, Williams Harvey Research Institute, Queen Mary and Westfield College, Charterhouse Square, London, UK

Introduction

Integrins are a family of heterodimeric cell-surface adhesion and signaling molecules each composed of an α and a β subunit. At present 18 α and 8 β subunits have been identified giving rise to 24 distinct integrin molecules. Many members of this family, primarily integrins containing a β_1 subunit, act as receptors for extracellular matrix proteins, *e.g.*, $\alpha_1\beta_1$ and $\alpha_2\beta_1$ are key receptors for collagen while $\alpha_3\beta_1$ and $\alpha_6\beta_1$ are key receptors for laminin. The integrin $\alpha_6\beta_1$ (VLA-6, CD49f/CD29) was originally identified on platelets but is now known to be expressed on numerous cell types including epithelial and endothelial cells and is indeed the key leukocyte laminin receptor. Its principal functional role relates to regulation of cell adhesion and motility, and, in line with this, $\alpha_6\beta_1$ interaction with its ligands leads to cellular cytoskeletal rearrangement and polarization. In the context of inflammatory events, the ability of $\alpha_6\beta_1$ to bind vascular laminins has led to much interest in the role of this integrin in regulation of leukocyte migration through the vascular basement membrane (BM). In blood vessels, the BM forms a thin protein sheet that underlies the endothelium and encases the pericytes/smooth muscle cells. Major constituents of this BM are laminins (laminin 8 and laminin 10) and collagen IV, which form two independent networks that are interconnected by molecules such as nidogens (nidogen-1 and nidogen-2) and the large heparan sulfate proteoglycan, perlecan [1]. In contrast to our growing knowledge of the molecules and mechanisms that mediate leukocyte migration through the endothelium [2], less is known about the mechanisms that mediate and regulate leukocyte migration through the vascular BM. This chapter briefly addresses the general biology of $\alpha_6\beta_1$ integrin and reviews the existing evidence implicating this integrin in regulation of leukocyte adhesion and transmigration especially at the level of the vascular BM.

Adhesion Molecules: Function and Inhibition, edited by Klaus Ley
© 2007 Birkhäuser Verlag Basel/Switzerland

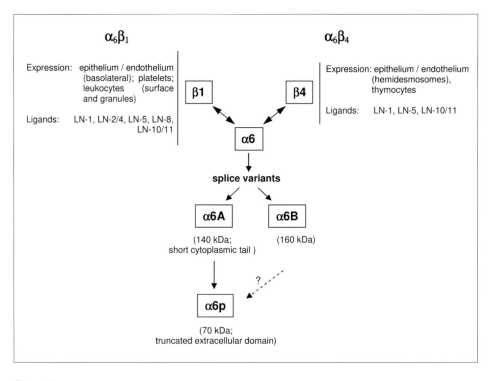

Figure 1

α_6 *integrin subunit structure and associations.*

The integrin α_6 subunit can pair with both β_1 and β_4 subunits to form two distinct laminin-binding integrins. The α_6 gene (chromosome 2 with 26 exons in both mouse and human) encodes for two spliced variants named α_{6A} and α_{6B} with a variability in the length of their intracellular regions, 36 and 54 amino acids, respectively. A truncated version of the α_6 subunit, called α_{6p} (found in carcinoma cells), results from post-translational cleavage of the β-propeller ligand-binding region of the integrin within its extracellular domain (LN, laminin).

$\alpha_6\beta_1$ structure, variants and expression profile

The α_6 integrin chain is found associated with only two β integrin chains giving rise to two distinct molecules $\alpha_6\beta_1$ and $\alpha_6\beta_4$, molecules that have different cellular expression profiles and laminin specificities (Fig. 1). The α_6 subunit is formed by a heavy and a light chain linked by disulfide bonds [3] with the heavy chain (875 amino acid residues containing the β-propeller domain, 110 kDa) forming the N-terminal extracellular region that interacts with both β_1 and β_4 subunits and ligands [4, 5]. The light chain of the α_6 subunit (170 amino acids, 30 kDa) includes a short extracellular sequence, a transmembrane region and an intracellular sequence that is devoid of any direct enzymatic properties. Interestingly, the α_6 subunit has two splice variants, α_{6A} and α_{6B} [4] that can both pair with β_1 and β_4 subunits. The two α_6 subunits only differ in their cytoplasmic domains, with α_{6A} and α_{6B} expressing 36 and 54 amino acid cytoplasmic domains, respectively (Fig. 1). While this structural difference does not appear to have a bearing to the ligand binding profiles of the molecules, it may be relevant in terms of their signaling profiles through altered ability to interact with different intracellular proteins and the cytoskeleton [4]. In addition, and more recently, a novel variant of the α_6 integrin subunit, termed α_{6p}, was identified in prostate carcinoma cell lines [6]. This variant can also pair with both β_1 and β_4 integrin subunits and the principal structural alteration is a shortened extracellular heavy chain as a result of missing the putative ligand-binding region of the integrin α subunit within the β-propeller domain. The functional implications of α_{6p} variant expression in terms of cell adhesion and proliferation *in vivo* remain to be fully determined [7].

$\alpha_6\beta_1$ is expressed on platelets [8] as well as several other cell types such as epithelial cells, neural stem cells, Schwann cells, synovial fibroblasts, endothelial cells and leukocytes. On endothelial and epithelial cells, it is primarily localized to the basolateral surface in direct contact with the BM, where it can contribute to maintaining the integrity and stabilization of the cell layer [9, 10]. In endothelial cells, this expression level can be up-regulated following stimulation with estradiol [11]. With respect to leukocytes, $\alpha_6\beta_1$ expression has been reported on lymphocytes [12–14], monocytes/macrophages [15–17] and granulocytes, *i.e.*, neutrophils [18, 19] and eosinophils [20]. Lack of specific monoclonal antibodies (mAbs) recognizing different variants of $\alpha_6\beta_1$ has restricted the studies addressing their cellular/tissue expression but biochemical studies have shown that both α_{6A} and α_{6B} are expressed by thymocytes and mature lymphocytes [21, 22]. To date very little is known about the relative expression, distribution or function of different $\alpha_6\beta_1$ variants in different leukocyte subpopulations and on endothelial cells in different vascular beds *in vivo*.

Ligand binding and signaling pathways

$\alpha_6\beta_1$ like all other integrins can exhibit outside-in signaling post binding to its ligands and also inside-out signaling resulting in regulated affinity and/or avidity of the molecule for its ligands. With respect to the former, the ligands for $\alpha_6\beta_1$ are all laminin isoforms, molecules that are major constituents of BMs [1, 23]. Laminins are cruciform-shaped heterotrimeric glycoproteins composed of three genetically distinct α, β and γ chains. To date, 5 α, 4 β and 3 γ laminin chains have been identified that combine to form 15 distinct laminin isoforms [24]. The major laminin ligands of the $\alpha_6\beta_1$ integrin are laminin 10 ($\alpha5\beta1\gamma1$) and laminin 11 ($\alpha5\beta2\gamma1$) and to a lesser extent laminin 1 ($\alpha 1\beta 1\gamma 1$), laminin 2/4 ($\alpha 2\beta 1\gamma 1/\alpha 2\beta 2\gamma 1$), laminin 5 ($\alpha 3\beta 3\gamma 2$) and laminin 8 ($\alpha 4\beta 1\gamma 1$) [25–27]. In general, the binding of β_1 integrins to their ligands can lead to regulated gene expression, cytoskeletal re-organization and enhanced inflammatory cellular responses [28, 29]. Details of the signaling pathways associated with $\alpha_6\beta_1$ ligation are not fully understood, but it appears that amongst the first responses is the recruitment of signaling molecules and localization of $\alpha_6\beta_1$ to focal adhesion sites, an event that acts to link the extracellular matrix to the actin cytoskeleton. This response is mediated *via* the cytoplasmic domain of the molecule in terms of both signal transduction mechanisms and mechanical stability of cells during their interaction with laminin. Of interest, recent studies have described PDZ recognition sequences in the α_6 intracellular tail that may play a role in recruitment of proteins such as members of the cytoskeleton, paxillin and vinculin [30–32], as well as several kinases such as focal adhesion kinase (FAK), the src-family kinase and the extracellular signal-regulated kinase (ERK) [28, 33–35]. In leukocytes $\alpha_6\beta_1$ ligation has also been shown to lead to activation of the MAPK pathway and to induce increased levels of cytoplasmic calcium *via* mobilization of intracellular stores and channel-dependent influx from the extracellular space [36, 37].

The ability of $\alpha_6\beta_1$ to interact with its ligands can also be regulated by inside-out signaling. This can be achieved through changes in integrin mobility and localization on the cell surface that result in integrin clustering and consequently enhanced integrin avidity and/or changes in conformation leading to increased affinity of the integrin [38]. Very little is known about the molecular interactions that mediate these events in the context of $\alpha_6\beta_1$ but a number of key regulators of $\alpha_6\beta_1$ function have now been identified. Specifically cell surface $\alpha_6\beta_1$ has been shown to interact with numerous proteins confined to lipid rafts in the cell membrane such as members of the tetraspanin family [39–41] and urokinase-type plasminogen activator receptor [7, 42]. Co-localization of $\alpha_6\beta_1$ integrin with these molecules that possess transductional properties appears to contribute to its localization on the cell surface and to play a role in regulating the adhesive property of the integrin. For example, Lammerding et al. [43] demonstrated that the tetraspanin molecule CD151 plays a key role in regulating the time-dependent gain of the adhesion strength of $\alpha_6\beta_1$ to laminin, with the C-terminal cytoplasmic tail of CD151 being particularly important. Collectively the findings of this study indicated an important role for the tetraspanin CD151 as a modulator of cytoskeletal engagements in $\alpha_6\beta_1$-dependent adhesion strengthening.

While certain cell types can adhere to laminin under non-activating conditions (*e.g.*, platelets, eosinophils), others require cellular stimulation (*e.g.*, neutrophils, macrophages and T cells), suggesting that the ability of $\alpha_6\beta_1$ to interact with its ligands under basal conditions is cell specific (see Tab. 1 and text below for more details). Furthermore, with respect to the latter, since cellular stimulation does not appear to lead to increased expression of $\alpha_6\beta_1$ [19], it is considered that stimuli such

Table 1 - Effect of anti-α_6 antibody blockade on cell adhesion and migration in vitro

Cell type	Responses inhibited by anti-α_6 mAb	Species	Refs.
Platelets	Adhesion and spreading on LN-coated plates	Human/mouse	[8, 64]
Platelets	Adhesion to LN-1- and LN-8-coated plates	Human	[66]
B lymphocytes	Adhesion to LN-coated plates following ligation of BCR, CD40 or stimulation with IL-4	Mouse	[14]
Monocyte/macrophage	Adhesion and spreading of PMA-stimulated cells on LN-coated plates	Mouse	[30, 15]
Monocytes	Adhesion and/or migration on LN-10/11- and LN-8-coated plates	Human	[17]
Eosinophils	Adhesion to LN-coated plates	Human	[20]
Neutrophils	Adhesion of PMA-stimulated cells to LN-coated plates	Human	[18]
Neutrophils	Adhesion of unstimulated and PMA-stimulated cells to LN-10-coated plates and transmigration through LN-10-coated filters in response to fMLP	Mouse	[51]
HSC/HPC	Adhesion of un-stimulated and PMA-stimulated cells to LN-8- and LN-10-coated plates	Mouse	[57]
Neutrophils, monocytes, lymphocytes	Tethering and adhesion on LN-coated plates under flow condition	Human	[44]
Neutrophils	Transmigration through fibroblast monolayers	Human	[48, 50]
Neutrophils	Transmigration through TNFα-stimulated HUVECs grown on LN-coated filters under flow conditions	Human	[49]

HSC/HPC, hematopoietic stem cells/hematopoietic progenitor cells; BCR, B cell receptor; LN, laminin (all the studies used crude extracts of laminin unless stated otherwise).

as PMA induce cell adhesion to laminin *via* enhanced $\alpha_6\beta_1$ affinity for its ligand, as has been suggested for numerous β_1 integrins on T cells [12]. Of interest, using a flow model, the ability of neutrophils to tether on laminin was reduced under conditions of neutrophil stimulation with IL-8 [44]. Since $\alpha_6\beta_1$ has been reported to be expressed on microvilli [45], it is possible that chemokine stimulation may induce morphological changes in neutrophils in terms of microvilli expression that could reduce the efficiency with which $\alpha_6\beta_1$ interacts with its laminin substrate under flow. Overall, little is known about the physiological mechanisms that regulate the expression/activation of $\alpha_6\beta_1$ on leukocytes *in vivo*. In this context, we have found that IL-1β-induced transmigrated neutrophils express a significantly elevated level of $\alpha_6\beta_1$ on their cell surface as compared to blood neutrophils that express low levels of $\alpha_6\beta_1$ [46] (Fig. 2). This effect appeared to be mediated by a PECAM-1-PECAM-1 interaction between neutrophils and endothelial cells during the transmigration response [46, 47]. The signaling mechanisms associated with this PECAM-1-mediated event is at present unclear but the rapid nature of the response suggests that the cell surface expression of $\alpha_6\beta_1$ occurs *via* mobilization of the integrin from intracellular stores as has been suggested for human neutrophils [48]. Collectively these findings provide the first indications of how $\alpha_6\beta_1$ expression maybe regulated *in vivo*.

Role of $\alpha_6\beta_1$ in leukocyte adhesion and transmigration

In contrast to other β_1 integrins that can bind to multiple extracellular matrix protein ligands, $\alpha_6\beta_1$ appears to be relatively specific for laminin, suggesting a significant role for this integrin in interaction of cells with BMs. The majority of investigations into the functional role of $\alpha_6\beta_1$ have addressed the adhesive and migratory responses of cells expressing this integrin on laminin-coated plates *in vitro*, studies that have been performed using functional blocking mAbs directed against the different subunits of $\alpha_6\beta_1$ (Tab. 1).

Platelets and numerous other cell types involved in inflammatory/immune reactions, such as T cells, macrophages and neutrophils, exhibit $\alpha_6\beta_1$-dependent adhesion and/or spreading responses on laminin-coated surfaces *in vitro* (Tab. 1). On CD4+ T cells this response was enhanced following T cell activation and differentiation, with memory T cells expressing three- to fourfold more $\alpha_6\beta_1$ than naive cells [12]. With respect to macrophages, antibodies against α_6 and β_1 integrin subunits were found to inhibit PMA-stimulated adhesion of mouse peritoneal macrophages to laminin-coated plates [15]. As detailed above PMA-induced enhanced adhesion to laminin was not associated with an increased cell surface expression of the integrin on macrophages but was linked to anchorage of $\alpha_6\beta_1$ to the cytoskeleton. Similarly, PMA-stimulated adhesion of human neutrophils to laminin was found to be $\alpha_6\beta_1$ dependent [18], while $\alpha_6\beta_1$ mediated adhesion of eosinophils to laminin under basal conditions [20]. More recently $\alpha_6\beta_1$ has been shown to mediate leukocyte (neutrophils, eosinophils, lymphocytes and monocytes) interactions with laminin-coated plates under physiological flow conditions [44] and to mediate neutrophil migration through TNF-α-stimulated cultured endothelial cells grown on a laminin substrate under flow [49]. In the latter study, the findings were attributable to suppression of neutrophil adhesion to the subendothelial cell laminin and not suppression of neutrophil/endothelial cell interactions. IL-8-induced neutrophil migration through human synovial fibroblasts has also been reported to be partially dependent

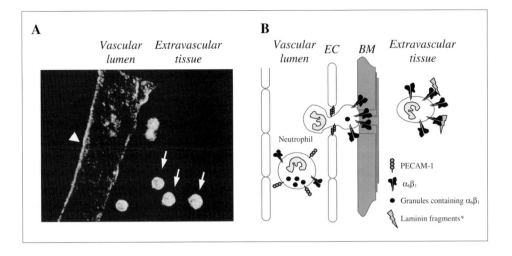

Figure 2

IL-1β-stimulated transmigrated neutrophils express α_6 integrin on their cell surface in vivo. Mice were injected via the intrascrotal route with IL-1β (30 ng) and 4 h later the cremaster muscle was dissected away from the animal. The tissue was fixed, immunostained for α_6 integrin expression using mAb GoH3 and analyzed by confocal microscopy. (A) The image shows transmigrated neutrophils next to a post-capillary venule (25-μm diameter). Both transmigrated leukocytes (arrows) and the venular endothelium are positive for α_6 (arrowhead). In contrast, leukocytes within the vascular lumen only exhibit a grainy expression of α_6, indicative of an intracellular granular expression profile of the molecule. (B) The diagram is a schematic illustration of the proposed mechanism by which $\alpha_6\beta_1$ is mobilized to the cell surface of transmigrated neutrophils following homophilic interaction of PECAM-1 at endothelial cell (EC) junctions. Briefly, it is hypothesized that during neutrophil transendothelial cell migration, interaction of leukocyte PECAM-1 with endothelial cell PECAM-1 triggers signaling pathways within the leukocyte that leads to mobilization of $\alpha_6\beta_1$ from intracellular granular stores to the cell surface of emigrating leukocytes. This increased expression of $\alpha_6\beta_1$ then facilitates migration of neutrophils through the venular basement membrane (BM) via interactions with its laminin constituents. *In line with this, in IL-1β-stimulated cremaster tissues, ~15% of transmigrated neutrophils express fragments of laminin, co-localized with $\alpha_6\beta_1$ integrin, on their cell surface (see text for details).

on $\alpha_6\beta_1$, a functional role that was particularly noticeable under conditions of β_2 integrin blockade [50]. Although the studies performed above were all conducted using laminin that is not found at sites of inflammation (mostly purified laminin-1), there is now evidence for the ability of $\alpha_6\beta_1$ to mediate stimulated neutrophil adhesion to and transmigration through filters coated with the vascular laminin isoforms, laminin 8 and laminin 10 [51].

Investigations into the functional differences of $\alpha_{6A}\beta_1$ and $\alpha_{6B}\beta_1$ have been hampered by lack of antibodies that distinguish between the two molecules but the functional roles of these molecules have been investigated by other methods. For example, using the macrophage cell line P388D1, macrophages that expressed $\alpha_{6A}\beta_1$ exhibited considerably more pseudopodia when plated onto laminin and were more migratory than cells that expressed the $\alpha_{6B}\beta_1$ integrin, supporting the concept that these molecules differ in their signaling properties as detailed above [16]. Furthermore, whereas deletion of α_6 causes perinatal lethality [52], a mouse line that expresses α_{6B} but not α_{6A}, is viable and largely phenotypically normal apart from a selective defect in lymphocyte distribution. Specifically, α_{6A}-deficient mice show normal lymphocyte homing to secondary lymphoid organs *in vivo*, but, interestingly, isolated T cells display a reduction in their ability to migrate through laminin-1-coated filters (but not fibronectin-coated filters) in response to the chemokines SDF-1/CXCL12 and MCP-1/CCL2 [53]. These findings suggest that, although expression of α_{6A} is necessary for the optimal migration of T cells on laminin-1 *in vitro*, it is not critical for *in vivo* lymphocyte homing to secondary lymphoid organs [53].

Despite the above, few studies have addressed the functional role of $\alpha_6\beta_1$ in regulation of inflammatory and immune responses *in vivo* (Tab. 2). In this context it has been shown that α_6 integrins play a role in homing of T cell precursors into the thymus [21] and retention/responses of germinal center B cells [14, 54]. Furthermore, the expression of $\alpha_6\beta_1$ is associated with an enhanced migratory and invasiveness phenotype of carcinoma cells *in vitro* and *in vivo* and reduced survival of patients [7, 55, 56]. More recently, using a passive cell transfer technique, it was shown that an anti-α_6 integrin mAb can compromise the homing of hematopoietic progenitor/stem cells (HPC/HSC) to the bone marrow *in vivo* [57].

With respect to acute inflammatory events, anti-α_6 and anti-β_1 mAbs have been shown to inhibit neutrophil migration into mouse airways following intratracheal administration of LPS and the chemokine KC [58]. Analysis of tissues by electron microscopy suggested that the inhibitory effect of the anti-β_1 mAb in this model was largely on migration of neutrophils from the interstitium into the alveoli, with only a minor effect on migration of neutrophils across the capillary wall [58]. Studies from our group have shown that treatment of mice with an anti-α_6 integrin mAb inhibits neutrophil transmigration through IL-1β-stimulated cremasteric venules as observed by intravital microscopy [46, 59]. Analysis of tissues by electron microscopy indicated that this inhibitory effect was at the level of the vascular BM in that the neutrophils were trapped within the vessel wall. Of interest, the ability of the anti-α_6 integrin mAb to suppress neutrophil transmigration in this model was stimulus specific as transmigration induced by TNF-α was unaffected [59]. Since work from our laboratory had previously shown that PECAM-1 can also mediate neutrophil migration through the vascular BM in a stimulus-specific manner [60, 61], we sought to investigate the possibility that PECAM-1-mediated neutrophil

Table 2 - Effect of anti-α_6 antibody blockade on cell migration in vivo

Cell type	Responses inhibited by anti-α_6 mAb	Species	Refs.
HSC/HPC	Homing to the bone marrow	Mouse	[57]
T lymphocytes	Homing to the thymus	Mouse	[21]
Neutrophils	Migration into the airways in response to intratracheal instillation of LPS and KC	Mouse	[58]
Neutrophils	Transmigration in response to IL-1β in a peritonitis model and through cremasteric venules	Mouse	[46, 47, 59]

HSC/HPC, hematopoietic stem cells/hematopoietic progenitor cells

migration through the BM was associated with regulated expression of the integrin $\alpha_6\beta_1$ on the cell surface of transmigrated neutrophils. Collectively, our findings to date suggest that PECAM-1-PECAM-1 interaction between neutrophils and endothelial cells at endothelial cell junctions elicits expression of $\alpha_6\beta_1$ on the cell surface of emigrating neutrophils that can then aid their migration through the vascular BM ([46, 47, 62]; Fig. 2). The precise mechanism by which $\alpha_6\beta_1$ mediates neutrophil migration through the BM is at present unclear but a recent study from our group has identified regions within the vascular BM where the deposition of certain BM constituents, such as laminin 10, is lower than the average level [63]. Neutrophils seem to use these vulnerable regions for penetrating the BM, a response that leads to a transient increase in the size of the regions. Of interest, ~15% of the transmigrated neutrophils expressed laminin 10 on their cell surface in association with $\alpha_6\beta_1$. Hence, a potential mechanism for the observed enhanced size of the laminin 10 low expression regions may be carriage/cleavage of laminin/laminin fragments post adhesion to $\alpha_6\beta_1$. Whether laminin 10 is cleaved or physically detached from the BM structure is at present unknown. Finally, since both PECAM-1-dependent and $\alpha_6\beta_1$-dependent neutrophil migration is stimulus specific, the molecular basis of pathways independent of these molecules needs to be clarified. It is possible that this may be linked to the ability of the inflammatory trigger to directly stimulate leukocytes and thus activate alternative mechanisms involved in leukocyte transmigration.

Although it is hypothesized that blockers of $\alpha_6\beta_1$ suppress cell migration *in vivo* *via* suppressing the physical ability of cells to migrate through laminin-containing barriers, it is also possible that inhibiting the stimulatory effects of laminin on cellular effector functions (*e.g.*, chemotaxis) may contribute to the inhibitory effects of anti-$\alpha_6\beta_1$ reagents. Furthermore, leukocyte/vessel wall interactions *in vivo* could also be potentially regulated by platelet-expressed $\alpha_6\beta_1$. Since platelets can spread on and adhere to laminin-coated plates in an $\alpha_6\beta_1$-dependent manner [8, 64], $\alpha_6\beta_1$-

mediated adhesion of platelets to exposed vascular BM at sites of vascular injury could potentially support leukocyte attachment and infiltration into sites of vascular damage [65].

Conclusions

In conclusion, as the principal leukocyte laminin receptor, the integrin $\alpha_6\beta_1$ plays a key role in mediating leukocyte migration through laminin-rich barriers such as the vascular BM. While there is increasing *in vivo* evidence to implicate this integrin in key inflammatory and immune events, there is a need for better understanding of the mechanisms associated with both $\alpha_6\beta_1$-dependent and $\alpha_6\beta_1$-independent modes of leukocyte migration *in vivo*. A better understanding of the signaling pathways that regulate $\alpha_6\beta_1$ expression and affinity in different leukocyte subtypes and endothelial cells of different vascular beds would also be of value in identifying the potential role of this integrin in different inflammatory scenarios and pathologies.

Acknowledgments
The authors are grateful to the British Heart Foundation and The Wellcome Trust for the generous support of their research.

References

1   Timpl R (1996) Macromolecular organization of basement membranes. *Curr Opin Cell Biol* 8: 618–624
2   Muller WA (2003) Leukocyte-endothelial-cell interactions in leukocyte transmigration and the inflammatory response. *Trends Immunol* 4: 327–334
3   Tamura RN, Rozzo C, Starr L, Chambers J, Reichardt LF, Cooper HM, Quaranta V (1990) Epithelial integrin α6β4: complete primary structure of α6 and variant forms of β4. *J Cell Biol* 111: 1593–1604
4   Hogervorst F, Kuikman I, van Kessel AG, Sonnenberg A (1991) Molecular cloning of the human α6 integrin subunit. Alternative splicing of α6 mRNA and chromosomal localization of the α6 and β4 genes. *Eur J Biochem* 199: 425–433
5   Hogervorst F, Admiraal LG, Niessen C, Kuikman I, Janssen H, Daams H, Sonnenberg A (1993) Biochemical characterization and tissue distribution of the A and B variants of the integrin α6 subunit. *J Cell Biol* 121: 179–191
6   Davis TL, Rabinovitz I, Futscher BW, Schnolzer M, Burger F, Liu Y, Kulesz-Martin M, Cress AE (2001) Identification of a novel structural variant of the α6 integrin. *J Biol Chem* 276: 26099–26106

7   Demetriou MC, Cress AE (2004) Integrin clipping: a novel adhesion switch? *J Cell Biochem* 91: 26–35

8   Sonnenberg A, Modderman PW, Hogervorst F (1988) Laminin receptor on platelets is the integrin VLA-6. *Nature* 336: 487–489

9   Yannariello-Brown J, Wewer U, Liotta L, Madri JA (1988) Distribution of a 69-kD laminin-binding protein in aortic and microvascular endothelial cells: modulation during cell attachment, spreading, and migration. *J Cell Biol* 106: 1773–1786

10 Schaapveld RQ, Borradori L, Geerts D, van Leusden MR, Kuikman I, Nievers MG, Niessen CM, Steenbergen RD, Snijders PJ, Sonnenberg A (1998) Hemidesmosome formation is initiated by the β4 integrin subunit, requires complex formation of β4 and HD1/plectin, and involves a direct interaction between β4 and the bullous pemphigoid antigen 180. *J Cell Biol* 142: 271–284

11 Cid MC, Esparza J, Schnaper HW, Juan M, Yague J, Grant DS, Urbano-Marquez A, Hoffman GS, Kleinman HK (1999) Estradiol enhances endothelial cell interactions with extracellular matrix proteins *via* an increase in integrin expression and function. *Angiogenesis* 3: 271–280

12 Shimizu Y, van Seventer GA, Horgan KJ, Shaw S (1990) Costimulation of proliferative responses of resting CD4$^+$ T cells by the interaction of VLA-4 and VLA-5 with fibronectin or VLA-6 with laminin. *J Immunol* 145: 59–67

13 Schweighoffer T, Luce GE, Tanaka Y, Shaw S (1994) Differential expression of integrins α6 and α4 determines pathways in human peripheral CD4$^+$ T cell differentiation. *Cell Adhes Commun* 2: 403–415

14 Ambrose HE, Wagner SD (2004) α6-integrin is expressed on germinal centre B cells and modifies growth of a B-cell line. *Immunology* 111: 400–406

15 Shaw LM, Messier JM, Mercurio AM (1990) The activation dependent adhesion of macrophages to laminin involves cytoskeletal anchoring and phosphorylation of the α6β1 integrin. *J Cell Biol* 110: 2167–2174

16 Wei J, Shaw LM, Mercurio AM (1998) Regulation of mitogen-activated protein kinase activation by the cytoplasmic domain of the α6 integrin subunit. *J Biol Chem* 273: 5903–5907

17 Pedraza C, Geberhiwot T, Ingerpuu S, Assefa D, Wondimu Z, Kortesmaa J, Tryggvason K, Virtanen I, Patarroyo M (2000) Monocytic cells synthesize, adhere to, and migrate on laminin-8 (α4β1γ1). *J Immunol* 165: 5831–5838

18 Bohnsack JF (1992) CD11/CD18-independent neutrophil adherence to laminin is mediated by the integrin VLA-6. *Blood* 79: 1545–1552

19 Rieu P, Lesavre P, Halbwachs-Mecarelli L (1993) Evidence for integrins other than β2 on polymorphonuclear neutrophils: expression of α6β1 heterodimer. *J Leukoc Biol* 53: 576–582

20 Georas SN, McIntyre BW, Ebisawa M, Bednarczyk JL, Sterbinsky SA, Schleimer RP, Bochner BS (1993) Expression of a functional laminin receptor α6β1 (very late activation antigen-6) on human eosinophils. *Blood* 82: 2872–2879

21   Ruiz P, Wiles MV, Imhof BA (1995) α6 integrins participate in pro-T cell homing to the thymus. *Eur J Immunol* 25: 2034–2041

22   Chang AC, Salomon DR, Wadsworth S, Hong MJ, Mojcik CF, Otto S, Shevach EM, Coligan JE (1995) α3β1 and α6β1 integrins mediate laminin/merosin binding and function as costimulatory molecules for human thymocyte proliferation. *J Immunol* 154: 500–510

23   Timpl R (1989) Structure and biological activity of basement membrane proteins. *Eur J Biochem* 180: 487–502

24   Hallmann R, Horn N, Selg M, Wendler O, Pausch F, Sorokin LM (2005) Expression and function of laminins in the embryonic and mature vasculature. *Physiol Rev* 85: 979–1000

25   Kikkawa Y, Sanzen N, Fujiwara H, Sonnenberg A, Sekiguchi K (2000) Integrin binding specificity of laminin-10/11: laminin-10/11 are recognized by α3β1, α6β1, and α6β4 integrins. *J Cell Sci* 113: 869–876

26   Gonzalez AM, Gonzales M, Herron GS, Nagavarapu U, Hopkinson SB, Tsuruta D, Jones JC (2002) Complex interactions between the laminin α4 subunit and integrins regulate endothelial cell behaviour *in vitro* and angiogenesis *in vivo*. *Proc Natl Acad Sci USA* 99: 16075–16080

27   Nishiuchi R, Murayama O, Fujiwara H, Gu J, Kawakami T, Aimoto S, Wada Y, Sekiguchi K (2003) Characterization of the ligand-binding specificities of integrin α3β1 and α6β1 using a panel of purified laminin isoforms containing distinct α chains. *J Biochem* 134: 497–504

28   Wei J, Shaw LM, Mercurio AM (1997) Integrin signalling in leukocytes: lessons from the α6β1 integrin. *J Leukoc Biol* 61: 397–407

29   Hynes RO (2002) Integrins: bidirectional, allosteric signaling machines. *Cell* 110: 673–687

30   Shaw LM, Mercurio AM (1994) Regulation of cellular interactions with laminin by integrin cytoplasmic domains: the A and B structural variants of the α6β1 integrin differentially modulate the adhesive strength, morphology, and migration of macrophages. *Mol Biol Cell* 5: 679–690

31   Shaw LM, Turner CE, Mercurio AM (1995) The α6Aβ1 and α6Bβ1 integrin variants signal differences in the tyrosine phosphorylation of paxillin and other proteins. *J Biol Chem* 270: 23648–23652

32   Sastry SK, Lakonishok M, Wu S, Truong TQ, Huttenlocher A, Turner CE, Horwitz AF (1999) Quantitative changes in integrin and focal adhesion signalling regulate myoblast cell cycle withdrawal. *J Cell Biol* 144: 1295–1309

33   Tani TT, Mercurio AM (2001) PDZ interaction sites in integrin a subunits. T14853, TIP/GIPC binds to a type I recognition sequence in α6A/α5 and a novel sequence in α6B. *J Biol Chem* 276: 36535–36542

34   El Mourabit H, Poinat P, Koster J, Sondermann H, Wixler V, Wegener E, Laplantine E, Geerts D, Georges-Labouesse E, Sonnenberg A, Aumailley M (2002) The PDZ domain

of TIP–2/GIPC interacts with the C-terminus of the integrin α5 and α6 subunits. *Matrix Biol* 21: 207–214

35 Ferletta M, Kikkawa Y, Yu H, Talts JF, Durbeej M, Sonnenberg A, Timpl R, Campbell KP, Ekblom P, Gensersch E (2003) Opposing roles of integrin α6Aβ1 and dystroglycan in laminin-mediated extracellular signal-regulated kinase activation. *Mol Biol Cell* 14: 2088–2103

36 Simms H, D'Amico R (1995) Regulation of polymorphonuclear neutrophil CD16 and CD11b/CD18 expression by matrix proteins during hypoxia is VLA-5, VLA-6 dependent. *J Immunol* 155: 4979–4990

37 Schottelndreier H, Potter BV, Mayr GW, Guse AH (2001) Mechanisms involved in α6β1-integrin-mediated Ca2⁺ signalling. *Cell Signal* 13: 895–899

38 Woods ML, Shimizu Y (2001) Signalling networks regulating β1 integrin-mediated adhesion of T lymphocytes to extracellular matrix. *J Leukoc Biol* 69: 874–880.

39 Berditchevski F, Zutter MM, Hemler ME (1996) Characterization of novel complexes on the cell surface between integrins and proteins with 4 transmembrane domains (TM4 proteins). *Mol Biol Cell* 7: 193–207

40 Sterk LM, Geuijen CA, van den Berg JG, Claessen N, Weening JJ, Sonnenberg A (2002) Association of the tetraspanin CD151 with the laminin-binding integrins α3β1, α6β1, α6β4 and α7β1 in cells in culture and *in vivo*. *J Cell Sci* 115: 1161–1173

41 Brown EJ (2002) Integrin-associated proteins. *Curr Opin Cell Biol* 14: 603–607

42 Demetriou MC, Pennington ME, Nagle RB, Cress AE (2004) Extracellular α6 integrin cleavage by urokinase-type plasminogen activator in human prostate cancer. *Exp Cell Res* 294: 550–558

43 Lammerding J, Kazarov AR, Huang H, Lee RT, Hemler ME (2003) Tetraspanin CD151 regulates α6β1 integrin adhesion strengthening. *Proc Natl Acad Sci USA* 100: 7616–7621

44 Kitayama J, Ikeda S, Kumagai K, Saito H, Nagawa H (2000) α6β1 integrin (VLA-6) mediates leukocyte tether and arrest on laminin under physiological shear flow. *Cell Immunol* 199: 97–103

45 Abitorabi MA, Pachynski RK, Ferrando RE, Tidswell M, Erle DJ (1997) Presentation of integrins on leukocyte microvilli: a role for the extracellular domain in determining membrane localization. *J Cell Biol* 139: 563–571

46 Dangerfield J, Larbi KY, Huang MT, Dewar A, Nourshargh S (2002) PECAM-1 (CD31) homophilic interaction up-regulates α6β1 on transmigrated neutrophils *in vivo* and plays a functional role in the ability of alpha6 integrins to mediate leukocyte migration through the perivascular basement membrane. *J Exp Med* 196: 1201–1211

47 Wang S, Dangerfield JP, Young RE, Nourshargh S (2005) PECAM-1, a6 integrins and neutrophil elastase cooperate in mediating neutrophil transmigration. *J Cell Sci* 118: 2067–2076

48 Roussel E, Gingras MC (1997) Transendothelial migration induces rapid expression on neutrophils of granule-release VLA6 used for tissue infiltration. *J Leukoc Biol* 62: 356–362

49 Kitayama J, Hidemura A, Saito H, Nagawa H (2000) Shear stress affects migration behavior of polymorphonuclear cells arrested on endothelium. *Cell Immunol* 203: 39–46

50 Gao JX, Wilkins J, Issekutz AC (1995) Migration of human polymorphonuclear leukocytes through a synovial fibroblast barrier is mediated by both beta2 (CD11/CD18) integrins and the beta 1 (CD29) integrins VLA-5 and VLA-6. *Cell Immunol* 163: 178–186

51 Sixt M, Hallmann R, Wendler O, Scharffetter-Kochanek K, Sorokin LM (2001) Cell adhesion and migration properties of β2-integrin negative polymorphonuclear granulocytes on defined extracellular matrix molecules. Relevance for leukocyte extravasation. *J Biol Chem* 276: 18878–18887

52 Georges-Labouesse E, Messaddeq N, Yehia G, Cadalbert L, Dierich A, Le Meur M (1996) Absence of integrin α6 leads to epidermolysis bullosa and neonatal death in mice. *Nat Genet* 13: 370–373

53 Gimond C, Baudoin C, van der Neut R, Kramer D, Calafat J, Sonnenberg A (1998) Cre-loxP-mediated inactivation of the α6A integrin splice variant *in vivo*: evidence for a specific functional role of α6A in lymphocyte migration but not in heart development. *J Cell Biol* 143: 253–266

54 Borland G, Cushley W (2004) Positioning the immune system: unexpected roles for α6-integrins. *Immunology* 111: 381–383

55 Friedrichs K, Ruiz P, Franke F, Gille I, Terpe HJ, Imhof BA (1995) High expression level of α6 integrin in human breast carcinoma is correlated with reduced survival. *Cancer Res* 55: 901–906

56 Rabinovitz I, Nagle RB, Cress AE (1995) Integrin α6 expression in human prostate carcinoma cells is associated with a migratory and invasive phenotype *in vitro* and *in vivo*. *Clin Exp Metastasis* 13: 481–491

57 Qian H, Tryggvason K, Jacobsen SE, Ekblom M (2006) Contribution of α6 integrins to hematopoietic stem and progenitor cell homing to bone marrow and collaboration with α4 integrins. *Blood* 107: 3503–3510

58 Ridger VC, Wagner BE, Wallace WA, Hellewell PG (2001) Differential effects of CD18, CD29, and CD49 integrin subunit inhibition on neutrophil migration in pulmonary inflammation. *J Immunol* 166: 3484–3490

59 Dangerfield JP, Wang S, Nourshargh S (2005) Blockade of α6 integrin inhibits IL-1β- but not TNFα-induced neutrophil transmigration *in vivo*. *J Leukoc Biol* 77: 159–165

60 Wakelin MW, Sanz MJ, Dewar A, Albelda SM, Larkin SW, Boughton-Smith N, Williams TJ, Nourshargh S (1996) An anti-platelet-endothelial cell adhesion molecule-1 antibody inhibits leukocyte extravasation from mesenteric microvessels *in vivo* by blocking the passage through the basement membrane. *J Exp Med* 184: 229–239

61 Thompson RD, Noble KE, Larbi KY, Dewar A, Duncan GS, Mak TW, Nourshargh S (2001) Platelet-endothelial cell adhesion molecule-1 (PECAM-1)-deficient mice demonstrate a transient and cytokine-specific role for PECAM-1 in leukocyte migration through the perivascular basement membrane. *Blood* 97: 1854–1860

62   Nourshargh S, Marelli-Berg FM (2005) Transmigration through venular walls: a key regulator of leukocyte phenotype and function. *Trends Immunol* 26: 157–165

63   Wang S, Voisin MB, Larbi KY, Dangerfield J, Scheiermann C, Tran M, Maxwell PH, Sorokin L, Nourshargh S (2006) Venular basement membranes contain specific matrix protein low expression regions that act as exit points for emigrating neutrophils. *J Exp Med* 203: 1519–1532

64   Inoue O, Suzuki-Inoue K, McCarty OJ, Moroi M, Ruggeri ZM, Kunicki TJ, Ozaki Y, Watson SP (2006) Laminin stimulates spreading of platelets through integrin $\alpha6\beta1$-dependent activation of GPVI. *Blood* 107: 1405–1412

65   Huo Y, Ley KF (2004) Role of platelets in the development of atherosclerosis. *Trends Cardiovasc Med* 14: 18–22

66   Geberhiwot T, Ingerpuu S, Pedraza C, Neira M, Lehto U, Virtanen I, Kortesmaa J, Tryggvason K, Engvall E, Patarroyo M (1999) Blood platelets contain and secrete laminin-8 ($\alpha4\beta1\gamma1$) and adhere to laminin-8 *via* $\alpha6\beta1$ integrin. *Exp Cell Res* 253: 723–732

Vascular adhesion protein-1 (VAP-1)

Marko Salmi[1] and Sirpa Jalkanen[2]

[1]MediCity Research Laboratory and Department of Microbiology and Immunology, University of Turku, 20520 Turku, Finland; [2]Department of Bacterial and Inflammatory Diseases, National Public Health Institute, 20520 Turku, Finland

Discovery of vascular adhesion protein-1

In 1980s the leukocyte adhesion molecules and their ligands on the vascular endo-thelium were thought to explain tissue-selective, or even tissue-specific, leukocyte traffic. At the same time it became apparent that vessels in inflamed joints displayed binding characteristics clearly distinct from those in peripheral lymph nodes, gut and skin. In search of joint-selective endothelial adhesion molecules we therefore isolated vascular fragments from inflamed human synovial samples to raise mono-clonal antibodies (mAbs) against endothelial antigens [1]. From this screen, we identified an mAb, 1B2, that readily stained synovial blood vessels and inhibited lymphocyte adhesion in classical frozen section binding assays. Differential screen-ing against other known endothelial adhesion molecules allowed us to conclude that we had identified a new adhesion molecule. This antigen was named vascular adhesion protein-1 (VAP-1; nowadays also known by the gene name amine oxidase copper containing -3, AOC3).

VAP-1 is an ecto-enzyme

It became immediately clear that VAP-1, like any other endothelial molecule involved in the adhesion cascade, is neither a tissue-specific nor an endothelium-specific addressin [1, 2]. Prominent synthesis of VAP-1 in smooth muscle cells was, in fact, helpful for large-scale immunoaffinity purifications needed for protein sequencing and cDNA cloning. The sequence of VAP-1 strikingly revealed that it did not belong to any of the adhesion molecule superfamilies known at that time [3]. Instead the sequence unequivocally identified VAP-1 as a semicarbazide-sensi-tive amine oxidase (SSAO).

Based on the chemical nature of the attached co-factor, amine oxidases are sub-divided into several groups (Fig. 1) [4]. The FAD-containing enzymes, monoamine oxidase-A (MAO-A), MAO-B and polyamine oxidase, are intracellular enzymes involved in the metabolism of neurotransmitters, spermine and spermidine. Cop-

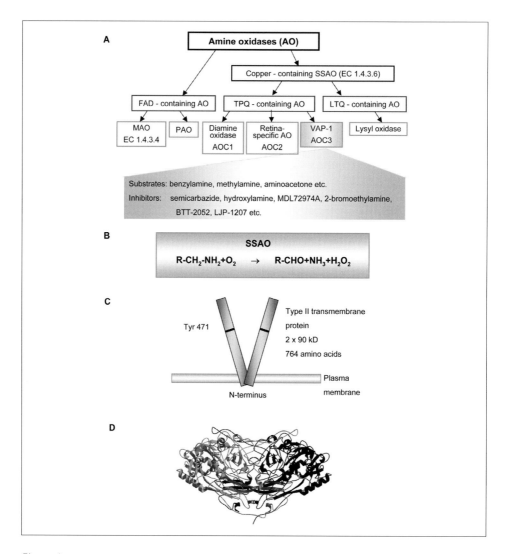

Figure 1

Classification and structure of VAP-1.

(A) Based on the sequence VAP-1 belongs to semicarbazide sensitive amine oxidases (SSAO). FAD, flavin adenine dinucleotide; TPQ, topaquinone; LTQ, lysine tyrosylquinone; MAO, monoamine oxidase; PAO, polyamine oxidase; AOC, amine oxidase, copper containing; EC, enzyme classification. The structures of several SSAO inhibitors can be found in [11, 18, 51] (B) The enzymatic reaction catalyzed by VAP-1. (C) VAP-1 is a homodimeric transmembrane protein. The tyrosine residue 471 that is modified into topaquinone is highlighted. (D) The overall fold of VAP-1 structure. One subunit of the dimer is shown in light gray and the other in dark gray (courtesy of Dr. Tiina Salminen, Åbo Akademi, Turku, Finland).

per-containing amine oxidases, on the other hand, include diamine oxidase (AOC1), retina-specific amine oxidase (AOC2) and VAP-1 (AOC3), which was the first AOC to be cloned in man. All AOC enzymes contain topaquinone, a unique, autocatalytically formed modification of tyrosine, at their active site [5]. The function of all AOCs is inhibited by carbonyl-reactive substances such as semicarbazide.

AOC catalyze oxidative deamination of primary amines in a two-step reaction that results in the production of the corresponding aldehyde, ammonium and hydrogen peroxide (Fig. 1) [4, 5]. During the reductive-half reaction the primary amine interacts with the topaquinone, and a covalent Schiff-base intermediate occurs between the enzyme and the substrate before an aldehyde is formed. During the second, oxidative half-reaction, the enzyme is reoxidized by molecular oxygen, and ammonium and hydrogen peroxide are released.

VAP-1 as an adhesion molecule

Anti-VAP-1 mAbs inhibit leukocyte-endothelial interactions *in vitro*

In the original report, the anti-VAP-1 mAb 1B2 was shown to inhibit lymphocyte binding to high endothelial venules in peripheral lymph nodes and tonsil *in vitro* by approximately 50% [1]. This and other anti-VAP-1 mAbs also efficiently block lymphocyte adhesion to flat-walled venules in inflamed tissues such as synovium, skin, and lamina propria of the gut [2, 6, 7]. In liver, VAP-1 plays a dominant role in sinusoidal endothelial cells, which lack many other adhesion molecules (*e.g.*, endothelial selectins) [8]. Granulocytes and tissue-derived macrophages also utilize VAP-1 in frozen section binding assays, indicating that VAP-1 can support adhesion of many different leukocyte subclasses [7, 9].

Anti-VAP-1 mAbs inhibit leukocyte rolling on, firm adhesion to and, most profoundly, transmigration through monolayers of endothelial cells under conditions of defined laminar shear stress [10–12]. These *in vitro* flow chamber studies have been performed with constitutively VAP-1-positive endothelial cells from the heart of rabbits, human umbilical vein endothelial cells (HUVECs) that have been transfected with VAP-1 and cultured human liver sinusoidal endothelial cells.

SSAO enzyme activity is involved in leukocyte binding *in vitro*

The unique molecular nature of VAP-1 has raised the obvious question whether the enzymatic activity of VAP-1 is also involved in its adhesive function. Use of semicarbazide, hydroxylamine and second-generation SSAO inhibitors has unequivocally shown that blocking of the catalytic activity of VAP-1 results in impaired rolling, firm adhesion and transmigration under flow conditions [10–12]. Also, when HUVECs are transfected with VAP-1 cDNA rendered enzymatically inactive through a single point mutation, they support less binding than cells transfected with the wild-type VAP-1 [11]. Moreover, HUVECs expressing the mutant VAP-1 no longer show any VAP-1-inhibitable granulocyte binding. Since the SSAO inhibitors or the point mutation do not affect the antibody-defined epitopes on VAP-1, and since anti-VAP-1 antibodies do not inhibit its catalytic activity, these data strongly suggest that VAP-1 can support leukocyte binding through both enzyme activity-dependent and enzyme activity-independent pathways.

Anti-VAP-1 mAbs and SSAO inhibitors interfere with leukocyte traffic *in vivo*

The first animal studies with VAP-1 were done with two anti-human VAP-1 mAbs cross-reacting with rabbit VAP-1 [13]. The inhibition of VAP-1 led to about 70% reduction in the influx of neutrophils into the inflamed peritoneal cavity. The reduction of leukocyte infiltration was in line with intravital analyses, which revealed that anti-VAP-1 antibodies markedly increased the rolling velocity of granulocytes in inflamed mesenteric vessels. At the same time, the number of firmly adherent cells and the number of transmigrated cells was reduced by more than 50%.

With the advent of function-blocking anti-mouse and anti-rat VAP-1 antibodies, the role of VAP-1 in recruitment of inflammatory cells was verified in many rodent models (Tab. 1). These antibodies reduce the development of inflammatory infiltrates in acute granulocyte-dependent (peritonitis), monocyte-dependent (CCL21-driven air-pouch inflammation) and lymphocyte-dependent (allograft rejection) models [14, 15]. Notably, a half-year long treatment with anti-VAP-1 mAbs targeting the VAP-1-positive microvessels in the pancreatic islets significantly decreased the incidence of diabetes in NOD mice without any adverse effects [14]. Interestingly, VAP-1 plays a very dominant role in migration of Th2-type, but not in that of Th1-type, lymphocytes to inflamed liver by ablating the leukocyte-endothelial contacts in liver sinusoids [16]. Real-time imaging has shown that the anti-VAP-1 mAbs also significantly increase the rolling velocity of granulocytes in inflamed vessels in cremaster [17].

SSAO inhibition alters leukocyte traffic *in vivo*. This was first shown in a carrageenan-induced air-pouch inflammation model in rats, in which administration of a hydrazine-based SSAO inhibitor led to significant reduction in the number of inflammatory cells [11]. Thereafter, this and many other SSAO inhibitors have been shown to reduce inflammation in many different models (Table 1). One of the best examples is a compound known as LJP1207 [N'-(2-phenyl-allyl)-hydrazine hydrochloride], which is orally bioavailable [18, 19]. In oxazolone-induced colitis it effectively blunts weight loss and mortality, and leads to lowered colonic cytokine levels, less inflammatory cell infiltrate, and diminished ulceration [18]. LJP1207

Table 1 - VAP-1 inhibition alleviates inflammation in vivo[a]

Model	Species	Treatment	Major effect	Ref
Peritonitis				
PP+IL-1	Rabbit	mAb	Leukocyte infiltration 70% ↓	[13]
PP+IL-1	Mouse	mAb	Granulocyte infiltration 50% ↓	[14]
TNF-α	Mouse	KO	Granulocyte infiltration 40% ↓	[17]
Arthritis				
adjuvant	Rat	inh	Clinical inflammation score 40% ↓	[20]
mAb-induced[b]	Mouse	inh	Clinical inflammation score 80% ↓	[20]
mAb-induced[b]	Mouse	KO	Clinical inflammation score 40% ↓	[20]
Liver inflammation				
allograft	Rat	mAb	Lymphocytes in allografts 60% ↓	[15]
ConA	Mouse	mAb	ALT level 90% ↓	[16]
Insulitis				
NOD	Mouse	mAb	Incidence of diabetes 50% ↓	[14]
Skin inflammation				
carrageenan	Rat	inh	Paw edema 70% ↓	[18]
carrageenan	Rat	inh	Leukocyte infiltration 50% ↓	[11]
CCL21-induced	Mouse	mAb	Monocyte infiltration 60% ↓	[14]
Lung inflammation				
LPS inhalation	Mouse	inh	Leukocytes in BAL lavage 40% ↓	[51]
Colitis				
oxazolone-induced	Mouse	inh	Survival 50% ↑	[18]
Endotoxemia				
LPS-induced	Mouse	inh	Mortality 50% ↓	[18]
Ischemia-reperfusion injury				
stroke	Rat	inh	Neurological ability 80% ↑	[19]

[a]*PP, proteose peptone, mAb, monoclonal anti-VAP-1 antibody; KO, VAP-1-knockout; inh, VAP-1 enzyme inhibitor.*
[b]*Anti-collagen type II antibody cocktail.*

also prolongs survival in systemic inflammation in LPS-induced endotoxemic shock [18]. Notably, the SSAO inhibition almost completely reverses the leukocyte extravasation in post-ischemic brain, even if the treatment is started several hours after the triggering of the ischemia [19].

VAP-1-deficient mice show abnormal leukocyte traffic

Formal proof for the involvement of VAP-1 in leukocyte traffic *in vivo* has been obtained through the generation of gene-deficient mice [17]. Deletion of VAP-1 results in the disappearance of all detectable VAP-1 protein and SSAO activity. VAP-1 does not seem to be essential for mouse development, since VAP-1-deficient mice are born at the expected Mendelian frequency, and are macroscopically healthy. Moreover, the histology of all studied organs (such as heart, lungs, liver, kidney, thymus, spleen, lymph nodes and gut) is microscopically normal. However, the leukocyte extravasation cascade is clearly affected in the absence of VAP-1. The rolling velocity of granulocytes is about four times faster and the number of firmly adherent cells and transmigrated cells is significantly diminished. In *in vivo* disease models the influx of granulocytes into inflamed peritoneum and into inflamed joints is also significantly impaired in VAP-1-deficient mice [17, 20]. So far similar findings have been observed in VAP-1-deficient mice both in 129S6 and C57BL/6 backgrounds.

Deletion of VAP-1 also affects constitutive lymphocyte homing to a minor degree [17]. Lymphocytes roll much faster and fewer cells firmly adhere in Peyer's patches in the absence of VAP-1. Lymphocyte traffic into mesenteric lymph nodes and spleen is reduced by about 20% in short-term homing assays, although migration to peripheral lymph nodes or Peyer's patches is not statistically significantly inhibited. Moreover, the homing of OVA-transgenic T cells into pancreata is impaired in the absence of VAP-1. Hence, studies on VAP-1-deficient mice have opened a new paradigm of ecto-enzyme regulated control of leukocyte traffic. This concept is now supported by genetically modified mice that lack other ecto-enzymes, such as nucleotidases (*e.g.*, CD73 and CD39), ADP-ribosyl cyclases (CD38) and peptidases (*e.g.*, CD26 and sheddases), and have profound defects in leukocyte migration [21].

Other amine oxidases and cell adhesion

It is noteworthy that both diamine oxidase (AOC1) and lysyl oxidase (LOX) are also involved in the control of inflammation and cell movement. A secreted form of diamine oxidase is bound to endothelial cells, and it oxidizes and inactivates histamine [22]. LOX, on the other hand, is involved in directional movement of smooth muscle cells and in invasion and matrix adhesion of metastatic cells [23, 24].

Inflammation regulates VAP-1 expression

In normal conditions VAP-1 is present in practically all human organs, but its expression is restricted to vascular endothelial cells, follicular dendritic cells, adipocytes and smooth muscle cells [2]. In mouse and rat, the expression pattern is somewhat different (*e.g.*, only low expression in the sinusoidal endothelium in liver) [25]. In endothelial cells, VAP-1 is mostly stored within intracellular vesicles in an enzymatically inactive form. However, at sites of inflammation, both in experimental animals and in man, it is rapidly translocated to the endothelial cell surface, where it is fully functional as an adhesin and enzyme [26, 27]. In organ culture systems, IL-1, IL-4, TNF-α, IFN-γ and LPS stimulations up-regulate endothelial VAP-1, but the substance(s) capable of directly inducing VAP-1 translocation or protein synthesis has not been identified [28].

During human ontogeny, VAP-1 is found in the smooth muscle already at week 7 of gestation. VAP-1 expression shows time-dependent switches during the embryonic life in the vascular beds of liver, kidney, gut and heart. Fetal VAP-1 is enzymatically active and able to bind lymphocytes in *in vitro* adhesion assays, suggesting that VAP-1 may support lymphocyte traffic already *in utero* [29].

VAP-1 also exists in soluble form

That SSAO activity is present in serum has been known for more than four decades, and it has been shown to be elevated in certain diseases such as in diabetes, atherosclerosis, heart failure, obesity and certain liver diseases [30–34] (Tab. 2). Moreover, this enzyme may contribute to the development of late diabetic complications such as retinopathy and nephropathy [35, 36]. However, VAP-1 is not simply a general inflammation marker, since in many disorders the soluble VAP-1 levels stay normal or even decrease [37] (Table 2).

VAP-1 protein is responsible for practically all SSAO activity in the serum, as demonstrated by VAP-1 depletion [37]. Under normal conditions the source of soluble VAP-1 in serum is preferentially endothelium and particularly the sinusoidal endothelium of the liver [34], and its concentration inversely correlates to the insulin level [38]. However, soluble VAP-1 can also be produced by smooth muscle and fat cells [39, 40]. In adipocytes, VAP-1 shedding by a batimastat-sensitive metalloprotease is increased by TNF-α [41]. The cleavage site of VAP-1 is close to the membrane and, therefore, the soluble form is nearly as big as the membrane form. Increased concentrations of soluble VAP-1 may promote inflammation by generating highly reactive end products, such as hydrogen peroxide and aldehydes, through its catalytic activity. In fact, old transgenic mice over-expressing human VAP-1 on endothelium have increased serum concentrations of VAP-1 and they develop symptoms resembling those of the metabolic syndrome in man. These findings strongly support the idea that increased VAP-1 concentration is the cause and not only the consequence of vasculopathies [40].

Table 2 - Soluble VAP-1 levels in different human diseases[a]

Elevated	Diabetes	
	High blood pressure	
	Liver diseases	- primary biliary cirrhosis
		- alcoholic liver disease
		- hepatitis
	Skin diseases	- psoriasis
		- atopic eczema
	Congestive heart failure	
	Multiple sclerosis	
	Alzheimer's disease (only in severe)	
No change or elevated[b]	Atherosclerosis	
	Obesity	
No change	Liver diseases	- paracetamol poisoning
		- primary sclerosing cholangitis
	Rheumatoid arthritis	
	Stroke	
	ADHD[c]	
No change or decreased[b]	Inflammatory bowel diseases	
Decreased	Depression	

[a]Due to space limitations references to the individual papers have not been included. This information can be obtained from the authors.
[b]Conflicting reports
[c]Attention deficit hyperactivity disorder

VAP-1 forms homodimers

The crystal structure of human VAP-1 has been solved by two groups [42, 43]. VAP-1 has a fold typical for a copper-containing amine oxidase family, although it is distinguished from the other family members in being membrane bound. The topaquinone cofactor and a copper ion are present in the active site that is at the bottom of a deep and narrow substrate channel accessible from the surface of the molecule. An RGD motif is displayed on the surface, where it has potential to bind integrins, although experimental evidence for any VAP-1–integrin interactions is lacking. A VAP-1 monomer has six potential N-glycosylation sites, which all can be used on the basis of mutagenesis analyses [42, 43] and several potential O-glycosylation sites. The VAP-1 mutants lacking the N-linked oligosaccharides on the top of the molecule show reduced leukocyte binding, although their catalytic activity is increased in comparison to the wild-type VAP-1 [44].

Dualistic adhesive nature of VAP-1

Many adhesion assays have shown that VAP-1 has functionally important surface epitope(s) that can be recognized with mAbs. These mAbs block VAP-1-mediated binding to endothelial cells without affecting the enzymatic function of VAP-1. On the other hand, inhibitors of the enzyme activity of VAP-1 prevent VAP-1-dependent leukocyte adherence to vascular endothelium and enzymatically inactive VAP-1 mutants cannot function as adhesins to mediate leukocyte-endothelial cell interactions. Moreover, combined treatments with a function-blocking antibody and an inhibitor do not have additional or synergistic effects, suggesting that VAP-1 works in a two-step fashion [10, 11]. VAP-1 presumably first acts *via* the antibody-defined epitope and binds to its lymphocyte counter-receptor (the identity of which is currently unknown) (Fig. 2). This interaction guides the lymphocyte to the enzymatic step, during which a transient Schiff-base is formed attaching lymphocyte and endothelium together (the life-time of the Schiff-base has been measured to be approximately 0.3 s, Judith Klinman, unpublished observations). During the enzymatic reaction hydrogen peroxide and ammonium are released and the lymphocyte counter-receptor becomes modified into an aldehyde. Thus, anti-VAP-1 mAbs are proposed to inhibit the first, adhesion epitope-dependent function of VAP-1, whereas the SSAO inhibitors block the second, enzyme activity-dependent function of VAP-1. Both mAbs and SSAO inhibitors can therefore block VAP-1-dependent leukocyte endothelial-contacts. It should be also noted that inhibition of VAP-1 by either means normally leads to about 40% reduction in leukocyte binding, and the remaining interactions are mediated by VAP-1-independent mechanisms of adhesion. Finally, the potent inflammatory mediators released during the VAP-1-catalyzed reaction can activate both the endothelium and the leukocyte, for example by up-regulating other adhesion molecules.

Adhesion-independent functions of VAP-1

In adipocytes, VAP-1 is up-regulated during differentiation [45]. In these cells VAP-1 enzyme activity increases insulin-independent glucose uptake *via* the GLUT4 pathway [46]. The insulin-like effects of VAP-1 are probably also important *in vivo*, since administration of VAP-1 substrates improves glucose transport in rat models of both type 1 and type 2 diabetes [47, 48]. However, since most protocols involve co-administration of vanadate, some caution is warranted in the interpretation of these results, as reaction of VAP-1-derived hydrogen peroxide with this compound

245

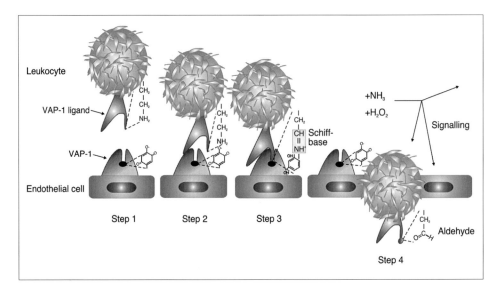

Figure 2
Putative model of VAP-1 function during the extravasation cascade.
Endothelial VAP-1 first makes contacts with its leukocyte counter-receptor using surface epitopes defined by the function blocking anti-VAP-1. Thereafter, the VAP-1 ligand, or possibly another substrate on the leukocyte surface, can penetrate into the enzymatic cavity of VAP-1 allowing triggering of the catalytic reaction. During the enzymatic reaction, the enzyme and the ligand become covalently, but transiently, bound to each other. The bioactive end-products of the reaction can also regulate the expression and/or function of other adhesion-related molecules. The black dot within VAP-1 represents the catalytic center of the molecule.

can lead to artificial formation peroxyvanadate, which has multiple effects on the cell signaling. The use of transgenic mice has shown that overexpression of VAP-1 in endothelial cells leads to a transient increase in glucose tolerance [40]. However, when the VAP-1 activity is increased chronically, diabetic complications such as glomerulosclerosis, increased blood pressure and atherosclerosis ensue.

The role of VAP-1 on smooth muscle cells remains more enigmatic. It could support leukocyte interactions with smooth muscle cells/pericytes during the diapedesis, but *ex vivo* studies have failed to support this concept [49]. Transgenic mice overexpressing VAP-1 in smooth muscle cells show altered elastin structure and decreased blood pressure, which suggests possible involvement of VAP-1 in the regulation of arterial tone [39]. Since VAP-1-deficient mice are, however, normotensive and have an apparently normal arterial tree (aorta, major arteries), the physiological function of smooth muscle VAP-1 remains to be determined [50].

VAP-1 as a diagnostic and therapeutic target

As presented above, several preclinical studies have demonstrated the efficacy of targeting VAP-1 in various disease models. Moreover, *in vitro* and *ex vivo* analyses using different patient materials and samples suggest that targeting of VAP-1 may also be feasible in clinical settings. From the drug design point of view, VAP-1 has several properties that make it attractive as an anti-inflammatory target. Most VAP-1 is stored within intracellular vesicles and rapidly released onto the endothelial cell surface upon inflammation. This property can be exploited in imaging of inflammation and most likely helps to minimize possible side effects of the anti-VAP-1 therapeutics [26]. VAP-1-knockout mice, which in normal non-challenged conditions are seemingly healthy, also suggest that targeting of VAP-1 may not markedly compromise the physiological functions of the body. Due to the unique functional mode of VAP-1, it can be blocked either by antibodies (or fragments thereof) or by enzyme inhibitors. Although antibody treatments against certain adhesion molecules have shown their efficacy and recently entered the market, they cannot be given orally. Small molecule SSAO inhibitors, in contrast, are orally bioavailable. The recently resolved crystal structure of VAP-1 will certainly facilitate the efforts to design novel SSAO/VAP-1 inhibitors to inhibit inappropriate inflammation.

References

1   Salmi M, Jalkanen S (1992) A 90-kilodalton endothelial cell molecule mediating lymphocyte binding in humans. *Science* 257: 1407–1409

2   Salmi M, Kalimo K, Jalkanen S (1993) Induction and function of vascular adhesion protein-1 at sites of inflammation. *J Exp Med* 178: 2255–2260

3   Smith DJ, Salmi M, Bono P, Hellman J, Leu T, Jalkanen S (1998) Cloning of vascular adhesion protein-1 reveals a novel multifunctional adhesion molecule. *J Exp Med* 188: 17–27

4   Jalkanen S, Salmi M (2001) Cell surface monoamine oxidases: enzymes in search of a function. *EMBO J* 20: 3893–3901

5   Klinman JP, Mu D (1994) Quinoenzymes in biology. *Annu Rev Biochem* 63: 299–344

6   Arvilommi A-M, Salmi M, Kalimo K, Jalkanen S (1996) Lymphocyte binding to vascular endothelium in inflamed skin revisited: a central role for vascular adhesion protein-1 (VAP-1). *Eur J Immunol* 26: 825–833

7   Salmi M, Rajala P, Jalkanen S (1997) Homing of mucosal leukocytes to joints. Distinct endothelial ligands in synovium mediate leukocyte-subtype specific adhesion. *J Clin Invest* 99: 2165–2172

8   McNab G, Reeves JL, Salmi M, Hubscher S, Jalkanen S, Adams DH (1996) Vascular adhesion protein 1 mediates binding of T cells to human hepatic endothelium. *Gastroenterology* 110: 522–528

9   Jaakkola K, Jalkanen S, Kaunismäki K, Vänttinen E, Saukko P, Alanen K, Kallajoki M, Voipio-Pulkki L-M, Salmi M (2000) Vascular adhesion protein-1, intercellular adhesion molecule-1 and P-selectin mediate leukocyte binding to ischemic heart in humans. *J Am Coll Cardiol* 36: 122–129

10   Salmi M, Yegutkin G, Lehvonen R, Koskinen K, Salminen T, Jalkanen S (2001) A cell surface amine oxidase directly controls lymphocyte migration. *Immunity* 14: 265–276

11   Koskinen K, Vainio PJ, Smith DJ, Pihlavisto M, Yla-Herttuala S, Jalkanen S, Salmi M (2004) Granulocyte transmigration through endothelium is regulated by the oxidase activity of vascular adhesion protein-1 (VAP-1). *Blood* 103: 3388–3395

12   Lalor PF, Edwards S, McNab G, Salmi M, Jalkanen S, Adams DH (2002) Vascular adhesion protein-1 mediates adhesion and transmigration of lymphocytes on human hepatic endothelial cells. *J Immunol* 169: 983–992

13   Tohka S, Laukkanen M-L, Jalkanen S, Salmi M (2001) Vascular adhesion protein 1 ions as a molecular brake during granulocyte rolling and mediates their recruitment *in vivo*. *FASEB J* 15: 373–382

14   Merinen M, Irjala H, Salmi M, Jaakkola I, Hanninen A, Jalkanen S (2005) Vascular adhesion protein-1 is involved in both acute and chronic inflammation in the mouse. *Am J Pathol* 166: 793–800

15   Martelius T, Salaspuro V, Salmi M, Krogerus L, Hockerstedt K, Jalkanen S, Lautenschlager I (2004) Blockade of vascular adhesion protein-1 inhibits lymphocyte at liver allograft rejection. *Am J Pathol* 165: 1993–2001

16   Bonder C, Swain MG, Zbytnuik LD, Norman MU, Yamanouchi J, Santamaria P, Ajuebor M, Salmi M, Jalkanen S, Kubes P (2005) Rules of recruitment of trafficking Th1 and Th2 cells in inflamed liver. *Immunity* 23: 153–163

17   Stolen CM, Marttila-Ichihara F, Koskinen K, Yegutkin GG, Turja R, Bono P, Skurnik M, Hanninen A, Jalkanen S, Salmi M (2005) Absence of the endothelial oxidase AOC3 leads to abnormal leukocyte traffic *in vivo*. *Immunity* 22: 105–115

18   Salter-Cid LM, Wang E, O'Rourke A M, Miller A, Gao H, Huang L, Garcia A, Linnik MD (2005) Anti-inflammatory effects of inhibiting the amine oxidase activity of SSAO. *J Pharmacol Exp Ther* 315:553–562.

19   Xu HL, Salter-Cid L, Linnik M, Wang E, Paisansathan C, Pelligrino D (2005) Vascular adhesion protein-1 plays an important role in post-ischemic inflammation and neuropathology in diabetic, estrogen-treated ovariectomized female rats subjected to transient forebrain ischemia. *J Pharmacol Exp Ther* 317:19–29

20   Marttila-Ichihara F, Smith DJ, Stolen C, Yegutkin GG, Elima K, Mercier N, Kiviranta R, Pihlavisto M, Alaranta S, Pentikainen U et al (2006) Vascular amine oxidases are needed for leukocyte extravasation into inflamed joints *in vivo*. *Arthritis Rheum* 54: 2852–2862

21   Salmi M, Jalkanen S (2005) Cell-surface enzymes in control of leukocyte trafficking. *Nat Rev Immunol* 5: 760–771

22   Haddock RC, Mack P, Leal S, Baenziger NL (1990) The histamine degradative uptake

pathway in human vascular endothelial cells and skin fibroblasts is dependent on extracellular Na⁺ and Cl⁻. *J Biol Chem* 265: 14395–14401

23 Li W, Liu G, Chou IN, Kagan HM (2000) Hydrogen peroxide-mediated, lysyl oxidase-dependent chemotaxis of vascular smooth muscle cells. *J Cell Biochem* 78: 550–557

24 Erler JT, Bennewith KL, Nicolau M, Dornhofer N, Kong C, Le QT, Chi JT, Jeffrey SS, Giaccia AJ (2006) Lysyl oxidase is essential for hypoxia-induced metastasis. *Nature* 440: 1222–1226

25 Bono P, Jalkanen S, Salmi M (1999) Mouse vascular adhesion protein-1 (VAP-1) is a sialoglycoprotein with enzymatic activity and is induced in diabetic insulitis. *Am J Pathol* 155: 1613–1624

26 Jaakkola K, Nikula T, Holopainen R, Vähäsilta T, Matikainen M-T, Laukkanen M-L, Huupponen R, Halkola L, Nieminen L, Hiltunen J et al (2000) *In vivo* detection of vascular adhesion protein-1 in experimental inflammation. *Am J Pathol* 157: 463–471

27 Vainio PJ, Kortekangas-Savolainen O, Mikkola JH, Jaakkola K, Kalimo K, Jalkanen S, Veromaa T (2005) Safety of blocking vascular adhesion protein-1 in patients with contact dermatitis. *Basic Clin Pharmacol Toxicol* 96: 429–435

28 Arvilommi A-M, Salmi M, Jalkanen S (1997) Organ-selective regulation of vascular adhesion protein-1 expression in man. *Eur J Immunol* 27: 1794–1800

29 Salmi M, Jalkanen S (2006) Developmental regulation of the adhesive and enzymatic activity of vascular adhesion protein-1 (VAP-1) in humans. *Blood* 108: 1555–1561

30 Boomsma F, van den Meiracker AH, Winkel S, Aanstoot HJ, Batstra MR, Man in ,t Veld AJ, Bruining GJ (1999) Circulating semicarbazide-sensitive amine oxidase is raised both in type I (insulin-dependent), in type II (non-insulin-dependent) diabetes mellitus and even in childhood type I diabetes at first clinical diagnosis. *Diabetologia* 42: 233–237

31 Karadi I, Meszaros Z, Csanyi A, Szombathy T, Hosszufalusi N, Romics L, Magyar K (2002) Serum semicarbazide-sensitive amine oxidase (SSAO) activity is an independent marker of carotid atherosclerosis. *Clin Chim Acta* 323: 139–146

32 Boomsma F, van Veldhuisen DJ, de Kam PJ, Man in't Veld AJ, Mosterd A, Lie KI, Schalekamp MA (1997) Plasma semicarbazide-sensitive amine oxidase is elevated in patients with congestive heart failure. *Cardiovasc Res* 33: 387–391

33 Meszaros Z, Szombathy T, Raimondi L, Karadi I, Romics L, Magyar K (1999) Elevated serum semicarbazide-sensitive amine oxidase activity in non-insulin dependent diabetes mellitus: correlation with body mass index and serum triglyceride. *Metabolism* 48: 113–117

34 Kurkijärvi R, Yegutkin GG, Gunson BK, Jalkanen S, Salmi M, Adams DH (2000) Circulating soluble vascular adhesion protein 1 accounts for the increased serum monoamine oxidase activity in chronic liver disease. *Gastroenterology* 119: 1096–1103

35 Gronvall-Nordquist JL, Backlund LB, Garpenstrand H, Ekblom J, Landin B, Yu PH, Oreland L, Rosenqvist U (2001) Follow-up of plasma semicarbazide-sensitive amine oxidase activity and retinopathy in Type 2 diabetes mellitus. *J Diabetes Complications* 15: 250–256

36 Boomsma F, Derkx FH, van den Meiracker AH, Man in ,t Veld AJ, Schalekamp MA

(1995) Plasma semicarbazide-sensitive amine oxidase activity is elevated in diabetes mellitus and correlates with glycosylated haemoglobin. *Clin Sci (Lond)* 88: 675–679

37 Kurkijärvi R, Adams DH, Leino R, Möttönen T, Jalkanen S, Salmi M (1998) Circulating form of human vascular adhesion protein-1 (VAP-1): increased serum levels in inflammatory liver diseases. *J Immunol* 161: 1549–1557

38 Salmi M, Stolen C, Jousilahti P, Yegutkin GG, Tapanainen P, Janatuinen T, Knip M, Jalkanen S, Salomaa V (2002) Insulin-regulated increase of soluble vascular adhesion protein-1 in diabetes. *Am J Pathol* 161: 2255–2262

39 Gokturk C, Nilsson J, Nordquist J, Kristensson M, Svensson K, Soderberg C, Israelson M, Garpenstrand H, Sjoquist M, Oreland L et al (2003) Overexpression of semicarbazide-sensitive amine oxidase in smooth muscle cells leads to an abnormal structure of the aortic elastic laminas. *Am J Pathol* 163: 1921–1928

40 Stolen CM, Madanat R, Marti L, Kari S, Yegutkin GG, Sariola H, Zorzano A, Jalkanen S (2004) Semicarbazide-sensitive amine oxidase overexpression has dual consequences: insulin mimicry and diabetes-like complications. *FASEB J* 18:702–704

41 Abella A, Garcia-Vicente S, Viguerie N, Ros-Baro A, Camps M, Palacin M, Zorzano A, Marti L (2004) Adipocytes release a soluble form of VAP-1/SSAO by a metalloprotease-dependent process and in a regulated manner. *Diabetologia* 47: 429–438

42 Airenne TT, Nymalm Y, Kidron H, Smith DJ, Pihlavisto M, Salmi M, Jalkanen S, Johnson MS, Salminen TA (2005) Crystal structure of the human vascular adhesion protein-1: unique structural features with functional implications. *Protein Sci* 14: 1964–1974

43 Jakobsson E, Nilsson J, Ogg D, Kleywegt GJ (2005) Structure of human semicarbazide-sensitive amine oxidase/vascular adhesion protein-1. *Acta Crystallogr D Biol Crystallogr* 61: 1550–1562

44 Maula SM, Salminen T, Kaitaniemi S, Nymalm Y, Smith DJ, Jalkanen S (2005) Carbohydrates located on the top of the "cap" contribute to the adhesive and enzymatic functions of vascular adhesion protein-1. *Eur J Immunol* 35: 2718–2727

45 Moldes M, Fève B, Pairault J (1999) Molecular cloning of a major mRNA species in murine 3T3 adipocyte lineage. differentiation-dependent expression, regulation, and identification as semicarbazide-sensitive amine oxidase. *J Biol Chem* 274: 9515–9523

46 Enrique-Tarancón G, Marti L, Morin N, Lizcano JM, Unzeta M, Sevilla L, Camps M, Palacín M, Testar X, Carpéné C et al (1998) Role of semicarbazide-sensitive amine oxidase on glucose transport and GLUT4 recruitment to the cell surface in adipose cells. *J Biol Chem* 273: 8025–8032

47 Abella A, Marti L, Camps M, Claret M, Fernandez-Alvarez J, Gomis R, Guma A, Viguerie N, Carpene C, Palacin M et al (2003) Semicarbazide-sensitive amine oxidase/vascular adhesion protein-1 activity exerts an antidiabetic action in Goto-Kakizaki rats. *Diabetes* 52: 1004–1013

48 Marti L, Abella A, Carpene C, Palacin M, Testar X, Zorzano A (2001) Combined treatment with benzylamine and low dosages of vanadate enhances glucose tolerance and reduces hyperglycemia in streptozotocin-induced diabetic rats. *Diabetes* 50: 2061–2068

49   Jaakkola K, Kaunismäki K, Tohka S, Yegutkin G, Vänttinen E, Havia T, Pelliniemi LJ, Virolainen M, Jalkanen S, Salmi M (1999) Human vascular adhesion protein-1 in smooth muscle cells. *Am J Pathol* 155: 1953–1965

50   Mercier N, Osborne-Pellegrin M, El Hadri K, Kakou A, Labat C, Loufrani L, Henrion D, Challande P, Jalkanen S, Feve B et al (2006) Carotid arterial stiffness, elastic fibre network and vasoreactivity in semicarbazide-sensitive amine-oxidase null mouse. *Cardiovasc Res* 72: 349–357

51   Kinemuchi H, Sugimoto H, Obata T, Satoh N, Ueda S (2004) Selective inhibitors of membrane-bound semicarbazide-sensitive amine oxidase (SSAO) activity in mammalian tissues. *Neurotoxicology* 25: 325–335

The role of endothelial cell-selective adhesion molecule (ESAM) in neutrophil emigration into inflamed tissues

Stefan Butz and Dietmar Vestweber

Max-Planck-Institute for Molecular Biomedicine, Röntgenstrasse 20, 48149 Münster, Germany

Introduction

Leukocyte emigration into inflamed tissues is among the most intensely pursued topics in the field of inflammation. Research focuses on the molecular factors activating endothelial cells and leukocytes, the adhesive molecules facilitating the contact between both cell types and the mechanisms allowing leukocytes to transmigrate through the blood vessel endothelium. In the last few years, studies have been intensified to understand how leukocytes, once captured to the vessel wall overcome the barrier made of endothelial cells linked to each other by interendothelial junctions. The mechanisms by which these leukocytes traverse the endothelial cell layer to reach the underlying tissue, a process called diapedesis, are largely unknown. Whereas convincing evidence has been published that polymorphonuclear leukocytes (PMN) can indeed migrate through endothelial cells in a transcellular fashion *in vivo* [1] as well as *in vitro* [2], careful quantitative analysis has demonstrated that at least *in vitro* the majority of PMNs and other leukocytes migrate *via* a paracellular route through the contact areas between endothelial cells [2, 3]. Consequently, a number of endothelial cell contact proteins such as PECAM-1, members of the junctional adhesion molecule family (JAM-A, -B and -C), CD99 and ICAM-2 have been reported to support leukocyte extravasation [4–7]. PECAM-1 was the first of these proteins that was identified in the context of leukocyte extravasation [8]. Its relevance for neutrophil extravasation is well established [9]. Although PECAM-1, JAM-A and ICAM-2 were shown by intravital microscopy to be involved in the transmigration process *in vivo*, the detailed molecular mechanism by which they participate in the process is still unknown. The only known endothelial cell contact protein that does not function as a supporter but rather as a barrier for neutrophil extravasation is VE-cadherin, a major adhesion molecule responsible for the stability of endothelial cell contacts [10].

Recently, two research teams have independently reported the identification of a novel endothelial cell contact protein, called endothelial cell-selective adhesion molecule (ESAM). Quertermous and his co-workers [11] discovered ESAM as a candidate molecule for mediating endothelial cell interactions during vascular development. Our group has identified the same molecule by searching for cell surface proteins that are potentially involved in the process of leukocyte emigration from the blood into a site of inflammation by regulating interendothelial cell contacts [12]. This chapter focuses on the functional role of ESAM in leukocyte emigration, which was established by analyzing ESAM-deficient mice in different inflammation models. These studies propose ESAM as a new player in this process, which is probably involved in the control of endothelial cell contacts.

Domain structure, subcellular localization and protein interactions of ESAM

ESAM is a member of the immunoglobulin (Ig) superfamily and belongs together with the three JAM proteins to the CTX subset of this family (see Fig. 1). This subset was named after the first member, the protein CTX (cortical thymocyte marker in Xenopus), a receptor specifically expressed on the surface of cortical thymocytes in Xenopus [13, 14]. Like all members of this family, ESAM contains two extracellular Ig domains, the membrane-distal one fits into the variable Ig domain subtype (V-like Ig domain), the membrane-proximal one into the constant domain subtype with C2 strand topology (C2-like Ig domain). The transmembrane domain is followed by a cytoplasmic tail of 120 amino acids containing proline-rich regions and a motif for a type I PDZ interaction at the very C terminus. ESAM is exclusively expressed on endothelial cells as well as by megakaryocytes and platelets [11, 12]. It is exposed on the surface of platelets upon activation by thrombin. ESAM is found throughout the vasculature on endothelium of arteries, veins and capillaries as well as in the endocardium. It is expressed in high endothelial venules (HEVs), a specialized endothelium for lymphocyte migration into lymph nodes [12]. On endothelial cells ESAM is associated with tight junctions as was demonstrated by immunogold labeling of brain and muscle capillaries [12]. Furthermore, ectopic expression in transfected epithelial cells revealed a strict limitation of its subcellular distribution to the apical ZO-1-containing areas of lateral contacts.

ESAM can mediate homotypic cell interactions. In a mixed cell culture assay with ESAM-transfected and mock-transfected CHO cells, cell contact staining for ESAM could only be observed when both neighboring cells expressed the protein [12]. This study strongly suggests that ESAM can bind to itself in a homophilic manner, although the possibility of indirect ESAM clustering at cell-cell contacts of ESAM-transfected CHO cells cannot formally be excluded. Furthermore, ESAM is capable of promoting calcium-independent cell adhesion in CHO cell-based aggregation assays [11, 12].

Although ESAM is related to the three JAM proteins, it belongs to another subgroup within the CTX-family. On the basis of sequence similarity, ESAM is

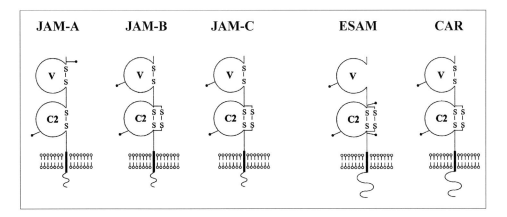

Figure 1
Schematic domain structure of the CTX family members ESAM, JAM-A, JAM-B, JAM-C, and CAR.
The CTX family represents a large subgroup of the Ig superfamily of cell surface receptors. CTX family proteins possess an extracellular region largely comprised of a membrane distal V-like Ig domain and a membrane proximal C2-like Ig domain. ESAM and CAR belong to different subset of CTX-family members than the JAMs (see text). The cytoplasmic domains of ESAM and CAR are 120 and 107 amino acids long, respectively (mouse sequences) and contain a type I PDZ recognition motif, whereas the cytoplasmic domains of JAM-A, JAM-B ad JAM-C are composed of around 45 amino acids and harbor a type II PDZ binding motif. Putative N-glycosylation sites are indicated by dots.

more closely related to the coxsackievirus-adenovirus receptor (CAR) [15, 16] (see Fig. 1), the brain- and testis-specific Ig superfamily protein (BT-IgSF) [17], and the junctional adhesion molecule 4 (JAM4) [18] and only distantly related to JAM-A [19, 20], JAM-B [21–23], and JAM-C [21, 24, 25]. In fact, ESAM, CAR, BT-IgSF, and JAM4 belong to the same subgroup of CTX-family proteins, whereas JAM-A, -B, and −C form another subgroup [24]. In line with this, the relatively long cytoplasmic tails of ESAM and CAR (120 and 107 amino acids, respectively; mouse sequences) contain a type I PDZ-domain binding motif, whereas the much shorter cytoplasmic domains of JAM-A, -B, and −C (45, 40, and 49 amino acids, respectively; mouse sequences) harbor a type II motif. Consequently, ESAM binds to different PDZ domain proteins than the three JAMs. JAM-A, -B, and -C directly bind to the cell polarity PDZ-domain protein PAR-3 [26–28], which is part of a ternary protein complex consisting of PAR-3, PAR-6 and the atypical protein kinase Cλ. This complex is essential for the establishment of cell polarity. ESAM does not bind to PAR-3, instead, it binds to the tight junction-associated PDZ domain protein MAGI-1 (membrane-associated guanylate kinase with inverted domain

255

structure) [24]. The cytoskeleton-associated binding partners such as synaptopodin and α-actinin-4 suggest that MAGI-1 might be implicated in actin cytoskeleton dynamics. These differences in the specificity for various scaffolding proteins strongly suggest different signaling potentials for ESAM and the JAMs. Further differences between the members of the two CTX subgroups become apparent when comparing the spectrum of extracellular ligands [29]. Like the JAMs, the members of the other subgroup (ESAM, CAR, JAM4, and BT-IgSF) can mediate homotypic cell interactions, but only the latter can promote cell aggregation. The JAMs are not capable of inducing the formation of stable aggregates by homophilic binding, when expressed in heterologous cells. Interestingly, a strong heterophilic interaction has been described between JAM-B and JAM-C [25, 30]. However, based on yet-unpublished data from our laboratory, ESAM does not interact with JAM-A, -B and -C or with CAR in CHO-based mixed cell culture assays. Each of the three JAMs can interact with distinct integrins on adjacent leukocytes [29]. So far, integrin binding or binding of ESAM to other cell surface proteins on leukocytes could not be shown. In yet-unpublished adhesion studies performed by our group, it was not possible to demonstrate robust cell adhesion of leukocytes to immobilized ESAM-Fc or to ESAM-transfected cells. Despite some similarities, like the overall structure and, at least for JAM-A and -C, the tight junction localization of these molecules, ESAM and the JAMs probably function in different ways.

ESAM participates in neutrophil, but not lymphocyte, extravasation at inflammatory sites

The localization of ESAM at endothelial tight junctions identifies it as an attractive candidate for a modulator of endothelial cell contact stability, or as a membrane protein potentially participating in paracellular leukocyte transmigration. To investigate this possibility, the ability of antibodies against ESAM or a soluble recombinant form of ESAM to interfere with endothelial cell contact integrity and leukocyte extravasation was tested. In addition, the consequences of gene disruption for each of the two functions were analyzed.

The direct blocking studies failed to reveal a function for ESAM [31]. Neither the two monoclonal antibodies nor the two polyclonal antisera against the extracellular domains of ESAM were able to inhibit the migration of neutrophils or of lymphocytes through the monolayer of cultured mouse endothelial cells *in vitro*. The same was found for a soluble ESAM-Fc fusion protein. In addition, none of the reagents inhibited the migration of activated lymphocytes into inflamed skin in a classical delayed-type hypersensitivity (DTH) reaction model [6] or the recruitment of neutrophils into thioglycolate- or IL-1β-stimulated peritoneum. However, cell aggregation assays with ESAM-transfected CHO cells put these findings into perspective: homotypic cell aggregation could not be blocked with anti-ESAM antibodies or with ESAM-Fc. Thus, neither the antibodies nor ESAM-Fc prevent ESAM from promoting homotypic cell adhesion. It could well be that the monoclonal antibodies do not bind at sites required for cell adhesion, and this is also possible for the polyclonal antibodies, although the latter is surprising. The fact that polyclonal antibodies against human ESAM do at least partially block the aggregation of CHO cells transfected with human ESAM (our unpublished observations) indeed suggests indirectly that our antibodies against mouse ESAM were just not able to precisely target the functionally relevant epitopes of mouse ESAM. The lack of any blocking activity of ESAM-Fc is less surprising, since it is not unusual that soluble forms of adhesion molecules bind much less efficiently to their ligands than the membrane anchored, possibly clustered form of a cell adhesion molecule.

Since the question of whether ESAM is involved in leukocyte extravasation could not be answered using the antibody-blocking experiments, gene disrupted mice were generated. Such mice were actually generated independently by the Quertermous lab and by our group [31, 32]. Both research teams showed that ESAM gene deficiency did not lead to any obvious abnormalities during embryonic development. The Quertermous group found that ESAM deficiency leads to defects in tumor angiogenesis, although embryonic angiogenesis was unaffected. Tumors derived from melanoma cells and from Lewis lung carcinoma cells grew to much smaller sizes in mice lacking ESAM than in wild-type mice. Vascular density inside the tumor tissue was also decreased. In addition, *in vitro* experiments with ESAM-deficient endothelial cells or cells overexpressing ESAM suggested that this protein plays a role in the migration and tube formation of endothelial cells [32]. This study provides valuable insights into a functional role of ESAM in angiogenic processes in the adult organism. Our group analyzed the ESAM gene-deficient mice for potential defects in leukocyte extravasation [31].

Examining the recruitment of activated lymphocytes into inflamed skin in a classical DTH model revealed no difference between ESAM gene-deficient and wild-type mice (Fig. 2A). These findings did not depend on the time point of analysis after T cell inoculation (5 or 15 h). Thus, ESAM is not required for the recruitment of activated lymphocytes into inflamed skin. To determine whether ESAM plays a role in neutrophil emigration, an experimental model of thioglycolate-induced peritonitis was used (Fig. 2B). Neutrophil accumulation into inflamed peritoneum was tested at 2, 4, and 6 h after thioglycolate injection. At the 2-h time point neutrophil extravasation was reduced by about 70% when compared with wild-type mice. Interestingly, neutrophil recruitment into the peritoneum was only transiently affected by the lack of ESAM. At later time points the accumulation of neutrophils in ESAM-deficient animals recovered (Fig. 2B), demonstrating that ESAM is required at an early step of the inflammatory process. Taken together, these results obtained by analyzing ESAM-deficient mice suggested that ESAM is important for the emigration of neutrophils into peritoneum, but not for the emigration of lymphocytes into skin [31].

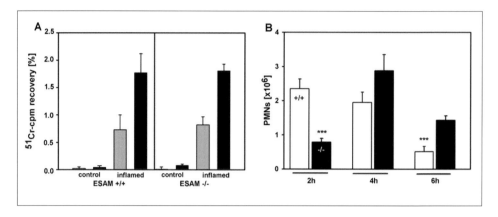

Figure 2
ESAM is involved in neutrophil extravasation but not in lymphocyte emigration into inflammatory sites.
*(A) Migration of lymphocytes into inflamed skin is unaffected by ESAM deficiency. Freshly isolated T cells from lymph nodes of 2,4-dinitrofluorobenzene (DNFB)-sensitized donor mice were labeled with ^{51}Cr and injected into the tail vein of sensitized and challenged ESAM-deficient (ESAM$^{-/-}$) or wild-type (ESAM$^{+/+}$) mice. After 5 h (gray bars) and after 15 h (black bars) mice were killed and radioactivity that accumulated in the non-inflamed ear (control) or in the DNFB-challenged ear (inflamed) was calculated as percent of total injected radioactivity. (B) Lack of ESAM impairs neutrophil recruitment into inflamed peritoneum. Wild-type (+/+) mice or ESAM-deficient (−/−) mice were intraperitoneally stimulated with thioglycolate and peritoneal neutrophils were removed by lavage 2, 4 or 6 h later, as indicated, and counted. The percentage of neutrophils was determined by staining for Gr-1 and FACS analysis. The data represent the mean ± SD of four or more mice in each group. The results are from one of three independent experiments with similar results. * * * p<0.001. Reproduced from [31] with permission (Copyright 2006, The Rockefeller University Press).*

Two important differences between the peritonitis and the skin DTH models prevented the more general conclusion to be drawn that ESAM is relevant for neutrophil but not for lymphocyte extravasation. First, the kinetics of neutrophil and lymphocyte extravasation differs considerably, with neutrophils extravasating much quicker and earlier than lymphocytes. Second, the two inflammation models deal with different tissues. To test whether lymphocytes do indeed extravasate independently from ESAM, while neutrophils do require ESAM, at the same time point and within the same tissues, the peritoneum was stimulated with a mixture of IL-1β and the chemokine CCL19 (MIP-3β, ELC). This treatment allowed neutrophil and lymphocyte recruitment into the peritoneum to be analyzed at the same time point, 2 h after stimulation. As illustrated in Figure 3, only neutrophil accumulation was

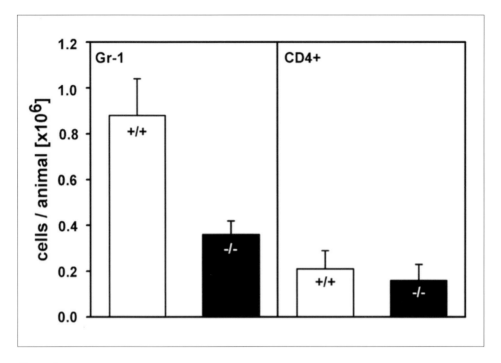

Figure 3
ESAM deficiency affects only neutrophil, and not lymphocyte, emigration into the perito-
neum after simultaneous stimulation with IL-1β and CCL19.
To allow a direct comparison of neutrophil and lymphocyte emigration in the same tissue at
the same time, an artificial situation was set up in which mice were stimulated intraperitone-
ally with IL-1β as well as with the chemokine CCL19 (MIP-3β). Peritoneal leukocytes were
removed by lavage 2 h later. Neutrophils and CD4+ T cells were counted by FACS analysis.
The data represent the mean ± SD of four or more mice in each group. The results are from
one of three independent experiments with similar results. Reproduced from [31] with per-
mission (Copyright 2006, The Rockefeller University Press).

affected by the lack of ESAM, whereas the recruitment of CCL19-responsive T cells (here: CD4+ T cells) was not perturbed. Thus, under conditions that allowed the analysis of neutrophil and lymphocyte emigration in the same tissue at the same time, only neutrophils required ESAM [31].

Whether lymphocytes migrate through endothelial barriers in the same way or differently from neutrophils is still an open question. Some of the adhesion mol-ecules at the endothelial cell contacts seem to support diapedesis of both types of leukocytes, such as JAM-A and ICAM-2 [4, 5, 7], although the evidence for a func-tion of JAM-A in lymphocyte migration through endothelial barriers so far is only based on *in vitro* studies [33]. In contrast, PECAM-1 only mediates the extravasation of myeloid cells [4]. Likewise, ESAM selectively supports neutrophil and not lymphocyte extravasation. This could suggest that PECAM-1 and ESAM are acting in a similar pathway, only used by neutrophils. However, analyzing the ability of anti-PECAM-1 antibodies to affect neutrophil extravasation in ESAM gene-deficient mice revealed that these antibodies reduced neutrophil recruitment despite the absence of ESAM [31]. Thus, PECAM-1 participates in neutrophil diapedesis in a manner that is independent of the presence of ESAM. Despite the differences between the transmigration steps that require ESAM and PECAM-1, each of these steps seem to be dispensable for lymphocyte extravasation.

ESAM is not only expressed on endothelial cells but also on platelets. Since platelets have been described to participate in leukocyte extravasation, and since ESAM can promote cell adhesion, it was an obvious possibility that ESAM could be involved in platelet-mediated leukocyte endothelial interactions that would contribute to the recruitment of leukocytes into inflamed tissue. This was tested by depleting peripheral platelet counts with antibodies against GP1b. Platelet depletion was done in ESAM-deficient and in wild-type mice and the effects on neutrophil recruitment into inflamed peritoneum were compared. Indeed, platelet depletion reduced neutrophil recruitment quite significantly in wild-type mice; however, this treatment had the same inhibitory effect on neutrophil recruitment in ESAM-deficient mice [31]. Thus, the platelet contribution was independent of ESAM. This defines endothelial tight junctions as the location where ESAM participates in neutrophil extravasation.

ESAM is involved in the transmigration step of neutrophil extravasation

The localization of ESAM at endothelial tight junctions makes it a candidate for a membrane protein participating in the diapedesis process, rather than in the docking and adherence of neutrophils to the inner surface of the blood vessel wall. To test this *in vivo*, leukocyte extravasation in the cytokine-stimulated cremaster muscle was visualized using reflected light oblique transillumination video microscopy [34]. This microscopic method significantly enhances the image contrast and therefore allows a highly reliable quantification of extravasated leukocytes. The results obtained from these intravital microscopy studies clearly demonstrated that the initial docking and adhesion of leukocytes to the inner surface of the blood vessel wall in cytokine-stimulated tissue was unaffected by the absence of ESAM. In contrast, the number of extravasating leukocytes within the first 4 h after cytokine stimulation was clearly reduced in ESAM-deficient mice, demonstrating that it was the diapedesis process that was affected by the lack of ESAM [31].

PECAM-1, JAM-A, and ICAM-2 are the other endothelial cell contact proteins for which a role in diapedesis has been documented by intravital microscopy [7, 35,

36]. An interesting and obvious question that arises from these findings is whether these Ig-superfamily members act in concert during this process, simultaneously or consecutively. Although it is not clear how these proteins cooperate in the diapedesis process, different responsiveness to inflammatory cytokines indicates some diversity of pathways. Interestingly, reduced extravasation of neutrophils in the cremaster of ESAM-deficient mice was observed to be independent of the type of cytokine used to trigger inflammation. The effect was observed with TNF-α as well as with IL-1β. In contrast, disrupting the genes for PECAM-1 or ICAM-2 (on the C57BL/6 genetic background) reduced neutrophil diapedesis in the cremaster only if the tissue was stimulated with IL-1β, but not if TNF-α was used for stimulation [7, 35]. Similarly, JAM-A-dependent leukocyte transmigration was observed upon IL-1β stimulation, but not if the leukocyte-activating chemoattractants LTB$_4$ or PAF were used for stimulation (cited as unpublished result in [7]). Since IL-1β is known to selectively activate only endothelial cells, whereas TNF-α, in addition, also activates leukocytes, it has been suggested that stimuli that directly activate leukocytes appear to by-pass the need for endothelial cell contact proteins such as PECAM-1, ICAM-2 and JAM-A. In contrast, ESAM is required even if TNF-α or thioglycolate are used as inflammatory stimulators. Thus, with respect to stimulus specificity, ESAM belongs to a different class of adhesion mechanisms involved in diapedesis than the one that comprises PECAM-1, ICAM-2 and JAM-A.

Role of ESAM in the regulation of endothelial cell permeability

The clear restriction of ESAM to endothelial tight junctions prompts the question of whether it would be involved in mechanisms that affect the integrity of endothelial cell contacts and vascular permeability. A major role as a stabilizing adhesion molecule in the maintenance of cell contact integrity in quiescent endothelium could be ruled out since, in the absence of any challenge, vascular permeability for the dye Evans Blue was not affected by ESAM gene deficiency [31]. Upon stimulation of the peritoneum with cytokine or thioglycolate, not only leukocyte extravasation, but also leakiness for the plasma protein-adsorbed dye was delayed in ESAM-deficient mice [31]. Thus, ESAM deficiency retards the increase of leakiness upon inflammatory stimulation in a similar way as it retards the extravasation of neutrophils. Since leakiness was probably just accompanying leukocyte extravasation, this was not a surprising finding. Much more important were experiments analyzing a role for ESAM in the regulation of endothelial permeability independent of leukocyte extravasation. Vascular permeability was stimulated in the skin with the growth factor VEGF. It was found that the VEGF-triggered increase of vascular permeability was dramatically delayed in ESAM$^{-/-}$ mice (Fig. 4), suggesting that ESAM is involved in the signaling pathway that connects VEGF signaling to the opening of endothelial junctions [31]. Importantly, these studies indeed measured the leak

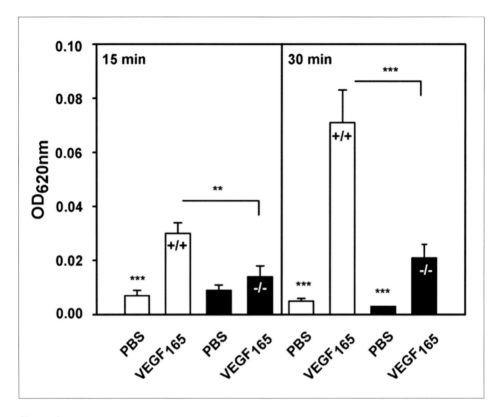

Figure 4

Lack of ESAM causes attenuation of VEGF-induced vascular permeability.

Wild-type mice (+/+, white bars) or ESAM-deficient mice (−/−, black bars) were intravenously injected with Evans blue. After 10 min, the mice were intradermally injected with VEGF or PBS, and killed after an additional 15 min or 30 min (as indicated). Skin samples were excised and the dye was extracted with formamide and quantified. The data represent the mean ± SD of three mice in each group. ** p<0.01, *** p<0.001. Reproduced from [31] with permission (Copyright 2006, The Rockefeller University Press).

of Evans Blue and were not affected by changes in the blood flow in the analyzed VEGF-injected skin area, since measuring the quantity of hemoglobin in the analyzed skin area by non-invasive white light spectroscopy showed no difference between VEGF injected and control injected skin areas [31]. Thus, ESAM must be somehow involved in signaling mechanisms that transmit the VEGF stimulus to the opening of endothelial junctions (Fig. 5).

Since the lack of ESAM slows down neutrophil extravasation as well as leukocyte independent opening of endothelial cell contacts, we propose that ESAM is

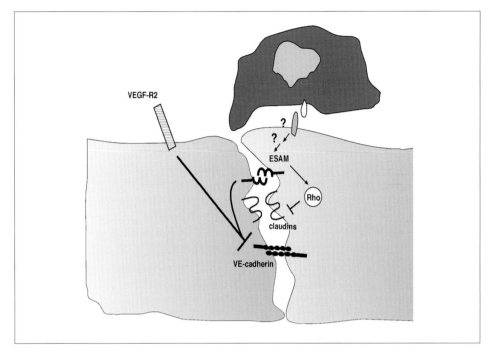

Figure 5
Speculative model for a role of ESAM in the control of endothelial cell contacts.
Deletion of the ESAM gene impairs two processes: VEGF-R2-triggered increase of vascular permeability and inflammation-induced extravasation of leukocytes. Since VEGF is thought to stimulate endothelial permeability by inhibiting the adhesive activity of VE-cadherin, ESAM may be involved in this process. In addition, knocking down the expression of ESAM leads to reduced levels of activated Rho. Since activated Rho supports the opening of tight junctions, and transendothelial migration of monocytes and neutrophils require active Rho, it is possible that ESAM is involved in signaling mechanisms that allow leukocytes to open junctions during diapedesis.

needed for signaling mechanisms that allow leukocytes to trigger the opening of cell contacts during diapedesis. Furthermore, this conclusion implies that ESAM functions in the junctional diapedesis pathways (Fig. 5).

JAM-C, another tight junction-enriched protein in endothelial cells, may also affect the integrity of endothelial cell contacts. It was reported that overexpression of JAM-C in cultured endothelial cells counteracted the adhesive function of VE-cadherin and that down-regulation of its expression by RNA interference (RNAi)-enhanced VE-cadherin-mediated adhesion [37]. Whether this is related to leukocyte diapedesis is still speculative, but certainly worth analyzing in the future.

Interplay of ESAM and the small GTPase Rho

The identified role of ESAM for the loosening of endothelial cell junctions suggests that ESAM is involved in signaling mechanisms that destabilize junctions. A first hint towards mechanisms possibly related to this function of ESAM at endothelial cell contacts has been provided by analyzing the effect of knocking down ESAM expression by RNAi on the activation levels of various small GTPases. These proteins are well known to act as signal transducers that regulate a diverse array of cellular functions such as cell proliferation and differentiation, actin cytoskeleton, membrane trafficking or cell adhesion. Knocking down ESAM expression in endothelial cells clearly reduced the level of activated Rho, whereas the levels of Rac1-, Rap1-, and Cdc42-GTP were not altered. These findings establish ESAM as the first tight junction-associated membrane protein that affects Rho activation [31].

In this regard, it is interesting that reports from several laboratories have identified Rho as crucial for leukocyte extravasation and endothelial cell permeability. Active Rho and Rho kinase have been found to be important for transendothelial migration of monocytes and neutrophils [38, 39]. In addition, Rho was found to mediate the opening of endothelial tight junctions triggered by the chemokine MCP-1 [40]. The same group showed recently that Rho acted in this process *via* Rho kinase and PKCa and that increased endothelial permeability was accompanied by Ser/Thr phosphorylation of occludin and claudin-5 [41]. Phosphorylation of occludin and claudin-5 was also achieved by co-culturing endothelial cells with monocytes, and this phosphorylation as well as transendothelial migration of monocytes was blocked by inhibiting Rho and Rho kinase [42]. Collectively, these reports suggest that activation of Rho is linked to the opening and destabilization of endothelial tight junctions. This assumption, combined with the *in vitro* results demonstrating reduced levels of activated endothelial Rho upon suppressing the expression of ESAM, suggests that the inhibition of neutrophil extravasation, which was observed in ESAM-deficient mice, could at least partly be due to insufficient levels of activated Rho required for the opening of endothelial junctions. In this context, it is intriguing that the PDZ domain protein MAGI, which has been found to associate with ESAM [24], is able to bind the RhoA-specific nucleotide exchange factor mNET1 (mouse neuroepithelioma transforming gene 1) [43].

Rho activation can also be triggered by cross-linking ICAM-1 on the endothelial cell surface with antibodies, supposedly mimicking leukocyte attachment to endothelia [44]. The same group has shown that the Rho inhibitor C3 transferase strongly reduces transendothelial migration but not adhesion of lymphocytes [45]. It is intriguing to speculate that ESAM is another cell surface molecule involved in the signaling cascade that connects the event of leukocyte docking to endothelial cells with the diapedesis step of leukocyte emigration (Fig. 5). In contrast to ICAM-1, which represents an endothelial adhesion molecule of major importance for the binding of leukocytes, a direct binding of leukocytes to immobilized recombinant

ESAM or to ESAM-transfected cells has not yet been demonstrated. Thus, if heterophilic ESAM ligands exist on neutrophils they would have to bind with affinities too low to support robust cell adhesion. Alternatively, ESAM could be involved in a signaling pathway in which it could serve as a transmitter of signals from other endothelial receptors, which would trigger the destabilization of endothelial contacts (Fig. 5).

It will be important to elucidate the signaling pathway that links the VEGF-stimulated increase of vascular permeability to ESAM. VEGF has been reported to trigger tyrosine phosphorylation of VE-cadherin and associated catenins [46], as well as phosphorylation of the tight junction components occludin and ZO-1 [47, 48]. These events were found to involve Src family kinases [48, 49]. More recently, a novel signaling pathway for VEGF-mediated vascular permeability has been described, in which serine-phosphorylated VE-cadherin has been shown to recruit β-arrestin2 *via* the Vav2-Rac-PAK signaling module and thereby induce its endocytosis [50]. This pathway leading to the disruption of endothelial cell junctions also requires Src kinase activity. Interestingly, this requirement for Src distinguishes VEGF-induced edema formation from inflammation-induced vascular leakiness [51]. Despite these differences, the opening of endothelial junctions during neutrophil diapedesis and VEGF-induced vascular permeability seem to converge on signaling steps that require ESAM.

Conclusion

In the last few years leukocyte diapedesis through the blood vessel wall has moved into the center of interest in leukocyte trafficking. Several endothelial cell contact proteins have now been demonstrated to participate *in vivo*, although the detailed mechanisms how they act in concert are still unknown. ESAM has been recently identified as a new player in the process of neutrophil transmigration. Together with our finding that ESAM is involved in the VEGF-triggered increase of vascular permeability, these results suggest that ESAM participates in the opening of endothelial cell contacts stimulated during neutrophil diapedesis.

References

1   Feng D, Nagy JA, Pyne K, Dvorak HF, Dvorak AM (1998) Neutrophils emigrate from venules by a transendothelial cell pathway in response to FMLP. *J Exp Med* 187: 903–915

2   Carman CV, Springer TA (2004) A transmigratory cup in leukocyte diapedesis both through individual vascular endothelial cells and between them. *J Cell Biol* 167: 377–388

3   Millan JL, Hewlett L, Glyn M, Toomre D, Clark P, Ridley AJ (2006) Lymphocyte tran-
     scellular migration occurs through recruitment of endothelial ICAM-1 to caveola- and
     F-actin-rich domains. *Nat Cell Biol* 8: 113–123

4   Muller WA (2003) Leukocyte-endothelial-cell interactions in leukocyte transmigration
     and the inflammatory response. *Trends Immunol* 24: 326–333

5   Imhof BA, Aurrand-Lions M (2004) Adhesion mechanisms regulating the migration of
     monocytes. *Nat Rev Immunol* 4: 432–444

6   Bixel G, Kloep S, Butz S, Petri B, Engelhardt B, Vestweber D (2004) Mouse CD99 par-
     ticipates in T cell recruitment into inflamed skin. *Blood* 104: 3205–3213

7   Huang MT, Larbi KY, Scheiermann C, Woodfin A, Gerwin N, Haskard DO, Nour-
     shargh S (2006) ICAM-2 mediates neutrophil transmigration *in vivo*: Evidence for
     stimulus specificity and a role in PECAM-1-independent transmigration. *Blood* 107:
     4721–4727

8   Muller WA, Weigl SA, Deng X, Phillips DM (1993) PECAM-1 is required for transen-
     dothelial migration of leukocytes. *J Exp Med* 178: 449–460

9   Schenkel AR, Chew TW, Muller WA (2004) Platelet endothelial cell adhesion molecule
     deficiency or blockade significantly reduces leukocyte emigration in a majority of mouse
     strains. *J Immunol* 173: 6403–6408

10   Gotsch U, Borges E, Bosse R, Böggemeyer E, Simon M, Mossmann H, Vestweber D
     (1997) VE-cadherin antibody accelerates neutrophil recruitment *in vivo*. *J Cell Sci* 110:
     583–588

11   Hirata K, Ishida T, Penta K, Rezaee M, Yang E, Wohlgemuth J, Quertermous T (2001)
     Cloning of an immunoglobulin family adhesion molecule selectively expressed by endo-
     thelial cells. *J Biol Chem* 276: 16223–16231

12   Nasdala I, Wolburg-Buchholz K, Wolburg H, Kuhn A, Ebnet K, Brachtendorf G, Samu-
     lowitz U, Kuster B, Engelhardt B, Vestweber D et al (2002) A transmembrane tight junc-
     tion protein selectively expressed on endothelial cells and platelets. *J Biol Chem* 277:
     16294–16303

13   Chretien I, Robert J, Marcuz A, Garcia-Sanz JA, Courtet M, Du Pasquier L (1996)
     CTX, a novel molecule specifically expressed on the surface of cortical thymocytes in
     Xenopus. *Eur J Immunol* 26: 780–791

14   Chretien I, Marcuz A, Courtet M, Katevuo K, Vainio O, Heath JK, White SJ, Du Pasqui-
     er L (1998) CTX, a Xenopus thymocyte receptor, defines a molecular family conserved
     throughout vertebrates. *Eur J Immunol* 28: 4094–4104

15   Tomko RP, Xu R, Philipson L (1997) HCAR and MCAR: the human and mouse cellular
     receptors for subgroup C adenoviruses and group B coxsackieviruses. *Proc Natl Acad
     Sci USA* 94: 3352–3356

16   Bergelson JM, Cunningham JA, Droguett G, Kurt-Jones EA, Krithivas A, Hong JS, Hor-
     witz MS, Crowell RL, Finberg RW (1997) Isolation of a common receptor for Coxsackie
     B viruses and adenoviruses 2 and 5. *Science* 275: 1320–1323

17   Suzu S, Hayashi Y, Harumi T, Nomaguchi K, Yamada M, Hayasawa H, Motoyoshi K

(2002) Molecular cloning of a novel immunoglobulin superfamily gene preferentially expressed by brain and testis. *Biochem Biophys Res Commun* 296: 1215–1221

18 Hirabayashi S, Tajima M, Yao I, Nishimura W, Mori H, Hata Y (2003) JAM4, a junctional cell adhesion molecule interacting with a tight junction protein, MAGI-1. *Mol Cell Biol* 23:4267–4282

19 Martin-Padura I, Lostaglio S, Schneemann M, Williams L, Romano M, Fruscella P, Panzeri C, Stoppacciaro A, Ruco L, Villa A et al (1998) Junctional adhesion molecule, a novel member of the immunoglobulin superfamily that distributes at intercellular junctions and modulates monocyte transmigration. *J Cell Biol* 142: 117–127

20 Malergue F, Galland F, Martin F, Mansuelle P, Aurrand-Lions M, Naquet P (1998) A novel immunoglobulin superfamily junctional molecule expressed by antigen presenting cells, endothelial cells and platelets. *Mol Immunol* 35: 1111–1119

21 Aurrand-Lions MA, Duncan L, Du Pasquier L, Imhof BA (2000) Cloning of JAM-2 and JAM-3: an emerging junctional adhesion molecular family? *Curr Top Microbiol Immunol* 251: 91–98

22 Cunningham SA, Arrate MP, Rodriguez JM, Bjercke RJ, Vanderslice P, Morris AP, Brock TA (2000) A novel protein with homology to the junctional adhesion molecule. Characterization of leukocyte interactions. *J Biol Chem* 275: 34750–34756

23 Palmeri D, van Zante A, Huang CC, Hemmerich S, Rosen SD (2000) Vascular endothelial junction-associated molecule, a novel member of the immunoglobulin superfamily, is localized to intercellular boundaries of endothelial cells. *J Biol Chem* 275: 19139–19145

24 Wegmann F, Ebnet K, Du Pasquier L, Vestweber D, Butz S (2004) Endothelial adhesion molecule ESAM binds directly to the multidomain adaptor MAGI-1 and recruits it to cell contacts. *Exp Cell Res* 300: 121–133

25 Arrate MP, Rodriguez JM, Tran TM, Brock TA, Cunningham SA (2001) Cloning of human junctional adhesion molecule 3 (JAM3) and its identification as the JAM2 counter-receptor. *J Biol Chem* 276: 45826–45832

26 Itoh M, Sasaki H, Furuse M, Ozaki H, Kita T, Tsukita S (2001) Junctional adhesion molecule (JAM) binds to PAR-3: a possible mechanism for the recruitment of PAR-3 to tight junctions. *J Cell Biol* 154: 491–497

27 Ebnet K, Suzuki A, Horikoshi Y, Hirose T, Meyer-zu-Brickwedde MK, Ohno S, Vestweber D (2001) The cell polarity protein ASIP/PAR-3 directly associates with junctional adhesion molecule (JAM). *EMBO J* 20:3738–3748

28 Ebnet K, Aurrand-Lions M, Kuhn A, Kiefer F, Butz S, Zander K, Meyer zu Brickwedde MK, Suzuki A, Imhof BA, Vestweber D (2003) The junctional adhesion molecule (JAM) family members JAM-2 and JAM-3 associate with the cell polarity protein PAR-3: a possible role for JAMs in endothelial cell polarity. *J Cell Sci* 116: 3879–3891

29 Ebnet K, Suzuki A, Ohno S, Vestweber D (2004) Junctional adhesion molecules (JAMs): more molecules with dual functions? *J Cell Sci* 117: 19–29

30 Liang TW, Chiu HH, Gurney A, Sidle A, Tumas DB, Schow P, Foster J, Klassen T, Dennis K, DeMarco RA et al (2002) Vascular endothelial-junctional adhesion molecule

(VE-JAM)/JAM 2 interacts with T, NK, and dendritic cells through JAM 3. *J Immunol* 168: 1618–1626

31   Wegmann F, Petri J, Khandoga AG, Moser C, Khandoga A, Volkery S, Li H, Nasdala I, Brandau O, Fassler R et al (2006) ESAM supports neutrophil extravasation, activation of Rho and VEGF-induced vascular permeability. *J Exp Med* 203: 1671–1677

32   Ishida T, Kundu RK, Yang E, Hirata K, Ho YD, Quertermous T (2003) Targeted disruption of endothelial cell-selective adhesion molecule inhibits angiogenic processes *in vitro* and *in vivo*. *J Biol Chem* 278: 34598–34604

33   Ostermann G, Weber KS, Zernecke A, Schroder A, Weber C (2002) JAM-1 is a ligand of the beta(2) integrin LFA-1 involved in transendothelial migration of leukocytes. *Nat Immunol* 3: 151–158

34   Mempel TR, Moser C, Hutter J, Kuebler WM, Krombach F (2003) Visualization of leukocyte transendothelial and interstitial migration using reflected light oblique transillumination in intravital video microscopy. *J Vasc Res* 40: 435–441

35   Wakelin MW, Sanz MJ, Dewar A, Albelda SM, Larkin SW, Boughton-Smith N, Williams TJ, Nourshargh S (1996) An anti-platelet-endothelial cell adhesion molecule-1 antibody inhibits leukocyte extravasation from mesenteric microvessels *in vivo* by blocking the passage through the basement membrane. *J Exp Med* 184: 229–239

36   Khandoga A, Kessler JS, Meissner H, Hanschen M, Corada M, Motoike T, Enders G, Dejana E, Krombach F (2005) Junctional adhesion molecule-A deficiency increases hepatic ischemia-reperfusion injury despite reduction of neutrophil transendothelial migration. *Blood* 106: 725–733

37   Orlova VV, Economopoulou M, Lupu F, Santoso S, Chavakis T (2006) Junctional adhesion molecule-C regulates vascular endothelial permeability by modulating VE-cadherin-mediated cell-cell contacts. *J Exp Med* 203: 2703–2714

38   Strey A, Janning A, Barth H, Gerke V (2002) Endothelial Rho signaling is required for monocyte transendothelial migration. *FEBS Lett* 517: 261–266

39   Saito H, Minamiya Y, Saito S, Ogawa J (2002) Endothelial Rho and Rho kinase regulate neutrophil migration *via* endothelial myosin light chain phosphorylation. *J Leukoc Biol* 72: 829–836

40   Stamatovic SM, Keep RF, Kunkel SL, Andjelkovic AV (2003) Potential role of MCP-1 in endothelial cell tight junction 'opening': signaling *via* Rho and Rho kinase. *J Cell Sci* 116: 4615–4628

41   Stamatovic SM, Dimitrijevic OB, Keep RF, Andjelkovic AV (2006) Protein kinase C-alpha: RhoA cross talk in CCL2-induced alterations in brain endothelial permeability. *J Biol Chem* 281: 8379–8388

42   Persidsky Y, Heilman D, Haorah J, Zelivyanskaya M, Persidsky R, Weber GA, Shimokawa H, Kaibuchi K, Ikezu T (2006) Rho-mediated regulation of tight junctions during monocyte migration across blood-brain barrier in HIV-1 encephalitis (HIVE). *Blood* 107: 4770–4780

43   Dobrosotskaya IY (2001) Identification of mNET1 as a candidate ligand for the first PDZ domain of MAGI–1. *Biochem Biophys Res Commun* 283: 969–975

44   Etienne S, Adamson P, Greenwood J, Strosberg AD, Cazaubon S, Couraud PO (1998) ICAM-1 signaling pathways associated with Rho activation in microvascular brain endothelial cells. *J Immunol* 161: 5755–5761

45   Adamson P, Etienne S, Couraud PO, Calder V, Greenwood J (1999) Lymphocyte migration through brain endothelial cell monolayers involves signaling through endothelial ICAM-1 *via* a rho-dependent pathway. *J Immunol* 162: 2964–2973

46   Esser S, Lampugnani MG, Corada M, Dejana E, Risau W (1998) Vascular endothelial growth factor induces VE-cadherin tyrosine phosphorylation in endothelial cells. *J Cell Sci* 111: 1853–1865

47   Antonetti DA, Barber AJ, Hollinger LA, Wolpert EB, Gardner TW (1999) Vascular endothelial growth factor induces rapid phosphorylation of tight junction proteins occludin and zonula occluden 1. A potential mechanism for vascular permeability in diabetic retinopathy and tumors. *J Biol Chem* 274: 23463–23467

48   Pedram A, Razandi M, Levin ER (2002) Deciphering vascular endothelial cell growth factor/vascular permeability factor signaling to vascular permeability. Inhibition by atrial natriuretic peptide. *J Biol Chem* 277: 44385–44398

49   Weis S, Shintani S, Weber A, Kirchmair R, Wood M, Cravens A, McSharry H, Iwakura A, Yoon YS, Himes N et al (2004) Src blockade stabilizes a Flk/cadherin complex, reducing edema and tissue injury following myocardial infarction. *J Clin Invest* 113: 885–894

50   Gavard J, Gutkind JS (2006) VEGF controls endothelial-cell permeability by promoting the β-arrestin-dependent endocytosis of VE-cadherin. *Nat Cell Biol* 8: 1223–1234

51   Eliceiri BP, Paul R, Schwartzberg PL, Hood JD, Leng J, Cheresh DA (1999) Selective requirement for Src kinases during VEGF-induced angiogenesis and vascular permeability. *Mol Cell* 4: 915–924

Structure and function of JAM proteins

Eric A. Severson and Charles A. Parkos

Epithelial Pathobiology Research Unit, Department of Pathology and Laboratory Medicine, Emory University School of Medicine, Atlanta, GA 30322, USA

Nomenclature

The immunoglobulin superfamily (IgSF) is a large class of proteins that includes the junctional adhesion molecule (JAM) family of proteins. The current nomenclature for JAM members designates the first three described JAM proteins as JAM-A, JAM-B, JAM-C. Two other related proteins have been reported that have not been included in the standard nomenclature and are termed JAM-4 and JAM-L (AMICA). In earlier studies, numerical designations were used to define JAM proteins according to the timing of initial characterizations. However, this early nomenclature led to confusion in terminology for JAM-B and C due to the timing in which human and murine JAM-B and JAM-C were reported. To avoid confusion, this review uses current nomenclature exclusively, as proposed originally by Muller, regardless of the designation given in the original reports [1].

Relationship to other IgSF family members

JAM-A, JAM-B and JAM-C are the three most closely related family members, having 32–33% amino acid identity, as highlighted in the phylogram in Figure 1. JAM-4 and JAM-L are more closely related to each other than they are to either JAM-A, B, C or other IgSF proteins. Furthermore, JAM-A, B, and C have a C-terminal PSD-95/Discs-Large/ZO-1 (PDZ) type II binding motif and a conserved R-EWK dimerization motif, both of which are lacking in JAM-4 and JAM-L. JAM-L has been reported to contain a dimerization motif; however, this domain is more similar to the CAR dimerization motif than it is to those on JAM-A, -B and -C [2]. The dimerization motif for JAM-4 has not yet been identified. Thus, there are significant differences in the protein sequences of regions on JAML and JAM-4 that, as discussed below, likely have important functional consequences for JAM-A, -B and -C. This review highlights the putative functional relevance of conserved structural elements on JAMs and thus primarily focuses on JAM-A, JAM-B and JAM-C.

Adhesion Molecules: Function and Inhibition, edited by Klaus Ley
© 2007 Birkhäuser Verlag Basel/Switzerland

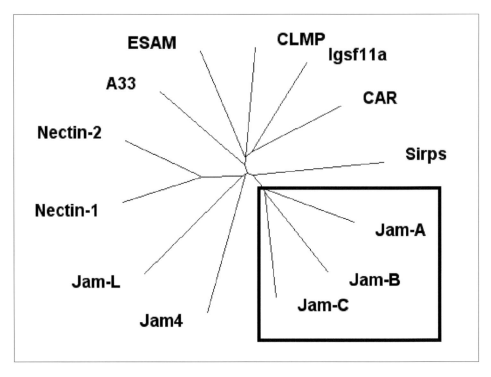

Figure 1
Phylogram tree generated with treeview from a CLUSTALW (http://www.ebi.ac.uk/ clustalw/) alignment of various IgSF members.
The relationship between JAM-A, JAM-B and JAM-C is highlighted in the box

Structure of JAM-A, JAM-B and JAM-C

The JAM-A, -B and -C pro-protein structures are similar, with each containing an N-terminal secretory signal peptide of 20–25 amino acids that targets the protein to the endoplasmic reticulum during synthesis after which cleavage occurs. As mentioned above, mature JAM proteins consist of two extracellular Ig-like domains followed by a single type I transmembrane domain and a cytoplasmic tail that terminates in a type II PDZ binding motif at the carboxy end. The conserved dimerization motif lies within the membrane-distal Ig loop. In addition, the cytoplasmic tails of JAM-A, -B, and -C have potential phosphorylation sites. These structural features are highlighted in Figure 2. Specific regions on JAMs have been linked to important protein functions. In particular there is a *cis*-dimerization interface on the membrane-distal Ig loop composed of an R-EWK motif (the asterisk in Fig. 2)

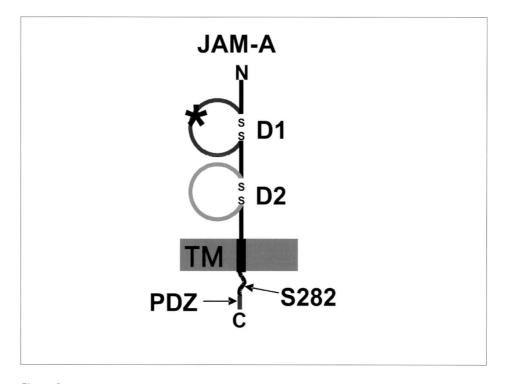

Figure 2
Structural schematic of the JAM-A protein.
*Key structural features include: D1, D2 - Ig loops; *, putative dimerization motif comprised of*
R-EWK in JAM-A, -B and -C; Tm, single pass transmembrane domain. ■, PDZ binding motif.
S282, Ser282, a phosphorylation site. Adapted from [55].

[3–5], a C-terminal PDZ binding motif [6, 7] and a phosphorylation site at Ser282 in both JAM-A [8] and JAM-C [9]. Other functionally important regions remain to be determined, such as additional phosphorylation sites and additional structural requirements that mediate interactions of JAM proteins between cells.

Tissue and cellular expression/localization

JAM proteins are expressed in a variety of tissues and cell types; however, the expression pattern for each family member differs. Such differences in expression patterns imply distinct functions for different JAMs. JAM-A is broadly expressed in endothelial cells [10], epithelial cells [10], fibroblasts [11] and hematopoietic cells [12, 13]; therefore, it is not surprising that JAM-A expression has been reported in nearly every organ. JAM-B is expressed exclusively in vascular and lymphatic endothelium and was originally termed vascular-endothelial JAM to reflect this restricted distribution [14]. Similar to JAM-A, JAM-C has a wider distribution with expression on endothelial cells [15], in lymphatics [15], leukocytes, [16], platelets [17], fibroblasts [11] and epithelial cells [18].

Cellular localization studies have revealed that JAM-A, -B and -C are expressed on the cell surface and concentrate at cell-cell junctions. In endothelial cells all three proteins localize to cell-cell junctions [10, 14, 15]. This localization suggests that JAM-A, -B and -C mediate adhesive interactions and could serve as potential ligands for migration of leukocytes. Interestingly, the distribution at endothelial cell-cell contacts is altered after treatment with inflammatory mediators. In particular, JAM-A is internalized from epithelial cell-cell contacts after treatment with cytokines such as TNF-α and IFN-γ [19], whereas JAM-C has been reported to redistribute from microvessel endothelial cell-cell contacts as a result of histamine and VEGF treatment [20]. Interestingly, in polarized epithelial cells, which have well-defined tight junctions, adherens junctions and desmosomes, JAM-A localizes primarily to the tight junction with some localization along the lateral cell border (see Fig. 3). JAM-C also localizes to intercellular junctions and has been reported to co-localize both with desmosomes and tight junctions in polarized intestinal epithelia [5, 9, 18].

From the above reports, it is apparent that JAM-A, -B and -C have overlapping but distinct expression and localization patterns. Indeed, these observations would predict that JAMs are functionally distinct *in vivo*. Currently, it is not known how expression of JAMs is regulated in different tissues. Furthermore, little is known about how JAM proteins are targeted to various regions of the cell.

Homophilic extracellular interactions

JAM-A, -B, and -C all contain an R-EWK motif in the membrane-distal Ig loop, which has been reported to mediate homophilic dimerization in *cis* [3–5, 14]. Evidence for *cis*-dimerization is based on crystal structure data, cross-linking studies, functional studies from JAM-A dimer mutants [3, 21, 22], purified protein-cell binding for JAM-B [14] and functional data for JAM-C dimerization mutants in the R-EWK motif [5]. Interestingly, there are no published reports demonstrating that the R-EWK motif is able to mediate interactions between cells (in *trans*).

Several lines of evidence indicate that interactions between cells (in *trans*) occur between JAMs. For instance, JAM-A overexpression in endothelial cells has been shown to mediate interactions with JAM-A on platelets [23]. JAM-C interactions between cells have also been reported to mediate interactions between an epithelial tumor cell line and endothelial cells [5]. In purified protein binding assays with JAM-A, -B, and -C, all mediated binding to CHO cells transfected with the corresponding JAM protein but did not mediate binding to untransfected CHO cells [5,

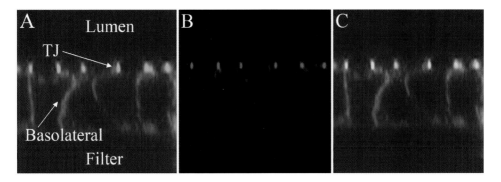

Figure 3
XZ reconstructed confocal immunofluorescence of JAM-A in T84 intestinal epithelial cells. JAM-A is shown (A) and ZO-1 (B). Both are merged in (C). Note the concentration of JAM-A at tight junctions, as demonstrated by colocalization with ZO-1. Some lateral staining of JAM-A can be visualized in (A) and (C). Adapted from [34].

24, 25]. Lastly, the murine JAM-A crystal structure [4] predicts that JAM *cis*-dimer structures may interact in *trans* forming tetramers and higher order structures.

Despite the fact that the crystal structures of both murine and human JAM-A predict that R-EWK mediates *cis*-dimerization between two JAM molecules [4, 21], if two JAM-A molecules are aligned to interact in *trans*, the positive charge from arginine and lysine and the negative charge from glutamic acid in the R-EWK domain would form repulsive electrostatic interactions to inhibit *trans*-dimerization. Thus, molecular mechanism(s) by which JAM molecules might interact in trans remains to be defined. Furthermore, the above observations do not provide the stoichiometry for *trans*-interactions or conclusively demonstrate the existence of homophilic *trans*-interactions in cells.

Heterophilic extracellular interactions

Currently, nearly all reports of extracellular binding interactions involving JAM-A, -B and -C have implicated the membrane-distal Ig loop with the exception of a single report describing involvement of the membrane-proximal Ig loop in ligand binding. Perhaps the best-documented reports of heterophilic interactions involve JAM-B and JAM-C [26]. This interaction is sufficient to mediate disruption of JAM-C dimers by soluble JAM-B [27]. Furthermore, immobilized JAM-B was actually used to purify JAM-C as an unknown ligand on leukocytes before the latter was identified. It is now apparent that JAM-B/JAM-C interactions are dependent upon the membrane-distal Ig loop for both binding partners. In addition, JAM-B/JAM-C dimers have been reported to interact with the leukocyte integrin $\alpha_m\beta_2$/Mac-I [17, 18, 27], while JAM-B has been reported to bind $\alpha_4\beta_1$/VLA-4 [26] integrins; an interaction blocked by the presence of soluble JAM-C. Furthermore, in the testis, JAM-C has been demonstrated to directly bind to CAR, presumably to maintain germ-line cell polarity [28]. Additionally, through a yeast two-hybrid screen, JAM-A was reported to bind to the leukocyte integrin $\alpha_L\beta_2$/LFA-1. The authors further demonstrated binding of the membrane-proximal Ig loop of JAM-A to the ligand binding I domain of $\alpha_L\beta_2$/LFA-1 [29, 30] and reported that such binding interactions mediate leukocyte transendothelial migration, which is discussed further below. There is also a report of association with $\alpha_v\beta_3$ integrins, but it is not known what domain of JAM-A is necessary for this interaction [31]. There are likely more heterophilic interactions for the JAM proteins that are yet to be reported, as the mechanisms for many JAM functions remain incompletely defined.

Intracellular protein-protein interactions

The intracellular domains of JAM-A, -B and -C have been shown to mediate functional responses through interactions that are mainly mediated by the PDZ binding motif at the C terminus (see Fig. 1). The PDZ binding motifs for JAM-A, -B and -C are SSFLV-COOH, KSFII-COOH, and SSFLI-COOH, respectively. However, not all of the intracellular protein interactions that have been reported appear to be dependent on these PDZ binding motifs, suggesting that there may be as-yet-undefined domains or phosphorylated residues in the cytoplasmic tail necessary for intracellular interactions. As shown in Table 1, many cytoplasmic proteins have been reported to interact either directly or indirectly with JAM-A, -B, or -C. Of note, most of the cytoplasmic protein-binding partners for JAMs are scaffolding proteins containing PDZ domains that serve to connect the plasma membrane to actin or microtubules, and thus mediate signaling through the assembly of protein complexes. Specifically, such events may play a role in the regulation of cell polarity, to mention one example.

Cellular function mediated by JAM-A, JAM-B and JAM-C

Determination of cell polarity

Transport of substances across epithelia and endothelia requires the differential compartmentalization of proteins in the basal and apical aspects of the cell membrane. These different membrane domains are separated by the junctional complex in epithelia and endothelia and are determined by the polarity of the cells. Cell polarity is important for separating these membrane regions (basal *versus* apical)

Table 1 - Interactions of the JAM family cytoplasmic tails: Interactions between the intracellular tail of JAM-A, JAM-B, and JAM-C with references

JAM-A		JAM-B		JAM-C	
Protein	Reference	Protein	Reference	Protein	Reference
ZO family	[6, 7]	PAR-3	[56]	CAR	[28]
Afadin	[6]	ZO-1	[56]	PAR-3	[56]
MUPP-1	[57]			ZO 1	[56]
PICK-1	[58]				
PAR-3	[35, 59]				
CASK/LIN-2	[60]				
Cingulin	[7]				
Occludin	[7]				

as well as determining the direction of cell migration in response to chemotactic gradients in leukocytes. Interestingly, JAM-A and JAM-C have been implicated in the regulation of cell polarity in a variety of cell types.

JAM-A has been shown to interact with PAR-3/PAR-6/aPKC through its carboxy PDZ binding motif and the PDZ domain of PAR-3. PAR-3 and PAR-6 are highly conserved polarity proteins first described in *C. elegans*. They both contain PDZ domains and function as scaffolding molecules that associate with atypical protein kinase C (aPKC). The proper cellular localization of this complex is involved in the establishment of cellular polarity through the action of aPKC. Recently, Rehder et al. [32] reported that epithelial cells transfected with JAM-A lacking the PDZ binding domain had impaired cyst formation in 3-D cell cultures, presumably due to a lack of interaction with PAR-3 and thus the PAR-3/PAR-6/aPKC complex.

JAM-C has been shown to interact with the PAR-3/PAR-6/aPKC complex in a similar manner, thus it is likely that JAM-C plays an important role in the maintenance of cell polarity as well. It has been reported that male JAM-C$^{-/-}$ mice are sterile due to the loss of polarization in spermatids, as defined by the absence of an acrosome and other polar structures. Presumably interaction of JAM-C in germ cells with JAM-B in sertoli cells is critical for development of polarity [33].

Barrier function

There are multiple lines of evidence that JAM-A plays a role in regulating barrier function as highlighted by the effects of its expression on transepithelial and transendothelial monolayer resistance to passive ion flow (TER). TER is determined using Ohms law by assessing the potential generated across monolayers of cells cultured on permeable supports during the passage of a constant electrical current using specialized commercially available current/voltage clamps. Initially, overexpression of JAM-A was reported to increase resistance across transfected CHO cells [10]. Furthermore, antibodies that bind near the *cis*-dimerization motif of JAM-A and inhibit JAM dimerization [34] as well as overexpression of the JAM-A intracellular domain [35] inhibit barrier recovery after disruption of intercellular junctions following transient calcium depletion [34, 35]. It is not known if antibodies that block cis-dimerization also block *trans*-dimerization. Interestingly, JAM-A dimerization-blocking antibodies also enhance corneal swelling in rabbit eye preparations, suggesting that disruption of JAM-A dimerization inhibits barrier recovery [36]. Based on the above evidence, it appears that *cis*-dimerization of JAM-A results in decreased in cell monolayer permeability/increased TER. Additionally, treatment with TNF-α and IFN-γ [19] redistribute JAM-A from cell-cell contacts and cause increased permeability. Further evidence for a role of JAM-A in regulating barrier function comes from siRNA studies in which down-regulation of JAM-A resulted in large increases in permeability of epithelial monolayers [37]. Since altered permeability associated with manipulation of JAMs is well documented, it is possible that some of the effects of JAM-A manipulation on leukocyte migration could be indirect and related to altered cell permeability. Increased cell permeability could potentially enhance leukocyte transmigration by decreasing the integrity of the barrier through which the leukocytes must pass or by facilitating the diffusion of a chemotactic gradient. Thus, while JAMs have been implicated in the regulation of leukocyte transmigration, the mechanisms remain unclear.

Cell adhesion, integrin regulation and cell migration

JAM proteins have been reported to be involved in cell-cell adhesion, both directly and indirectly through interactions with other proteins such as integrins. JAM-A was first described having a role in adhesion of platelets to endothelial cells. Investigators observed that platelet adherence to activated endothelial cells was blocked by soluble JAM-A or peptide mimetics of the membrane-distal Ig loop, suggesting direct binding of JAM-A [23]. JAM-C has also been reported to directly mediate adhesion of a tumor cell line to endothelial cells in culture [5]. Adhesion of lymphocytes to endothelial cells has been shown to be mediated by interactions between JAM-B and JAM-C [38]. The above reports serve as examples of direct binding of JAM proteins in mediating adhesive cell-based interactions; however, none define the exact stoichiometry of binding.

Indirectly, JAM proteins have been shown to modulate adhesion in a number of instances, usually through effects on integrin expression levels or activation. In endothelial cells, JAM-A siRNA has been reported to reduce adherence to vitronectin [39], and JAM-A$^{-/-}$ endothelial cells were shown to have decreased adhesion to fibronectin [40]. Overexpression of JAM-A in endothelial cells results in increased adhesion to fibronectin that is mediated by $\alpha_v\beta_3$ integrin, which has been reported to directly interact with JAM-A [41]. In epithelial cells, siRNA-mediated down-regulation of JAM-A causes decreased adhesion of cells to collagen I, collagen IV, and fibronectin and this decrease is mediated by diminished β_1 integrin protein expression [37]. Our own observations indicate that interfering with JAM-A function through expression of dimerization-defective mutants causes decreased cell adhesion to fibronectin, suggesting a role for the dimerization interface in the regulating integrin protein expression levels (unpublished observations). Further examples of interactions of JAM proteins and integrins are highlighted by a JAM-B/JAM-C/$\alpha_m\beta_2$ complex that forms between endothelial cells and leukocytes during leukocyte transmigration [27]. JAM-C has also recently been reported to regulate cell adhesion to fibronectin in an indirect fashion by activation of β_1 and β_3 integrins without changing integrin protein levels [9].

In some of the studies cited above, it was also observed that loss of JAM-A expression through siRNA-mediated down-regulation or gene knockout resulted in decreased endothelial cell motility. Since cellular adhesion is an initial event in a number of cellular processes, such as cell migration and angiogenesis, it is not surprising that JAM-mediated regulation of cell adhesion would also have an effect on cell migration. Indeed, depending upon the localization of active integrins, strong adhesive interactions could reduce cell migration, while strong interactions with an extracellular matrix substrate in cellular extensions could accelerate cell migration. The mechanisms by which JAMs regulate cell adhesion are largely unknown. It is possible that JAM *cis*-dimerization may be required to the formation of a signaling complex through its interactions with scaffolding proteins. *Cis*-dimerization is likely a necessary requirements for at least some of JAM-A functions since our unpublished studies indicate that interfering with JAM-A dimerization decreases cell migration. We propose that *cis*-dimerization is important for the formation of a signaling complex through increased spatial proximity of scaffolding proteins such as ZO-1 and Afadin by interactions with the JAM PDZ motif. Clues as to signaling events downstream of JAM-A come from studies that implicate the small GTPase, Rap1 [9, 37], a closely related ras homologue that has previously been reported to regulate integrin levels [9, 37]. Specifically, our data indicate that Rap1a is specifically activated through interactions with scaffolding proteins in the presence of JAM-A and that active Rap1a protects β_1 integrin from proteolysis. Other candidate elements of the JAM signaling cascade include members of the MEK pathway, where it has been reported that siRNA-mediated down-regulation of JAM-A leads to decreases in pERK1/2 [31, 39]. From these observations we can hypothesize that JAM-A dimerization may be necessary to assemble a signaling complex that includes multi-function scaffold proteins such as ZO-1 or Afadin, which in turn lead to activation of Rap1 and ERK. Such signaling events downstream of JAM-A,

-B and -C are just beginning to be examined and may be the key to understanding mechanisms for the diversity of JAM functions detailed in this review.

Angiogenesis

Since endothelial cell migration and permeability are critical determinants of angiogenesis, it is not surprising that the JAM family members are involved in angiogenesis. It is likely that JAM regulation of angiogenesis is secondary to altered cell migration and adhesion. Alternatively, angiogenesis could be altered due to signaling events initiated by JAM proteins. There are several reports implicating JAM-A in angiogenesis. In studies with JAM-A$^{-/-}$ endothelial cells, it has been reported that bFGF-mediated endothelial cell migration and angiogenesis are attenuated. Such effects may involve the MEK pathway, as pERK1/2 is decreased after bFGF treatment in cells with down-regulated JAM-A [31, 39]. JAM-C also has been reported to play a role in angiogenesis, as antibodies specific to JAM-C have been observed to decrease angiogenesis and size of tumors in nude mice under conditions of high VEGF levels [42]. Interestingly, there are no reports investigating the role of JAM-B in angiogenesis. Decreased or blocked JAM-A and JAM-C thus appears to inhibit angiogenesis in the above-reported settings; however, it remains to be determined if such inhibition is due to activation of various signaling pathways, changes in cell adhesion/migration or other currently undefined mechanisms.

Role of JAM-A in leukocyte transmigration

There are numerous reports suggesting a role for JAM-A in regulating leukocyte transmigration. In an early report on JAM-A, it was noted that the anti-JAM-A antibody BV12 inhibited transendothelial migration of murine monocytes *in vivo* [10]. Paradoxically, JAM-A-deficient dendritic cells in the skin are reported to have increased migration and random motility compared to wild-type dendritic cells, suggesting impaired polarization of dendritic cells with enhanced migratory ability [43]. It is now apparent that JAM-A$^{-/-}$ neutrophils have a defect in polarization that most likely results in diminished capacity to migrate [44]. In ischemia/reperfusion mouse models, there is decreased neutrophil transmigration in hepatic tissues in the absence of JAM-A [45]. Finally, in a model of atherosclerotic endothelium, JAM-A$^{-/-}$, ApoE$^{-/-}$ mice have impaired macrophage recruitment compared to JAM-A$^{+/+}$, ApoE$^{-/-}$ mice.

The above studies are highly suggestive of JAM-A playing a key role in leukocyte transmigration. However, they do not mechanistically define the connection between JAM-A expression and leukocyte transmigration. There are a number of possibilities that may provide insight into how JAM-A regulates leukocyte transmigration. In particular, three well-defined effects of JAM-A expression are decreased paracellular permeability, increased adhesion of leukocytes to endothelial cells and leukocyte polarization. JAM-mediated changes in paracellular permeability could alter leukocyte transmigration across endothelial and epithelial monolayers simply by tightening or loosening cell-cell contact through which the leukocytes must pass or by regulating the diffusion of chemotactic gradients. Likewise, increased cell adhesion is an early step in leukocyte transmigration and changes in both homophilic JAM interactions followed by altered integrin levels could lead to increases or decreases in leukocyte transmigration. Lastly, determination of cell polarity in leukocytes in response to a chemotactic gradient is crucial for directed migration across epithelial and endothelial monolayers. As has been reported for JAM-A$^{-/-}$ neutrophils and dendritic cells, absence of JAM-A expression presumably leads to abnormal PAR-3/PAR-6/aPKC complex localization and hence defective cell polarization. There could be additional signaling events influenced by the absence or alteration of JAM-A protein levels that could influence leukocyte transmigration through as-yet-undescribed mechanisms. A final possibility for the connection between JAM-A and leukocyte transmigration would be a JAM-A-leukocyte ligand interaction.

There are conflicting reports on the existence of a JAM-A-leukocyte ligand interaction. Initially, JAM-A encoded by a full-length cDNA was identified as binding to the leukocyte integrin LFA-1 ($\alpha_1\beta_2$) in a yeast two-hybrid assay using the α_1 subunit as bait. It was reported that the membrane-proximal extracellular domain of JAM-A is a ligand for LFA-1 and migration of Jurkat T cells across activated endothelial cells was inhibited by both anti-JAM-A mAb and anti-LFA-1 mAbs [30]. A second report published by the same group indicated that JAM-A binds to the I domain of LFA-1 through the membrane-proximal Ig loop of JAM-A [29, 30]. In contrast, another early report indicated that inhibition of JAM-A with multiple dimerization-blocking antibodies did not inhibit leukocyte transmigration across endothelial cells *in vitro* [46]. Furthermore, while JAM-A is localized to cell-cell contacts where neutrophils cross the endothelial monolayer, it does not colocalize with LFA-1 as would be expected if CD11a/CD18 was a ligand for JAM-A [47]. From these studies, it is clear that more studies are necessary to understand the *in vivo* relevance of JAM-A binding to LFA-1.

Role of JAM-B/JAM-C in leukocyte transmigration

JAM-B has been demonstrated to mediate interactions with lymphocytes through a heterophilic interaction with JAM-C [38]. In addition, inhibition of JAM-C interaction with JAM-B through a blocking antibody (H33) resulted in redistribution of JAM-C to the luminal aspect of endothelial cells monolayers that correlated with increased leukocyte transmigration [27]. The authors hypothesized that interactions with junction-associated JAM-B results in sequestration of JAM-C to cell-cell contacts. Under this scenario, inhibition of binding between JAM-B and JAM-C would "release" JAM-C to the cell surface for subsequent participation in endothelial-leukocyte interactions.

There are two reports demonstrating direct binding of JAM-C to the leukocyte adhesive integrin CD11b/CD18 ($\alpha_m\beta_2$) [17, 18]. Leukocyte migration across endothelial and epithelial monolayers was inhibited with JAM-C inhibiting antibody treatment (Gi-l1 or luca14) or soluble JAM-C [17, 18]. Furthermore, in vivo overexpression of JAM-C in endothelial cells has been reported to reduce circulating white blood cell levels, presumably due to increased JAM-C-dependent migration into tissues [18, 48, 49]; however, it is not known if JAM-C plays a role in hematopoiesis or egress of leukocytes from the bone marrow. Finally, it has been reported that JAM-C expression is increased in endothelial cells during cerulean-induced pancreatits and the severity of disease is attenuated in parallel with reduction of leukocyte infiltration by anti-JAM-C antibodies [50]. The above evidence suggests that JAM-C directly binds to CD11b/CD18 on leukocytes, and this interaction is important for leukocyte transmigration in vitro and in vivo. Details of specific regions on JAM-C that mediate such interactions remain to be determined. It will be intriguing to see if specific mutations of JAM-C result in attenuated transmigration through altered ligand binding or if altered transmigration is secondary to other JAM-C/JAM-B-mediated interactions.

The relationship of JAM function and human diseases: therapeutic implications

JAM proteins have been linked to pathophysiology of several disorders in addition to being co-opted as a virus receptor. Since JAM-A and JAM-C have crucial roles in cellular adhesion and angiogenesis, their expression may be important in the growth of primary tumors, tumor metastasis and survival of cancer cells. Additionally, given the roles of the JAM proteins in cell permeability, cell polarization and heterophilic ligand binding, they are likely to play key roles in a number of inflammatory conditions such as pancreatits (for JAM-C [50]) and ischemia/reperfusion (for JAM-A [44]). Expression of JAM-A and JAM-C has also been implicated in atherosclerosis, as wild-type endothelial cells lacking either protein have more severe atherosclerosis compared to endothelial cells deficient for JAM-A or JAM-C [51, 52]. Furthermore, the role of JAM-C as a polarity molecule makes it a potentially crucial determinant in male fertility, as disruptions in JAM-C inhibits spermatid polarization and cause infertility in mice [33]. If further studies reveal that JAM-C deficiency is a cause of human infertility, gene therapy with JAM-C might be worth considering as a way to produce mature spermatids. Finally, the localization of JAM-A at intercellular junctions in mucosae and the vasculature is possibly a feature that viral agents have evolved to exploit to gain entry into the host. In the phylogram of Figure 1, an interesting feature of several JAM-related IgSF proteins is that they are used as a receptor for a number of viral pathogens. In particular, CAR is a receptor for adenovirus and nectin is a receptor for herpes virus. Similarly, JAM-A has been shown to serve as receptor for reovirus [53] and calcivirus [54].

Given the binding interactions with CD11b/CD18 that have been reported, JAM-C would be a potential anti-inflammatory target, as inhibition of such binding could potentially decrease pathological leukocyte transmigration. Along similar lines, small molecule-mediated inhibition of dimerization of JAM-A could possibly be exploited to increase vascular or mucosal permeability for a number of applications in addition to inhibition of leukocyte migration.

Conclusions

JAM proteins are important mediators of cell polarity, paracellular permeability/barrier function, cell adhesion/migration and angiogenesis. The multiple functions of JAMs appear to be linked to key structural features shared by various family members. In particular, the dimerization motif within the membrane distal Ig loop might be an attractive target for development of therapeutics. Evidence discussed above indicates that disruption of this motif may attenuate a variety of pathological conditions while also providing an avenue to enhance drug delivery across cellular barriers. Despite the growing list of reports on function of JAM-A, -B and -C, the precise mechanism(s) governing JAM function remain a mystery. A better understanding of signaling pathways influenced by JAM protein expression and ligand interactions will provide important insights into how this important protein family contributes to health and disease.

Acknowledgements
Funding for studies in the Parkos laboratory was provided for by the NIH grants DK61379, DK72564 and HL72124.

References

1   Muller WA (2003) Leukocyte-endothelial-cell interactions in leukocyte transmigration and the inflammatory response. *Trends Immunol* 24: 327–334

2   Moog-Lutz C, Cave-Riant F, Guibal FC, Breau MA, Di Gioia Y, Couraud PO, Cayre YE, Bourdoulous S, Lutz PG (2003) JAML, a novel protein with characteristics of a junctional adhesion molecule, is induced during differentiation of myeloid leukemia cells. *Blood* 102: 3371–3378

3   Bazzoni G, Martinez-Estrada OM, Mueller F, Nelboeck P, Schmid G, Bartfai T, Dejana

E, Brockhaus M (2000) Homophilic interaction of junctional adhesion molecule. *J Biol Chem* 275: 30970–30976

4   Kostrewa D, Brockhaus M, D'Arcy A, Dale GE, Nelboeck P, Schmid G, Mueller F, Bazzoni G, Dejana E, Bartfai T et al (2001) X-ray structure of junctional adhesion molecule: structural basis for homophilic adhesion *via* a novel dimerization motif. *EMBO J* 20: 4391–4398

5   Santoso S, Orlova VV, Song K, Sachs UJ, Andrei-Selmer CL, Chavakis T (2005) The homophilic binding of junctional adhesion molecule-C mediates tumor cell-endothelial cell interactions. *J Biol Chem* 280: 36326–36333

6   Ebnet K, Schulz CU, Meyer Zu Brickwedde MK, Pendl GG, Vestweber D (2000) Junctional adhesion molecule interacts with the PDZ domain-containing proteins AF-6 and ZO-1. *J Biol Chem* 275: 27979–27988

7   Bazzoni G, Martinez-Estrada OM, Orsenigo F, Cordenonsi M, Citi S, Dejana E (2000) Interaction of junctional adhesion molecule with the tight junction components ZO-1, cingulin, and occludin. *J Biol Chem* 275: 20520–20526

8   Ozaki H, Ishii K, Arai H, Horiuchi H, Kawamoto T, Suzuki H, Kita T (2000) Junctional adhesion molecule (JAM) is phosphorylated by protein kinase C upon platelet activation. *Biochem Biophys Res Commun* 276: 873–878

9   Mandicourt G, Iden S, Ebnet K, Aurrand-Lions M, Imhof BA (2006) JAM-C regulates tight junctions and integrin-mediated cell adhesion and migration. *J Biol Chem* 282: 1830–1837

10   Martin-Padura I, Lostaglio S, Schneemann M, Williams L, Romano M, Fruscella P, Panzeri C, Stoppacciaro A, Ruco L, Villa A et al (1998) Junctional adhesion molecule, a novel member of the immunoglobulin superfamily that distributes at intercellular junctions and modulates monocyte transmigration. *J Cell Biol* 142: 117–127

11   Morris AP, Tawil A, Berkova Z, Wible L, Smith CW, Cunningham SA (2006) Junctional adhesion molecules (JAMs) are differentially expressed in fibroblasts and co-localize with ZO-1 to adherens-like junctions. *Cell Commun Adhes* 13: 233–247

12   Gupta SK, Pillarisetti K, Ohlstein EH (2000) Platelet agonist F11 receptor is a member of the immunoglobulin superfamily and identical with junctional adhesion molecule (JAM): regulation of expression in human endothelial cells and macrophages. *IUBMB Life* 50: 51–56

13   Naik UP, Ehrlich YH, Kornecki E (1995) Mechanisms of platelet activation by a stimulatory antibody: cross-linking of a novel platelet receptor for monoclonal antibody F11 with the Fc gamma RII receptor. *Biochem J* 310: 155–162

14   Palmeri D, van Zante A, Huang CC, Hemmerich S, Rosen SD (2000) Vascular endothelial junction-associated molecule, a novel member of the immunoglobulin superfamily, is localized to intercellular boundaries of endothelial cells. *J Biol Chem* 275: 19139–19145

15   Aurrand-Lions M, Duncan L, Ballestrem C, Imhof BA (2001) JAM-2, a novel immunoglobulin superfamily molecule, expressed by endothelial and lymphatic cells. *J Biol Chem* 276: 2733–2741

16 Arrate MP, Rodriguez JM, Tran TM, Brock TA, Cunningham SA (2001) Cloning of human junctional adhesion molecule 3 (JAM3) and its identification as the JAM2 counter-receptor. *J Biol Chem* 276: 45826–45832

17 Santoso S, Sachs UJ, Kroll H, Linder M, Ruf A, Preissner KT, Chavakis T (2002) The junctional adhesion molecule 3 (JAM–3) on human platelets is a counterreceptor for the leukocyte integrin Mac-1. *J Exp Med* 196: 679–691

18 Zen K, Babbin BA, Liu Y, Whelan JB, Nusrat A, Parkos CA (2004) JAM-C is a component of desmosomes and a ligand for CD11b/CD18-mediated neutrophil transepithelial migration. *Mol Biol Cell* 15: 3926–3937

19 Ozaki H, Ishii K, Horiuchi H, Arai H, Kawamoto T, Okawa K, Iwamatsu A, Kita T (1999) Cutting edge: combined treatment of TNF-alpha and IFN-gamma causes redistribution of junctional adhesion molecule in human endothelial cells. *J Immunol* 163: 553–557

20 Orlova VV, Economopoulou M, Lupu F, Santoso S, Chavakis T (2006) Junctional adhesion molecule-C regulates vascular endothelial permeability by modulating VE-cadherin-mediated cell-cell contacts. *J Exp Med* 203: 2703–2714

21 Prota AE, Campbell JA, Schelling P, Forrest JC, Watson MJ, Peters TR, Aurrand-Lions M, Imhof BA, Dermody TS, Stehle T (2003) Crystal structure of human junctional adhesion molecule 1: implications for reovirus binding. *Proc Natl Acad Sci USA* 100: 5366–5371

22 Mandell KJ, McCall IC, Parkos CA (2004) Involvement of the junctional adhesion molecule-1 (JAM1) homodimer interface in regulation of epithelial barrier function. *J Biol Chem* 279: 16254–16262

23 Babinska A, Kedees MH, Athar H, Ahmed T, Batuman O, Ehrlich YH, Hussain MM, Kornecki E (2002) F11-receptor (F11R/JAM) mediates platelet adhesion to endothelial cells: role in inflammatory thrombosis. *Thromb Haemost* 88: 843–850

24 Liang TW, DeMarco RA, Mrsny RJ, Gurney A, Gray A, Hooley J, Aaron HL, Huang A, Klassen T, Tumas DB, Fong S (2000) Characterization of huJAM: evidence for involvement in cell-cell contact and tight junction regulation. *Am J Physiol Cell Physiol* 279: C1733–1743

25 Cunningham SA, Arrate MP, Rodriguez JM, Bjercke RJ, Vanderslice P, Morris AP, Brock TA (2000) A novel protein with homology to the junctional adhesion molecule. Characterization of leukocyte interactions. *J Biol Chem* 275: 34750–34756

26 Cunningham SA, Rodriguez JM, Arrate MP, Tran TM, Brock TA (2002) JAM2 interacts with alpha4beta1. Facilitation by JAM3. *J Biol Chem* 277: 27589–27592

27 Lamagna C, Meda P, Mandicourt G, Brown J, Gilbert RJ, Jones EY, Kiefer F, Ruga P, Imhof BA, Aurrand-Lions M (2005) Dual interaction of JAM-C with JAM-B and alpha(M)beta2 integrin: function in junctional complexes and leukocyte adhesion. *Mol Biol Cell* 16: 4992–5003

28 Mirza M, Hreinsson J, Strand ML, Hovatta O, Soder O, Philipson L, Pettersson RF, Sollerbrant K (2006) Coxsackievirus and adenovirus receptor (CAR) is expressed in

male germ cells and forms a complex with the differentiation factor JAM-C in mouse testis. *Exp Cell Res* 312: 817–830

29 Fraemohs L, Koenen RR, Ostermann G, Heinemann B, Weber C (2004) The functional interaction of the beta 2 integrin lymphocyte function-associated antigen-1 with junctional adhesion molecule-A is mediated by the I domain. *J Immunol* 173: 6259–6264

30 Ostermann G, Weber KS, Zernecke A, Schroder A, Weber C (2002) JAM-1 is a ligand of the beta(2) integrin LFA-1 involved in transendothelial migration of leukocytes. *Nat Immunol* 3: 151–158

31 Naik MU, Mousa SA, Parkos CA, Naik UP (2003). Signaling through JAM-1 and alphavbeta3 is required for the angiogenic action of bFGF: dissociation of the JAM-1 and alphavbeta3 complex. *Blood* 102: 2108–2114

32 Rehder D, Iden S, Nasdala I, Wegener J, Brickwedde MK, Vestweber D, Ebnet K (2006) Junctional adhesion molecule-a participates in the formation of apico-basal polarity through different domains. *Exp Cell Res* 312: 3389–3403

33 Gliki G, Ebnet K, Aurrand-Lions M, Imhof BA, Adams RH (2004) Spermatid differentiation requires the assembly of a cell polarity complex downstream of junctional adhesion molecule-C. *Nature* 431: 320–324

34 Liu Y, Nusrat A, Schnell FJ, Reaves TA, Walsh S, Pochet M, Parkos CA (2000) Human junction adhesion molecule regulates tight junction resealing in epithelia. *J Cell Sci* 113: 2363–2374

35 Ebnet K, Suzuki A, Horikoshi Y, Hirose T, Meyer Zu Brickwedde MK, Ohno S, Vestweber D (2001) The cell polarity protein ASIP/PAR-3 directly associates with junctional adhesion molecule (JAM). *EMBO J* 20: 3738–3748

36 Mandell KJ, Holley GP, Parkos CA, Edelhauser HF (2006) Antibody blockade of junctional adhesion molecule-A in rabbit corneal endothelial tight junctions produces corneal swelling. *Invest Ophthalmol Vis Sci* 47: 2408–2416

37 Mandell KJ, Babbin BA, Nusrat A, Parkos CA (2005) Junctional adhesion molecule 1 regulates epithelial cell morphology through effects on beta1 integrins and Rap1 activity. *J Biol Chem* 280: 11665–11674

38 Liang TW, Chiu HH, Gurney A, Sidle A, Tumas DB, Schow P, Foster J, Klassen T, Dennis K, DeMarco RA et al (2002) Vascular endothelial-junctional adhesion molecule (VE-JAM)/JAM 2 interacts with T, NK, and dendritic cells through JAM 3. *J Immunol* 168: 1618–1626

39 Naik MU, Vuppalanchi D, Naik UP (2003) Essential role of junctional adhesion molecule-1 in basic fibroblast growth factor-induced endothelial cell migration. *Arterioscler Thromb Vasc Biol* 23: 2165–2171

40 Bazzoni G, Tonetti P, Manzi L, Cera MR, Balconi G, Dejana E (2005) Expression of junctional adhesion molecule-A prevents spontaneous and random motility. *J Cell Sci* 118: 623–632

41 Naik MU, Naik UP (2006) Junctional adhesion molecule-A-induced endothelial cell migration on vitronectin is integrin alpha v beta 3 specific. *J Cell Sci* 119: 490–499

42 Lamagna C, Hodivala-Dilke KM, Imhof BA, Aurrand-Lions M (2005) Antibody against

junctional adhesion molecule-C inhibits angiogenesis and tumor growth. *Cancer Res 65*: 5703–5710

43 Cera MR, Del Prete A, Vecchi A, Corada M, Martin-Padura I, Motoike T, Tonetti P, Bazzoni G, Vermi W, Gentili F et al (2004) Increased DC trafficking to lymph nodes and contact hypersensitivity in junctional adhesion molecule-A-deficient mice. *J Clin Invest* 114: 729–738

44 Corada M, Chimenti S, Cera MR, Vinci M, Salio M, Fiordaliso F, De Angelis N, Villa A, Bossi M, Staszewsky LI et al (2005) Junctional adhesion molecule-A-deficient polymorphonuclear cells show reduced diapedesis in peritonitis and heart ischemia-reperfusion injury. *Proc Natl Acad Sci USA* 102: 10634–10639

45 Khandoga A, Kessler JS, Meissner H, Hanschen M, Corada M, Motoike T, Enders G, Dejana E, Krombach F (2005) Junctional adhesion molecule-A deficiency increases hepatic ischemia-reperfusion injury despite reduction of neutrophil transendothelial migration. *Blood* 106: 725–733

46 Shaw SK, Perkins BN, Lim YC, Liu Y, Nusrat A, Schnell FJ, Parkos CA, Luscinskas FW (2001) Reduced expression of junctional adhesion molecule and platelet/endothelial cell adhesion molecule-1 (CD31) at human vascular endothelial junctions by cytokines tumor necrosis factor-alpha plus interferon-gamma Does not reduce leukocyte transmigration under flow. *Am J Pathol* 159: 2281–2291

47 Shaw SK, Ma S, Kim MB, Rao RM, Hartman CU, Froio RM, Yang L, Jones T, Liu Y, Nusrat A et al (2004) Coordinated redistribution of leukocyte LFA-1 and endothelial cell ICAM-1 accompany neutrophil transmigration. *J Exp Med* 200: 1571–1580

48 Aurrand-Lions M, Lamagna C, Dangerfield JP, Wang S, Herrera P, Nourshargh S, Imhof BA (2005) Junctional adhesion molecule-C regulates the early influx of leukocytes into tissues during inflammation. *J Immunol* 174: 6406–6415

49 Chavakis T, Keiper T, Matz-Westphal R, Hersemeyer K, Sachs UJ, Nawroth PP, Preissner KT, Santoso S (2004) The junctional adhesion molecule-C promotes neutrophil transendothelial migration *in vitro* and *in vivo*. *J Biol Chem* 279: 55602–55608

50 Vonlaufen A, Aurrand-Lions M, Pastor CM, Lamagna C, Hadengue A, Imhof BA, Frossard JL (2006) The role of junctional adhesion molecule C (JAM-C) in acute pancreatitis. *J Pathol* 209: 540–548

51 Zernecke A, Liehn EA, Fraemohs L, von Hundelshausen P, Koenen RR, Corada M, Dejana E, Weber C (2006) Importance of junctional adhesion molecule-A for neointimal lesion formation and infiltration in atherosclerosis-prone mice. *Arterioscler Thromb Vasc Biol* 26: e10–13

52 Keiper T, Santoso S, Nawroth PP, Orlova V, Chavakis T (2005) The role of junctional adhesion molecules in cell-cell interactions. *Histol Histopathol* 20: 197–203

53 Barton ES, Forrest JC, Connolly JL, Chappell JD, Liu Y, Schnell FJ, Nusrat A, Parkos CA, Dermody TS (2001) Junction adhesion molecule is a receptor for reovirus. *Cell* 104: 441–451

54 Makino A, Shimojima M, Miyazawa T, Kato K, Tohya Y, Akashi H (2006) Junctional

adhesion molecule 1 is a functional receptor for feline calicivirus. *J Virol* 80: 4482–4490

55   Mandell KJ, Parkos CA (2005). The JAM family of proteins. *Adv Drug Deliv Rev* 57: 857–867

56   Ebnet K, Aurrand-Lions M, Kuhn A, Kiefer F, Butz S, Zander K, Meyer zu Brickwedde MK, Suzuki A, Imhof BA, Vestweber D (2003) The junctional adhesion molecule (JAM) family members JAM-2 and JAM-3 associate with the cell polarity protein PAR-3: a possible role for JAMs in endothelial cell polarity. *J Cell Sci* 116: 3879–3891

57   Hamazaki Y, Itoh M, Sasaki H, Furuse M, Tsukita S (2002) Multi-PDZ domain protein 1 (MUPP1) is concentrated at tight junctions through its possible interaction with claudin-1 and junctional adhesion molecule. *J Biol Chem* 277: 455–461

58   Reymond N, Garrido-Urbani S, Borg JP, Dubreuil P, Lopez M (2005) PICK-1: a scaffold protein that interacts with Nectins and JAMs at cell junctions. *FEBS Lett* 579: 2243–2249

59   Itoh M, Sasaki H, Furuse M, Ozaki H, Kita T, Tsukita S (2001) Junctional adhesion molecule (JAM) binds to PAR-3: a possible mechanism for the recruitment of PAR-3 to tight junctions. *J Cell Biol* 154: 491–497

60   Martinez-Estrada OM, Villa A, Breviario F, Orsenigo F, Dejana E, Bazzoni G (2001) Association of junctional adhesion molecule with calcium/calmodulin-dependent serine protein kinase (CASK/LIN-2) in human epithelial caco-2 cells. *J Biol Chem* 276: 9291–9296

Promises and limitations of targeting adhesion molecules for therapy

Karyn Yonekawa[1] and John M. Harlan[2]

[1]Division of Nephrology, Department of Pediatrics, University of Washington, Seattle, WA 98195, USA; [2]Division of Hematology, Department of Medicine, University of Washington, Seattle, WA 98195, USA

Introduction

Dramatic progress in elucidating the molecular basis of inflammation over the past two decades has led to the development new anti-inflammatory and immunomodulatory therapies. In particular, the adhesion molecules involved in leukocyte trafficking from the blood stream to tissue have emerged as important therapeutic targets. Extensive preclinical studies have shown that blockade of leukocyte or endothelial adhesion molecules is efficacious in diverse disease models, prompting many pharmaceutical and biotechnology companies to develop adhesion antagonists. However, because of the close relationship between inflammation and host defense and tissue repair, anti-adhesion therapy may also be a double-edged sword. This chapter reviews the promises and limitations of anti-adhesion therapy, focusing on those drugs that have completed clinical trials.

Approaches to targeting adhesion molecules

There are a number of approaches to anti-adhesion therapy. Blockade of receptor-ligand interactions by a monoclonal antibody (mAb), soluble receptor or peptide, or small-molecule ligand mimetic is the most direct, and most drugs that have been tested in clinical trials are of this category. Other potential therapeutics include small-molecule antagonists that interrupt the signaling pathways that regulate integrin receptor affinity or the expression of endothelial adhesion molecules, allosteric inhibitors that prevent the conformational changes necessary for increased integrin receptor affinity, agents that interfere with the biosynthesis of carbohydrate ligands for selectin receptors, and drugs that impair transendothelial migration.

Adhesion Molecules: Function and Inhibition, edited by Klaus Ley
© 2007 Birkhäuser Verlag Basel/Switzerland

Targeting adhesion molecules in human disease

In reviewing the clinical trials of anti-adhesion therapy, only therapies that have completed Phase II or Phase III studies are considered.[1] Table 1 lists the therapies by mechanism of action and clinical indication.

Asthma

The asthmatic response to allergen is characterized by airway hyper-responsiveness and inflammation with the accumulation of effector cells such as lymphocytes and eosinophils. The trafficking of these immune cells to the lung in asthma involves both selectins (reviewed in [1]) and integrins, particularly $\alpha_L\beta_2$ (LFA-1) and $\alpha_4\beta_1$ (VLA-4) (reviewed in [2]).

Given the efficacy observed in preclinical studies, it is surprising that only a few adhesion agents have completed Phase II testing. A small Phase IIa study was undertaken in asthmatic patients using Bio-1211, an aerosolized small-molecule inhibitor of VLA-4; however, it was determined that that particular compound did not warrant further development [3]. Also, a planned Phase II trial of a small-molecule dual inhibitor of $\alpha_4\beta_1$ and $\alpha_4\beta_7$ was discontinued [4]. Efalizumab, a humanized anti-α_L mAb, was tested in patients with mild asthma. Although the number of activated eosinophils was significantly decreased after 4 and 8 weeks of treatment, the early and late percent fall in forced expiratory volume at 1 s after allergen challenge did not reach statistical significance [5]. Bimosiamose, an E-, P-, and L-selectin antagonist, showed efficacy in a small Phase IIa trial of 12 patients, and it is currently being evaluated in a Phase II trial as an inhaled treatment for mild allergic asthma [6].

Atherosclerosis

Vascular cell adhesion molecule-I (VCAM-1) is an immunoglobulin gene super-family (IgSF) member that is the primary endothelial ligand for $\alpha_4\beta_1$. The interaction of $\alpha_4\beta_1$ on monocytes and lymphocytes with endothelial VCAM-1 is thought to play a pivotal role in atherogenesis [7]. AGI-1067 is an oral antioxidant that inhibits the transcription of VCAM-1 as well as several other cytokine-induced, redox-sensi-

---

1 Phase I clinical trials test a new therapy in a small group of people (20–80) to assess safety, establish safe dosage ranges and identify side effects. Phase II clinical trials determine efficacy and further evaluate safety in a larger group of people (several hundred). Phase III studies confirm efficacy in a larger population (several hundred to several thousand) comparing the intervention to standard treatments and continue to monitor adverse effects.

Table 1 - Clinical trials of anti-adhesion therapies

Target	Drug	Clinical indication	Company	Trial result	Ref
Integrins:					
α_L	Efalizumab	Asthma	Genentech/Xoma	Negative	[5]
		Psoriasis		Positive	[39, 40]
		Psoriatic arthritis		Negative	[41]
		Rheumatoid arthritis		Negative	[45]
	Odulimomab	Transplant	IMTIX/Pasteur Merieux Serums et Vaccins	Negative	[56, 57]
	25.3	GVHD	Immunotech	Positive	[61]
		Transplant		Positive	[54, 55]
α_4	Natalizumab	Crohn's disease	Biogen Idec/Elan Pharmaceuticals	Positive	[23–25]
		Multiple sclerosis		Positive	[32]
$\alpha_M\beta_2$	UK279,276	Stroke	Pfizer	Negative	[49]
$\alpha_4\beta_7$	MLN02	Crohn's disease	Millennium/Genentech	Negative	[20]
		Ulcerative colitis		Positive	[22]
β_2	Erlizumab	Myocardial infarction	Genentech	Negative	[36]
		Traumatic shock		Negative	[65]
	Rovelizumab	Multiple sclerosis	ICOS	Negative	[31]
		Myocardial infarction		Negative	[35]
		Stroke		Negative	[48]
		Traumatic shock		Negative	[66]
IgSF ligands:					
ICAM-1	Alicaforsen	Crohn's disease	ISIS Pharmaceuticals	Negative	[18, 19]
		Rheumatoid arthritis		Negative	[46]
		Ulcerative colitis		Negative	[21]
	Enlimomab	Burns	Boehringer Ingelheim	Negative	[12]
		Stroke		Negative	[50]
		Transplant		Negative	[58]
Selectins:					
L-selectin	Aselizumab	Traumatic shock	Scil Biomedicals GmbH	Negative	[67]
Sialyl Lewis X	CY-1503	Cardiopulmonary bypass	Cytel Corporation	Negative	[15]
		Thromboend-arterectomy		Negative	[16]
Other:					
	AGI-1067	Atherosclerosis	Atherogenics	Positive	[9]

tive genes [8]. AGI-1067 improved luminal dimensions at the intervention site and reduced restenosis in the Canadian antioxidant restenosis trial (CART-1) [9]. Currently, AGI-1067 is in Phase III testing for patients with established coronary artery disease [10].

Burns

Tissue damage following a burn can worsen as a result of progressive destruction from the acute inflammatory response [11]. A murine mAb to human ICAM-1, enlimomab, was tested in the treatment of partial thickness burns. There was improved healing at high-risk sites with enlimomab compared with placebo and trends toward significance for the primary endpoint of reduction in the total area of grafts. However, the scar ratings at long-term follow-up were not statistically different from control [12].

Cardiopulmonary surgery

Ischemia-reperfusion injury has been postulated to contribute to many pathological conditions, including postoperative cardiopulmonary bypass, myocardial infarction, stroke, and shock. Preclinical studies in animal models of ischemia-reperfusion disorders showed striking protection against reperfusion injury with blockade of leukocyte or endothelial adhesion molecules (reviewed in [13]).

There is a prominent inflammatory reaction to cardiopulmonary bypass that may contribute to the development of postoperative complications such as myocardial infarction, pulmonary injury, and renal and neurological dysfunction [14]. It has been postulated that a major component of the inflammatory damage results from a reperfusion injury. The selectin antagonist, CY-1503 (Cylexin®), is an oligosaccharide mimetic of sialyl Lewis X, a selectin ligand. CY-1503 was tested in a Phase II trial for prevention of reperfusion injury in neonates and infants undergoing hypothermic cardiopulmonary bypass. Treatment with CY-1503 did not significantly improve other early postoperative outcomes or decrease the occurrence of adverse events [15].

CY-1503 was also tested in a small Phase II trial in patients undergoing pulmonary thromboendarterectomy for chronic thromboembolic pulmonary hypertension [16]. Patients undergoing this procedure may suffer from reperfusion lung injury. Treatment with CY-1503 significantly reduced the incidence of lung injury compared with placebo but the number of days on ventilator, number of days in the intensive care unit and hospital, and mortality were not different.

Inflammatory bowel disease

Although distinct diseases, Crohn's disease and ulcerative colitis are both chronic inflammatory conditions, which are theorized to stem from repeated stimulation of the immune system by normal bowel flora [17]. Alicaforsen [18, 19], an antisense oligodeoxynucleotide to ICAM-1 and MLN02 [20], a humanized mAb against $\alpha_4\beta_7$, did not show efficacy in Phase II clinical trials for Crohn's disease. However, with MLN02, although the treatment group did not achieve superior clinical response when compared to placebo, the primary endpoint of the trial, the secondary endpoint of disease remission was achieved in patients receiving the higher dose treatment.

Alicaforsen and MLN02 have also undergone Phase II trials for ulcerative colitis. Alicaforsen, when administered as an enema during acute exacerbations of mild to moderate left-sided ulcerative colitis, failed to produce a significant difference in the study's primary endpoint (disease activity index at 6 weeks) when compared to placebo [21]. In contrast, treatment with MLN02 significantly improved ulcerative colitis clinical remission rates as compared to placebo [22].

Natalizumab (Tysabri®) is a recombinant, humanized mAb to α_4, which inhibits binding of $\alpha_4\beta_7$ to mucosal addressin cell adhesion molecule-1 (MAdCAM-1) as well as $\alpha_4\beta_1$ to VCAM-1. Natalizumab has shown promise as a treatment for moderately severe Crohn's disease [23–25]. Recently, data from an open-label extension study of patients with Crohn's disease who had achieved remission with natalizumab demonstrated that remission in these patients was maintained for 2 years [26].

In 2004, natalizumab was approved by the U.S. Federal Drug Administration (FDA) for the treatment of relapsing-remitting multiple sclerosis (discussed below). However, in February 2005, its manufacturers, Biogen Idec and Elan Corporation, voluntarily suspended all commercial distribution and clinical trials of natalizumab due to the occurrence of three cases of progressive multifocal leukoencephalopathy (PML). This rare, devastating neurological disorder occurs as an opportunistic infection in organ transplantation, hematological malignancies, and HIV disease. It is due to reactivation of latent JC polyoma virus infection in peripheral tissue, particularly kidney and lymphoid tissue, including spleen and bone marrow. Blood-borne lymphocytes harboring the JC virus are then thought to bring the virus to the brain with subsequent infection. Two of the PML patients had multiple sclerosis and were also taking interferon-beta, and the third, a patient with Crohn's disease, had lymphopenia and history of treatment with the immunosuppressant drugs, azathioprine and infliximab. However, PML has not been reported in multiple sclerosis or Crohn's disease, even in patients receiving immunosuppressant drugs, thus making it likely the cases were related to natalizumab treatment. There are two plausible mechanisms by which natalizumab could promote development of PML. First, the drug could prevent the trafficking of JC virus-specific cytotoxic lymphocytes to the brain, allowing progression of the infection. Second, natalizumab treatment results in increased circulating lymphocytes, and the mobilization of B lymphocytes from marrow and spleen B cells could result in JC viremia. Of some reassurance, a retrospective safety review was performed on over 3000 patients who had received natalizumab, and no new cases of PML were found [27].

Although natalizumab was approved for restricted marketing in multiple sclerosis in June 2006 (see below), it has not been approved for treatment of Crohn's disease. In considering its use in Crohn's disease it will be necessary to carefully weigh the risk, albeit quite small, of a devastating fatal disease in a non-life-threatening disorder for which there are other efficacious therapies.

Multiple sclerosis

Multiple sclerosis is an autoimmune central nervous system disorder characterized by discrete areas of demyelination and axonal injury due to autoreactive T cells [28]. The course of multiple sclerosis is variable with the majority of patients having a relapsing-remitting disease pattern at onset.

Anti-integrin therapy directed against $\alpha_4\beta_1$ [29] or $\alpha_L\beta_2$ and $\alpha_M\beta_2$ [30] was shown to be efficacious in experimental allergic encephalomyelitis, a model for multiple sclerosis. However, in a Phase II trial of patients with acute exacerbations of multiple sclerosis, blockade of β_2 integrin with the anti-β_2 mAb, rovelizumab, did not demonstrate significant clinical benefit [31].

In contrast, the anti-α_4 mAb, natalizumab, has been shown to successfully reduce the risk of sustained progression of disability and of clinical relapse over 2 years [32]. As discussed above, natalizumab, which was approved by the FDA for the treatment of relapsing multiple sclerosis in November 2004, was removed from the U.S. market in February 2005 following the reports of three cases of PML. After careful consideration of the risks and benefits, the FDA approved reintroduction of the natalizumab in June 2006 with revised safety warnings and a risk management plan addressing the potential risk of PML [33]. The drug is only to be used as monotherapy in patients who have not responded to, or cannot tolerate, other treatments for multiple sclerosis.

Myocardial infarction

In animal models of myocardial infarction, leukocytes, particularly neutrophils, are important mediators of reperfusion injury, and blockade or deficiency of the adhesion molecules involved in leukocyte recruitment attenuates myocardial damage (reviewed in [34]). However, clinical trials of adhesion antagonists in acute myocardial infarction have been disappointing. Two multi-center, randomized, placebo-controlled trials tested the efficacy of two different recombinant, humanized, anti-β_2

integrin mAbs in the setting of acute myocardial infarction. Both failed to meet their endpoints. Rovelizumab failed to reduce infarct size following angioplasty [35], and erlizumab did not modify coronary blood flow, infarct size, or the rate of ECG ST-segment elevation resolution [36]. In the PSALM (P-Selectin Antagonist Limiting Myonecrosis) trial, patients with ST-elevation acute myocardial infarction were treated with recombinant P-selectin glycoprotein ligand-immunoglobulin or placebo as an adjuvant to thrombolysis. The trial was prematurely stopped by the sponsor for lack of efficacy [37].

Psoriasis

Psoriasis is an inheritable, chronic inflammatory condition that primarily affects the skin with scaly papular or plaque-like lesions. Extracutaneous manifestations include psoriatic arthritis and inflammatory bowel disease. Psoriasis is theorized to be the result of keratinocyte stimulation and subsequent dendritic and T cell activation [38].

$\alpha_L\beta_2$ on resident T lymphocytes in lesions binds to ICAM-1 on the antigen-presenting cells, and it is also important for lymphocyte trafficking to the inflamed skin. Patients with moderate to severe psoriasis who received subcutaneous injections of the anti-α_L mAb efalizumab exhibited significant improvement in their psoriasis compared with controls [39, 40]. In October 2003, the FDA approved efalizumab (Raptiva®) for the treatment of chronic moderate to severe psoriasis. Efalizumab was not, however, found to be of clinical benefit to patients with psoriatic arthritis [41].

The E-, P-, and L-selectin antagonist bimosiamose is currently under Phase II development as it showed efficacy in an exploratory Phase IIa trial of 12 patients with stable psoriatic plaques [6]. After 10 days of treatment, the plaques treated with a bimosiamose topical microemulsion demonstrated a decrease in plaque thickness as compared to control [42].

Rheumatoid arthritis

Rheumatoid arthritis is a progressive systemic autoimmune disease, which can lead to severe disabilities due to chronic inflammation. Early Phase I/II trials in rheumatoid arthritis using the murine anti-ICAM-1 mAb, enlimomab, were encouraging [43, 44]. Unfortunately, more recent clinical trials have not been successful. A Phase II trial of the anti-α_L mAb efalizumab in moderate to severe rheumatoid arthritis was terminated as there was no overall benefit compared to control [45]. The antisense oligodeoxynucleotide to ICAM-1, alicaforsen, also failed to demonstrate significant efficacy in the treatment of severe rheumatoid arthritis [46].

Stroke

Cerebral ischemia is characterized by an acute inflammatory response with an accumulation of leukocytes, predominantly neutrophils initially. Adhesion molecule antagonists demonstrated significant neuroprotection in animal models of cerebral ischemia (reviewed in [47]). Unfortunately, as with myocardial infarction, the promising preclinical studies did not translate into clinical efficacy. A Phase III trial of the anti-β_2 mAb, rovelizumab, in acute stroke was terminated due to lack of efficacy [48]. A Phase II trial testing UK-279,276 (neutrophil inhibitory factor), a recombinant glycoprotein that inhibits $\alpha_M\beta_2$, was also discontinued due to an early determination that treatment had no effect on outcome [49]. A Phase III trial of the anti-ICAM-1 mAb enlimomab demonstrated no efficacy, and, in fact, worsened stroke outcome [50].

Transplantation

Although graft survival has improved as immunosuppressive therapy has evolved, rejection remains an important post-transplant complication. Transplanted organs and organs undergoing rejection show leukocyte accumulation and increased expression of IgSF ligands [51, 52], and adhesion antagonists have shown efficacy in preclinical models (reviewed in [53]).

In bone marrow transplantation, graft failure after human leukocyte antigen nonidentical bone marrow transplantation in children was reduced by treatment with the murine anti-α_L mAb 25.3 [54]. Additionally, the use of mAb 25.3 in conjunction with an anti-CD2 antibody in patients receiving partially incompatible bone marrow transplants resulted in a high rate of engraftment [55]. However, this was accompanied by a high incidence of lethal infections and relapses.

In renal transplantation, a Phase II trial of a murine anti-α_L mAb, odulimomab, demonstrated no difference in the incidence and severity of acute rejection in the first 3 months compared to anti-thymocyte globulin [56]. A Phase III trial of odulimomab also showed little efficacy in prevention of delayed graft function, another important post-transplant complication [57]. Short-term induction therapy with the anti-ICAM-1 mAb enlimomab after cadaveric renal transplantation also did not reduce the rate of delayed graft function or of acute rejection [58]. In a Phase I/II dose-finding study of the anti-α_L mAb efalizumab, there was a worrisome development of post-transplant lymphoproliferative disease in several patients who received the highest dose [59].

Emigrated leukocytes are also the effectors of graft *versus* host disease (GVHD) [60]. In a small clinical trial, the anti-α_L mAb 25.3 was effective in the treatment of steroid-resistant, acute GVHD [61].

Traumatic shock

Hemorrhagic shock due to trauma and resuscitation has been postulated to represent a type of whole body ischemia-reperfusion with leukocyte-mediated reperfusion injury producing multiple organ failure [62]. In animal models of hemorrhagic shock and resuscitation, adhesion molecule blockade reduced mortality as well as organ failure [63, 64]. Two Phase II trials tested blockade of β_2 integrin in the setting of hemorrhagic shock following trauma. Neither erlizumab nor rovelizumab demonstrated efficacy compared with placebo upon analysis of a number of clinical endpoints [65, 66]. In another Phase II trial treatment of severely injured trauma patients, the anti-L-selectin mAb aselizumab had no significant impact on multiple efficacy variables [67].

Conclusion

The elucidation of the molecular basis of leukocyte trafficking over the past two decades promised a new generation of therapeutics for diverse inflammatory and immune diseases. Anti-adhesion therapies have been shown to be remarkably effective in a variety of preclinical studies, and there are now two approved drugs, efalizumab in psoriasis and natalizumab in multiple sclerosis. Unfortunately, however, there are over 20 negative clinical trials of anti-adhesion therapy in multiple other indications. In some cases, the failure of anti-adhesion therapies in the clinic may have resulted from common problems in clinical trial design, such as the dosing and timing of treatment or the selection of patient populations and clinical endpoints. In the putative ischemia-reperfusion disorders (*i.e.*, cardiopulmonary bypass, myocardial infarction, stroke, and traumatic shock), the disappointing results with anti-adhesion therapies most likely reflect the inadequacy of the current experimental models in predicting the human diseases [68]. Additionally, although some compromise in host defense against microorganisms is an expected consequence of inhibiting leukocyte trafficking, the occurrence of PML with natalizumab treatment has raised concerns that perturbation of the immune system with anti-adhesion therapies in other chronic diseases will result in similar rare infections. Certainly, in the short-term this complication will postpone the development of $\alpha_4\beta_1$ antagonists in inflammatory bowel disease and asthma.

Despite the challenges, there remain reasons to be optimistic about prospects for new drugs targeting leukocyte and endothelial adhesion molecules. There is clinical proof of efficacy with approved drugs in psoriasis and multiple sclerosis, and other agents are advancing in the development pipelines of multiple pharmaceutical companies.

References

1   Romano SJ (2005) Selectin antagonists: therapeutic potential in asthma and COPD. *Treat Respir Med* 4: 85–94

2   Bochner B (2004) Adhesion molecules as therapeutic targets. *Immunol Allergy Clin North Am* 24: 615–630

3   Wallstreet Online. United States Securities and Exchange Commission Annual Report. Available at: http://www.wallstreet-online.de/community/thread/108514–1.html (accessed 26 November 2006)

4   APM Health Europe. Healthcare Business News for European Pharmaceutical Markets. Available at: http://www.apmhealtheurope.com/story.php?Numero=4243&profil=34 (accessed 26 November 2006)

5   Gauvreau GM, Becker AB, Boulet LP, Chakir J, Fick RB, Greene WL, Killian KJ, O'Byrne PM, Reid JK, Cockcroft DW (2003) The effects of an anti-CD11a mAb, efalizumab, on allergen-induced airway responses and airway inflammation in subjects with atopic asthma. *J Allergy Clin Immunol* 112: 331–338

6   Revotar Biopharmaceuticals AG. Product Pipeline. Available at: http://www.revotar.com (accessed 2 January 2007)

7   Cybulsky MI, Iiyama K, Li H, Zhu S, Chen M, Iiyama M, Davis V, Gutierrez-Ramos JC, Connelly PW, Milstone DS (2001) A major role for VCAM-1, but not ICAM-1, in early atherosclerosis. *J Clin Invest* 107: 1255–1262

8   Kunsch C, Luchoomun J, Grey JY, Olliff LK, Saint LB, Arrendale RF, Wasserman MA, Saxena U, Medford RM (2004) Selective inhibition of endothelial and monocyte redox-sensitive genes by AGI-1067: a novel antioxidant and anti-inflammatory agent. *J Pharmacol Exp Ther* 308: 820–829

9   Tardif JC, Gregoire J, Schwartz L, Title L, Laramee L, Reeves F, Lesperance J, Bourassa MG, L'Allier PL, Glass M et al (2003) Effects of AGI-1067 and probucol after percutaneous coronary interventions. *Circulation* 107: 552–558

10   AtheroGenics Inc. Science & Technology Product Pipeline. Available at: http://www.atherogenics.com/science/product.html (accessed 26 November 2006)

11   Mileski WJ, Rothlien R, Lipsky P (1994) Interference with the function of leukocyte adhesion molecules by monoclonal antibodies: a new approach to burn injury. *Eur J Pediatr Surg* 4: 225–230

12   Mileski WJ, Burkhart D, Hunt JL, Kagan RJ, Saffle JR, Herndon DN, Heimbach DM, Luterman A, Yurt RW, Goodwin CW et al (2003) Clinical effects of inhibiting leukocyte adhesion with monoclonal antibody to intercellular adhesion molecule-1 (enlimomab) in the treatment of partial-thickness burn injury. *J Trauma* 54: 950–958

13   Thiagarajan RR, Winn RK, Harlan JM (1997) The role of leukocyte and endothelial adhesion molecules in ischemia-reperfusion injury. *Thromb Haemost* 78: 310–314

14   Paparella D, Yau TM, Young E (2002) Cardiopulmonary bypass induced inflammation: pathophysiology and treatment. An update. *Eur J Cardiothorac Surg* 21: 232–244

15   Clinicaltrials.gov. Cylexin for reduction of reperfusion injury in infant heart surgery.

Available at: http://clinicaltrials.gov/ct/gui/show/NCT00226369 (accessed 26 November 2006)

16 Kerr KM, Auger WR, Marsh JJ, Comito RM, Fedullo RL, Smits GJ, Kapelanski DP, Fedullo PF, Channick RN, Jamieson SW et al (2000) The use of Cylexin (CY-1503) in prevention of reperfusion lung injury in patients undergoing pulmonary thromboendarterectomy. *Am J Respir Crit Care Med* 162: 14–20

17 Podolsky DK (2002) Inflammatory bowel disease. *N Engl J Med* 347: 417–429

18 Schreiber S, Nikolaus S, Malchow H, Kruis W, Lochs H, Raedler A, Hahn EG, Krummenerl T, Steinmann G, German ICAM Study Group (2001) Absence of efficacy of subcutaneous antisense ICAM-1 treatment of chronic active Crohn's disease. *Gastroenterology* 120: 1339–1346

19 Yacyshyn BR, Chey WY, Goff J, Salzberg B, Baerg R, Buchman AL, Tami J, Yu R, Gibiansky E, Shanahan WR (2002) Double blind, placebo controlled trial of the remission inducing and steroid sparing properties of an ICAM-1 antisense oligodeoxynucleotide, alicaforsen (ISIS 2302), in active steroid dependent Crohn's disease. *Gut* 51: 30–36

20 Millennium. Millennium announces Phase II data for MLN02 in Crohn's disease. Available at: http://investor.millennium.com/phoenix.zhtml?c=80159&p=irol-newsmediaArticle&ID=333736&highlight= (accessed 26 November 2006)

21 Van Deventer SJH, Wedel MK, Baker B, Xia S, Chuang E, Miner PB (2006) A Phase II dose ranging, double-blind, placebo-controlled study of alicaforsen enema in subjects with acute exacerbation of mild to moderate left-sided ulcerative colitis. *Aliment Pharmacol Ther* 23: 1415–1425

22 Millennium. In Phase II study, Millennium investigational drug MLN02 produced significant improvement in ulcerative colitis remission rates as compared to placebo. Available at: http://investor.millennium.com/phoenix.zhtml?c=80159&p=irol-newsmediaArticle&ID=415406&highlight= (accessed 26 November 2006)

23 Sandborn WJ, Colombel JF, Enns R, Feagan BG, Hanauer SB, Lawrance IC, Panaccione R, Sanders M, Schreiber S, Targan S et al (2005) Natalizumab induction and maintenance therapy for Crohn's disease. *N Engl J Med* 353: 1912–1925

24 Macdonald JK, McDonald JW (2006) Natalizumab for induction of remission in Crohn's disease. *Cochrane Database Syst Rev* 3: CD006097

25 Ghosh S, Goldin E, Gordon FH, Malchow HA, Rask-Madsen J, Rutgeerts P, Vyhnalek P, Zadorova Z, Palmer T, Donoghue S et al (2003) Natalizumab for active Crohn's disease. *N Engl J Med* 348: 24–32

26 Biogen Idec. Tysabri® maintained remission in patients with moderate-to-severe Crohn's disease treated for longer than two years according to data presented this week. Available at: http://www.biogenidec.com/site/019_0.html (accessed 26 November 2006)

27 Yousry TA, Major EO, Ryschkewitsch C, Fahle G, Fischer S, Hou J, Curfman B, Miszkiel K, Mueller-Lenke N, Sanchez E et al (2006) Evaluation of patients treated with natalizumab for progressive multifocal leukoencephalopathy. *N Engl J Med* 354: 924–933

28 Hafler DA (2004) Multiple sclerosis. *J Clin Invest* 113: 788–794

29 Yednock TA, Cannon C, Fritz LC, Sanchez-Madrid F, Steinman L, Karin N (1992) Prevention of experimental autoimmune encephalomyelitis by antibodies against alpha 4 beta 1 integrin. *Nature* 356: 63–66

30 Gordon EJ, Myers KJ, Dougherty JP, Rosen H, Ron Y (1995) Both anti-CD11a (LFA-l) and anti-CD11b (MAC-1) therapy delay the onset and diminish the severity of experimental autoimmune encephalomyelitis. *J Neuroimmunol* 62: 153–160

31 All about multiple sclerosis. Icos Corp Won't Study LeukArrest as chronic MS treatment. Available at: http://www.mult-sclerosis.org/news/Oct1999/LeukArrestStudy Stopped.html (accessed 26 November 2006)

32 Polman CH, O'Connor PW, Havrdova E, Hutchinson M, Kappos L, Miller DH, Phillips JT, Lublin FD, Giovannoni G, Wajgt A et al (2006) A randomized, placebo-controlled trial of natalizumab for relapsing multiple sclerosis. *N Engl J Med* 354: 899–910

33 U.S. Food and Drug Administration. FDA approves resumed marketing of Tysabri under a special distribution program. Available at: http://www.fda.gov/bbs/topics/ NEWS/2006/NEW01380.html (accessed 26 November 2006)

34 Kakkar AK, Lefer DJ (2004) Leukocyte and endothelial adhesion molecule studies in knockout mice. *Curr Opin Pharmacol* 4: 154–158

35 Faxon DP, Gibbons RJ, Chronos NAF, Gurbel PA, Sheehan F (2002) The effect of blockade of the CD11/CD18 integrin receptor on infarct size in patients with acute myocardial infarction treated with direct angioplasty: the results of the HALT-MI study. *J Am Coll Cardiol* 40: 1199–1204

36 Baran KW, Nguyen M, McKendall GR, Lambrew CT, Dykstra G, Palmeri ST, Gibbons RJ, Borzak S, Sobel BE, Gourlay SG et al (2001) Double-blind, randomized trial of an anti-CD18 antibody in conjunction with recombinant tissue plasminogen activator for acute myocardial infarction: limitation of myocardial infarction following thrombolysis in acute myocardial infarction (LIMIT AMI) study. *Circulation* 104: 2778–2783

37 Mertens P, Maes A, Nuyts J, Belmans A, Desmet W, Esplugas E, Charlier F, Figueras J, Sambuceti G, Schwaiger M (2006) Recombinant P-selectin glycoprotein ligand-immunoglobulin, a P-selectin antagonist, as an adjunct to thrombolysis in acute myocardial infarction. The P-Selectin Antagonist Limiting Myonecrosis (PSALM) trial. *Am Heart J* 152: 125.e1–125.e8

38 Nickoloff BJ, Nestle FO (2004) Recent insights into the immunopathogenesis of psoriasis provide new therapeutic opportunities. *J Clin Invest* 113: 1664–1675

39 Gordon KB, Papp KA, Hamilton TK, Walicke PA, Dummer W, Li N, Bresnahan BW, Menter A (2003) Efalizumab for patients with moderate to severe plaque psoriasis: a randomized controlled trial. *JAMA* 290: 3073–3080

40 Lebwohl M, Tyring SK, Hamilton TK, Toth D, Glazer S, Tawfik NH, Walicke P, Dummer W, Wang X, Garovoy MR et al (2003) A novel targeted T-cell modulator, efalizumab, for plaque psoriasis. *N Engl J Med* 349: 2004–2013

41 Genetech. Genentech and XOMA announce results of Phase II study of Raptiva in psoriatic arthritis patients. Available at: http://www.gene.com/gene/news/press-releases/ display.do?method=detail&id=7287 (accessed 26 November 2006)

42 Friedrich M, Vollhardt K, Zahlten R, Sterry W, Wolff G. Demonstration of anti-psoriatic efficacy of a new topical formulation of the small molecule selectin antagonist bimosiamose. Available at: http://mcic3.textor.com/cgi-bin/mc/printabs.pl?APP=PSORIASIS2004-abstract&TEMPLATE=&keyf=0099&showHide=show&client= (accessed 5 January 2007)

43 Kavanaugh AF, Davis LS, Nichols LA, Norris SH, Rothlein R, Scharschmidt LA, Lipsky PE (1994) Treatment of refractory rheumatoid arthritis with a monoclonal antibody to intercellular adhesion molecule 1. *Arthritis Rheum* 37: 992–999

44 Kavanaugh AF, Davis LS, Jain RI, Nichols LA, Norris SH, Lipsky PE (1996) A Phase I/II open label study of the safety and efficacy of an anti-ICAM-1 (intercellular adhesion molecule-1; CD54) monoclonal antibody in early rheumatoid arthritis. *J Rheumatol* 23: 1338–1344

45 Genetech. Genentech and XOMA discontinue rheumatoid arthritis trial. Available at: http://www.gene.com/gene/news/press-releases/display.do?method=detail&id=6107 (accessed 26 November 2006)

46 Maksymowych WP, Blackburn WD Jr, Tami JA, Shanahan WR Jr (2002) A randomized, placebo controlled trial of an antisense oligodeoxynucleotide to intercellular adhesion molecule-1 in the treatment of severe rheumatoid arthritis. *J Rheumatol* 29: 447–453

47 Sughrue ME, Mehra A, Connolly ES Jr, D'Ambrosio AL (2004) Anti-adhesion molecule strategies as potential neuroprotective agents in cerebral ischemia: a critical review of the literature. *Inflamm Res* 53: 497–508

48 Stroke Center at Barnes-Jewish Hospital and Washington University School of Medicine. Stroke Trials Registry. Available at: http://www.strokecenter.org/trials/trialDetail.aspx?tid=50 (accessed 26 November 2006)

49 Krams M, Lees KR, Hacke W, Grieve AP, Orgogozo JM, Ford GA (2003) Acute Stroke Therapy by Inhibition of Neutrophils (ASTIN): an adaptive dose-response study of UK–279,276 in acute ischemic stroke. *Stroke* 34: 2543–2548

50 Enlimomab Acute Stroke Trial Investigators (2001) Use of anti-ICAM-1 therapy in ischemic stroke: results of the enlimomab acute stroke trial. *Neurology* 57: 1428–1434

51 Fuggle SV, Koo DD (1998) Cell adhesion molecules in clinical renal transplantation. *Transplantation* 65: 763–769

52 Tanio JW, Basu CB, Albelda SM, Eisen HJ (1994) Differential expression of the cell adhesion molecules ICAM-1, VCAM-1, and E-selectin in normal and posttransplantation myocardium. *Cell* adhesion molecule expression in human cardiac allografts. *Circulation* 89: 1760–1768

53 Dedrick RL, Bodary S, Garovoy MR (2003) Adhesion molecules as therapeutic targets for autoimmune diseases and transplant rejection. *Expert Opin Biol Ther* 3: 85–95

54 Fischer A, Friedrich W, Fasth A, Blanche S, Le Deist F, Girault D, Veber F, Vossen J, Lopez M, Griscelli C et al (1991) Reduction of graft failure by a monoclonal antibody (anti-LFA-1 CD11a) after HLA nonidentical bone marrow transplantation in children with immunodeficiencies, osteopetrosis, and Fanconi's anemia: a European Group for

Immunodeficiency/European Group for Bone Marrow Transplantation report. *Blood* 77: 249–256

55 Cavazzana-Calvo M, Bordigoni P, Michel G, Esperou H, Souillet G, Leblanc T, Stephan JL, Vannier JP, Mechinaud F, Reiffers J et al (1996) A Phase II trial of partially incompatible bone marrow transplantation for high-risk acute lymphoblastic leukaemia in children: prevention of graft rejection with anti-LFA-1 and anti-CD2 antibodies. Societe Francaise de Greffe de Moelle Osseuse. *Br J Haematol* 93: 131–138

56 Hourmant M, Bedrossian J, Durand D, Lebranchu Y, Renoult E, Caudrelier P, Buffet R, Oulillou JP (1996) A randomized multicenter trial comparing leukocyte function-associated antigen-1 monoclonal antibody with rabbit antithymocyte globulin as induction treatment in first kidney transplantations. *Transplantation* 62: 1565–1570

57 Matthews JB, Ramos E, Bluestone JA (2003) Clinical trials of transplant tolerance: slow but steady progress. *Am J Transplant* 3: 794–803

58 Salmela K, Wramner L, Ekberg H, Hauser I, Bentdal O, Lins LE, Isoniemi H, Backman L, Persson N, Neumayer HH et al (1999) A randomized multicenter trial of the anti-ICAM-1 monoclonal antibody (enlimomab) for the prevention of acute rejection and delayed onset of graft function in cadaveric renal transplantation: a report of the European Anti-ICAM-1 Renal Transplant Study Group. *Transplantation* 67: 729–736

59 Kuypers DRJ, Vanrenterghem YFC (2004) Monoclonal antibodies in renal transplantation: old and new. *Nephrol Dial Transplant* 19: 297–300

60 Ferrara JL, Reddy P (2006) Pathophysiology of graft-*versus*-host disease. *Semin Hematol* 43: 3–10

61 Stoppa AM, Maraninchi D, Blaise D, Viens P, Hirn M, Olive D, Reiffers J, Milpied N, Gaspard MH, Mawas C (1991) Anti-LFA1 monoclonal antibody (25.3) for treatment of steroid-resistant grade III-IV acute graft-*versus*-host disease. *Transpl Int* 4: 3–7

62 Vedder NB, Fouty BW, Winn RK, Harlan JM, Rice CL (1989) Role of neutrophils in generalized reperfusion injury associated with resuscitation from shock. *Surgery* 106: 509–516

63 Vedder N, Winn RK, Rice CL, Chi EY, Arfors KE, Harlan JM (1988) A monoclonal antibody to the adherence-promoting leukocyte glycoprotein, CD18, reduces organ injury and improves survival from hemorrhagic shock and resuscitation in rabbits. *J Clin Invest* 81: 939–944

64 Mileski WJ, Winn RK, Vedder N, Pohlman TH, Harlan JM, Rice CL (1990) Inhibition of CD18-dependent neutrophil adherence reduces organ injury after hemorrhagic shock in primates. *Surgery* 108: 206–212

65 Rhee P, Morris J, Durham R, Hauser C, Cipolle M, Wilson R, Luchette F, McSwain N, Miller R (2000) Recombinant humanized monoclonal antibody against CD18 (rhuMAb CD18) in traumatic hemorrhagic shock: results of a phase II clinical trial. Traumatic Shock Group. *J Trauma* 49: 611–619

66 Vedder N, Harlan JM, Winn RK, Maier R, Hoyt D, Moore F, Fabian T, West M, Peitzman A, Moore E et al (2000) Immunomodulators: inhibitors of adhesion. *Shock* 13: 1

67 Seekamp A, van Griensven M, Dhondt E, Diefenbeck M, Demeyer I, Vundelinckx G, Haas N, Schaechinger U, Wolowicka L, Rammelt S et al (2004) The effect of anti-L-selectin (aselizumab) in multiple traumatized patients – Results of a Phase II clinical trial. *Crit Care Med* 32: 2021–2028

68 Iwata A, Harlan JM, Vedder NB, Winn RK (2002) The caspase inhibitor z-VAD is more effective than CD18 adhesion blockade in reducing muscle ischemia-reperfusion injury: implication for clinical trials. *Blood* 100: 2077–2080

Index

F-actin 102, 104
α-actinin 103
ADAM-17 156
adherens junction 274
adhesion, heterophilic 202, 211
adhesion, homophilic 202–204
AGI-1067 290
alicaforsen 293, 295
amine oxidase copper containing -3, AOC3 237
Anaplasma phagocytophilum 15
angiogenesis 280, 283
anti-α_L mAb 25.3 296
antigen-presenting cell (APC) 111
AOC1 242
Arg-Gly-Asp (RGD) 121
Arp2/3 105
arrest of leukocytes 151, 184
aselizumab 297
asthma, adhesion antagonists in 290
asthma, L-selectin in 47–49
atherogenesis 153
atherosclerosis 153, 282, 290
atherosclerosis, adhesion antagonists in 290

B lymphocytes, homing of 43
bimosiamose 82, 290, 295
burns, adhesion antagonists in 292

calcium 102, 104
cardiac development 124

cardiopulmonary surgery, adhesion antagonists in 292
catch bond 9, 10
CD11a 176, 184
CD11b (Mac-1α) 176, 184
CD18, β_2 subunit 176, 184
CD24 77
CD31 226, 253, 260, 261
CD44 77
CD44 in leukocyte-endothelial interactions 38
CD49f/CD29 221–230
CD54 36–38, 99–112, 187
CD99 214, 253
cell permeability 261, 278, 282
cell polarization 282
chemotaxis, L-selectin in 39
collagen IV 221
collagen-induced arthritis (CIA) 80
contact hypersensitivity 155
core2GlcNAcT-I 6
cortactin 102, 105
coxsackievirus-adenovirus receptor (CAR) 255
Crohn's disease, adhesion antagonists in 293
CY-1503 292

delayed-type contact hypersensitivity (DTH) 79
DTH reaction model 256, 257
diabetes 48
diabetes, L-selectin in 47, 48
diamine oxidase AOC1 242

The PIR-Series
Progress in Inflammation Research

Homepage: http://www.birkhauser.ch

Up-to-date information on the latest developments in the pathology, mechanisms and therapy of inflammatory disease are provided in this monograph series. Areas covered include vascular responses, skin inflammation, pain, neuroinflammation, arthritis cartilage and bone, airways inflammation and asthma, allergy, cytokines and inflammatory mediators, cell signalling, and recent advances in drug therapy. Each volume is edited by acknowledged experts providing succinct overviews on specific topics intended to inform and explain. The series is of interest to academic and industrial biomedical researchers, drug development personnel and rheumatologists, allergists, pathologists, dermatologists and other clinicians requiring regular scientific updates.

Available volumes:
T Cells in Arthritis, P. Miossec, W. van den Berg, G. Firestein (Editors), 1998
Chemokines and Skin, E. Kownatzki, J. Norgauer (Editors), 1998
Medicinal Fatty Acids, J. Kremer (Editor), 1998
Inducible Enzymes in the Inflammatory Response,
    D.A. Willoughby, A. Tomlinson (Editors), 1999
Cytokines in Severe Sepsis and Septic Shock, H. Redl, G. Schlag (Editors), 1999
Fatty Acids and Inflammatory Skin Diseases, J.-M. Schröder (Editor), 1999
Immunomodulatory Agents from Plants, H. Wagner (Editor), 1999
Cytokines and Pain, L. Watkins, S. Maier (Editors), 1999
In Vivo *Models of Inflammation*, D. Morgan, L. Marshall (Editors), 1999
Pain and Neurogenic Inflammation, S.D. Brain, P. Moore (Editors), 1999
Anti-Inflammatory Drugs in Asthma, A.P. Sampson, M.K. Church (Editors), 1999
Novel Inhibitors of Leukotrienes, G. Folco, B. Samuelsson, R.C. Murphy (Editors), 1999
Vascular Adhesion Molecules and Inflammation, J.D. Pearson (Editor), 1999
Metalloproteinases as Targets for Anti-Inflammatory Drugs,
    K.M.K. Bottomley, D. Bradshaw, J.S. Nixon (Editors), 1999
Free Radicals and Inflammation, P.G. Winyard, D.R. Blake, C.H. Evans (Editors), 1999
Gene Therapy in Inflammatory Diseases, C.H. Evans, P. Robbins (Editors), 2000
New Cytokines as Potential Drugs, S. K. Narula, R. Coffmann (Editors), 2000
High Throughput Screening for Novel Anti-inflammatories, M. Kahn (Editor), 2000
Immunology and Drug Therapy of Atopic Skin Diseases,
    C.A.F. Bruijnzeel-Komen, E.F. Knol (Editors), 2000
Novel Cytokine Inhibitors, G.A. Higgs, B. Henderson (Editors), 2000
Inflammatory Processes. Molecular Mechanisms and Therapeutic Opportunities,
    L.G. Letts, D.W. Morgan (Editors), 2000
Cellular Mechanisms in Airways Inflammation, C. Page, K. Banner, D. Spina (Editors), 2000
Inflammatory and Infectious Basis of Atherosclerosis, J.L. Mehta (Editor), 2001

Lymphocyte Trafficking in Health and Disease

Badolato, R., University of Brescia, Italy / Sozzani, S., University of Brescia, Italy (Eds)

2006. XII, 242 p. 21 illus., 1 in color. Hardcover
ISBN-13 978-3-7643-7308-5
PIR — Progress in Inflammation Research

Since the discovery of chemokines and of chemokine receptors it has become evident that expression of chemokines at the site of inflammation may regulate the composition of cellular infiltrate, thereby directing the type of immune response. Recently, the molecular characterization of inherited disorders of immune system, (e.g., Wiskott-Aldrich syndrome, WHIM syndrome, leukocyte adhesion deficiency), which are characterized by cytoskeleton/adhesion defects or by altered response of chemokine receptors has contributed to clarifying the key players of immune response in normal physiology and in disease. This book, which deals with the description of the role of chemokines in immune response and underlines potential targets of therapeutical intervention, offers a series of contributions of the most challenging aspects of lymphocyte migration in homeostasis and in disease.

From the contents:
Lymphocyte trafficking: from immunology paradigms to disease mechanisms.- Biology of chemokines.- Lymphocyte-endothelial cell interaction.- Chemokine receptor expression in effector and memory T cell subsets.- Migration of dendritic cell subsets.- Migration of NK cells.- Lymphocyte trafficking and chemokine receptors during pulmonary disease.- Lymphocyte migration to the brain.- Lymphocyte migration to the kidney.- Lymphocyte migration to pancreatic islets.- Controlling leukocyte trafficking in disease.- Leukocyte adhesion deficiency .- Wiskott-Aldrich syndrome as a model of cytoskeleton defects .- From CXCR4 mutations to WHIM syndrome

Birkhäuser Verlag AG
Viaduktstrasse 42
4051 Basel / Switzerland

Tel. +41 61 205 07 77
e-mail: sales@birkhauser.ch
www.birkhauser.ch

Regulatory T Cells in Inflammation

Taams, L.S., King's College London, UK / Akbar, A.N., Royal Free and University College Medical School, London, UK / Wauben, M.H.M., University of Utrecht, The Netherlands (Eds)

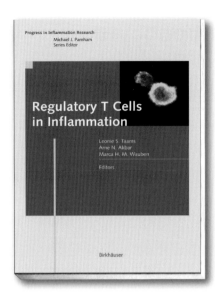

Progress in Inflammation Research
Michael J. Parnham
Series Editor

Regulatory T Cells in Inflammation

Leonie S. Taams
Arne N. Akbar
Marca H. M. Wauben

Editors

Birkhäuser

BIRKHÄUSER

2005. XI, 240 p. Hardcover
ISBN-13 978-3-7643-7088-6
PIR — Progress in Inflammation Research

Regulatory T-cells are essential components of the immune system, and several different subsets of regulatory T-cells have been described. Considerable regulatory function has been attributed to the CD4+CD25+ T-cell subset. These cells act by suppressing adaptive and possibly innate immune responses thereby maintaining or restoring the balance between immunity and tolerance. The suppressive effects of CD4+CD25+ regulatory T-cells are cell-contact dependent. Recent developments and viewpoints in the field of CD4+CD25+ regulatory T-cells as well as the potential use of regulatory T-cells in immunotherapy of inflammatory diseases are discussed in this volume. By linking data from experimental models with recent findings from the clinic, this book will be of interest to immunologists and other biomedical researchers as well as clinicians interested in the regulation and manipulation of the immune response during inflammatory disease.

From the contents:
Part I. Origin, function and distribution of regulatory T cells: History of CD4+CD25+ regulatory T cells. ‚Natural‘ and ‚induced‘ regulatory T cells - purpose and problems associated with an emerging distinction. The role of interleukin-10 in regulatory T-cell suppression: reconciling the discrepancies. Activation and distribution of regulatory T cells in naive and antigen-stimulated immune systems. Regulatory T cells and the innate immune system.- Part II. The potential use of regulatory T cells in immunotherapy: Exploiting the potential of regulatory T cells in the control of type 1 diabetes. Regulatory T cells in type 1 autoimmune diabetes. The potential for targeting CD4+CD25+ regulatory T cells in the treatment of multiple sclerosis in humans. Immunotherapy of rheumatoid arthritis using CD4+CD25+ regulatory T cells. Potential for manipulation of regulatory T cells in treatment or prevention of allergic disease. The role of regulatory T cells in cutaneous disorders.

Birkhäuser Verlag AG
Viaduktstrasse 42
4051 Basel / Switzerland

Tel. +41 61 205 07 77
e-mail: sales@birkhauser.ch
www.birkhauser.ch